Fighting America's hidden health threat

Ritchie C. Shoemaker, MD
with
James Schaller, MD
& Patti Schmidt

GATEWAY PRESS, INC.
Baltimore, MD 2005

Please direct all correspondence and book orders to:
Ritchie C. Shoemaker
P.O. Box 25
Pocomoke City, MD 21851

Library of Congress Control Number 2005923094
ISBN 0-9665535-3-5

Published for the author by
Gateway Press, Inc.
1001 N. Calvert Street
Baltimore, MD 21202-3897

www.gatewaypress.com

Printed in the United States of America

Dedications

To JoAnn Shoemaker who makes my life worthwhile.

To Sally Shoemaker who brings life to my dreams.

To Sara and Wells Shoemaker who gave me life.

With thanks to all the people who gave time and energy to this book: *Mold Warriors* will always be a part of your life.

A special thanks to Charles Schaefer, MD, otolaryngologist extraordinaire, Salisbury, Maryland for saving my life by stopping the hemorrhages caused by mold toxins when everything else had failed.

To mold patients everywhere: never let anyone stand in the way of your right to a healthful life.

Mold makes us sick. It is just that simple.

We can prove it.

Ritchie C. Shoemaker MD
Pocomoke, Maryland

CONTENTS

CONTENTS

Appendices

FOREWORD

By Richard Lipsey, PhD

Mold illness is real even when 80% of molds found indoors are non-pathogenic. When we start to eliminate the threat to health posed by water intrusion and a build-up of pathogenic molds and bacteria, especially in schools and other public buildings, we will all benefit. As a practicing toxicologist for the past 33 years, I've seen my share of medical and legal cases related to exposure to environmental agents. From chemicals in ground water possibly causing leukemia to asbestos exposure causing an unusual kind of lung cancer, we know that toxic chemicals that can hurt us are everywhere in our environment and that toxic chemicals under certain conditions threaten human life. Though I do not practice medicine, I share the feelings of physicians who search for effective treatments for these illnesses. We need to know the identity and levels of these chemicals, recognize when illness is due to exposure, stop the exposure, diagnose the injury and begin therapy. I do all of the above except diagnosis and treatment.

I've been extensively involved as an expert in legal cases that have involved people whose illnesses were acquired following exposure to toxin-forming organisms (including molds) growing in buildings with water damage. I appear in court regularly, equally for defendants and plaintiffs, and have always been permitted to present my opinions (Editors note: see Chapter 19, Lawyers, Lawyers, Lawyers). The same can't be said for other experts, including physicians. Identifying health effects caused by exposure to indoor molds has been a struggle across the nation for some time, medically and legally. Over the years I have seen a number of medical theories come and go.

Perhaps those prior theories are simply part of a natural progression of thought regarding human illness caused by exposure to indoor molds. I have talked to a number of patients who have been treated successfully by Dr. Ritchie Shoemaker for illness caused by mold exposure. I have read his opinions in court cases, his academic articles and his theories on how significant amounts of toxins made by indoor organisms, including fungi, initiate a series of reactions that each cause adverse health effects. "Significant amounts" vary, depending on the patient and his/her health at the time of exposure. Clearly documenting the genetic susceptibility to toxins brings a new dimension to how we think about mold illness.

These ideas are new ground for a classically trained toxicologist. In the past, we learned that, "The poison is in the dose," implying that the greater amount of toxin exposure, the greater the illness. Dr. Shoemaker's data show that idea doesn't apply to mold illness. When I inspect and sample a building where people have been exposed, I rarely know the dose.

It makes sense that the complicated illnesses I see in people who live or work in sick buildings is complex in its physiology. And although there is no one panacea for mold illness, following Dr. Shoemaker's Biotoxin Pathway makes sense because it outlines specific physiologic disturbances and how they arise in mold-exposed, ill people.

The treatment approach used by Dr. Shoemaker differs from that of other expert physicians with whom I have worked. I think it would be premature to say that only one group of ideas on therapy or one approach is valid: I have seen successes and failures with many approaches. But we must not eliminate any ideas about therapy when so many people are so ill from mold exposure.

The point I would make is that recognition of mold illness has gone through a rapid evolution. There's been a remarkable turnaround in academic attitudes in just two years. In late 2002, some consensus panels tried to tell us that mold illness didn't exist; yet the Institute of Medicine recently acknowledged that exposure to damp buildings can cause respiratory effects and that genetics and cytokines are involved in mold illness. And more recently, the data presented by Dr. Shoemaker and other practitioners shows that mold illness is multisystemic and that in addition to respiratory effects, chronic fatigue, joint problems, neurologic and cognitive effects are nearly always present in mold patients. Now we have good reason to look at physiological changes caused by exposures and to develop sequential therapies based on what mold illness does to us. The opinions of 2002 can no longer be considered to be "current" or accurate.

I, for one, am eager to see what the next two years brings in the identification and treatment of mold illness. The Biotoxin Pathway enables us to profile the illness based on what it does to us. With a logical framework that underlies further research, we are on the brink of bringing tremendous help to millions of people at risk for chronic illnesses.

Mold Warriors presents Dr. Shoemaker's ideas on many facets of mold, written from the perspective of a treating physician on the front lines. As he continues his own work, he calls for more research on mold growth and mold illness. I agree that the basic science that will look at each of his observations needs to be funded. But we need to pay careful attention to what he's observed as a treating physician for more than 2000 mold patients. We must learn from the knowledge he's gained from the bedside.

In this era of high tech medicine, we can often get caught up with what SPECT, MRI and CT scans and sophisticated assays show. We shouldn't forget that important advances in medicine, such as Dr. Shoemaker's work, usually begin with carefully recorded clinical observations that lead to further research.

What's important in Dr. Shoemaker's work is that he uses the time-honored scientific method in his work. He relies on history, physical exam, laboratory testing, differential diagnosis and treatment to establish his opinions. Some "experts," don't have the same sound processes to rely on. I testify about 90 times a year and witness the testimony of many "experts" who have yet to treat their first case of mold illness. We need to listen to opinion based on observable clinical reality before we listen to opinion based on opinion.

I recommend that you read *Mold Warriors*. Mark the time you do and return to it two years from now. Remember what you read and compare that to what we know then.

Richard Lipsey, PhD
Toxicologist
Jacksonville, Florida

FOREWORD

by Bianca Jagger

For most people, toxic mold is an alien subject that has no bearing on their daily lives. After all, there's so much to think about in today's world—armed conflicts, terrorism, world poverty and hunger, or the epidemics that constantly sweep our troubled globe—and toxic mold doesn't seem to rank up there with those critical issues. However, you may think differently after reading this book.

Indeed, as a human rights advocate who has spent most of my life campaigning for those in need, I have to admit that before I became personally affected, I too, didn't pay the subject of toxic mold much attention. Then a few years ago when my apartment in New York became contaminated with toxic mold, my life took on a drastic turn for the worse. Despite repeated pleas for help to remediate the problem, my landlord refused to take the necessary measures to repair the water intrusion that was causing the toxic mold to spread. As a result, I became extremely ill and was forced to vacate my home of twenty years and leave behind all my contaminated personal property.

Some may regard mold illness as a controversial issue; but it's not controversial for those living in conditions similar or worse than mine, their lives torn apart by this terrible scourge, ill, homeless and often facing bankruptcy. Many of the families affected have children; scientific evidence has revealed that young children's lives are most at risk when exposed to toxic mold. Last year Two-year-old Neveah Lair died in a Coventry apartment complex in Bakersfield, California, her mother contends that her death was caused by toxic mold and is presently in the mist of a lawsuit.

The toxic mold that grew in my apartment made me sick and I remain ill today. I've spent much of my life as an advocate for human rights, struggling for social and economic justice and equal treatment for men and women in all walks of life, all over the world. Over the years, I've had to face many dangers and obstacles in my work as a human rights defender—death squads in Central America, corrupt politicians in many parts of the world and the perpetrators of genocide in the former Yugoslavia. Yet, little did I know that the insidious cancer growing within the walls of my apartment as a result of mold infestation would rob from me the quality of my life.

Today, more than three years later, I remain ill, am unable to occupy my home and unable to have access to my personal property; I live in constant pain, battling chronic fatigue and multiple health problems, because my home became contaminated with the growth of molds that make toxins.

I'm not alone in my illness or in my desire for the return to health. I am a prisoner of what toxins do, and so too are countless others. By writing a foreword to Dr. Ritchie Shoemaker's book, *Mold Warriors*, and supporting him in his quest to find relief for those suffering from neurotoxin-based illnesses like mine, I'm continuing my own work to end violations of human rights in the U.S. The Federal and State Governments must focus their attention on the biotoxins epidemic and the damaging inflammation biotoxins cause. If members of Congress and State politicians fail to enact legislations that will establish Federal and State guidelines to protect the innocent and punish the violators there will continue to be no justice for the hundred of thousands of victims throughout the U.S.

Asbestos was the health nightmare of the 1980s and 90s, now buildings contaminated with biotoxins are the scourge of the early 21st Century.

I've consulted other specialists, and I've found them to be caring individuals who wanted to help me. But my illness remains with me. Mold hurts us in ways modern medicine has only begun to understand. I'm aware that some claim toxic mold doesn't cause anything more than upper respiratory illnesses. They're wrong; toxic mold has seriously affected my life, my health and my work. In fact, toxic mold can impinge on one's life forever. I know it.

I see my experience at my apartment in New York and the resulting illness as a violation of my fundamental rights to live in a "habitable living condition." On September 21 2003, New York City (HPD) House and Preservation Department issued 8 violations in my apartment. This included 3 "C" violations, considered immediately hazardous, 4 "B" violations consider hazardous and 1 "A" violation. Despite the gravity of the situation my landlord denied the existence of toxic mold and ignored the violations, which indicated that the procedures in place are not enough to compel landlords to address the hazardous situation cause by toxic mold.

That's because there's no Federal or State legislation requiring landlords to maintain adequate environmental living conditions in their apartment buildings. Toxic mold contamination resulted in my illness and ultimately, it's reflected on the pages of the laboratory results that Dr. Shoemaker has sent me. I know my enemy is more than just the various toxic molds that were growing indoors in my home—*Stachybotrys, Aspergillus, Penicillium*. I see my enemy with faces of letters, like masks from a Mardi Gras parade. Ghostly letters, like MSH, MMP9, ADH, VEGF, IL-1B and C3a, rise from the black and white lab copies to pull at every fiber of my soul and haunt my attempts at sleep.

I didn't know these letters or what they meant before April 2004, when my research led me to Dr. Shoemaker's office in tiny Pocomoke, Maryland. He says I'm one of the worst mold illness sufferers among the thousands he's seen. His testing—all those acronyms—and the results from them show what's wrong with me physiologically. I know he's researching new approaches for those with illness as severe as mine, and I hope one day I will benefit from those new treatments. As his work proceeds, other physicians are looking at new ways to help mold patients. In time, with the help of my doctors, including Dr. Shoemaker, I hope to regain my strength.

What's missing from this new frontier? The list is long, but it includes Federal and State Government legislation that would establish guidelines to protect home owners and tenants from toxic mold, including mandatory inspections for homes, apartments and public buildings. We need an organized governmental approach, we need doctors to be informed as to how to recognize and treat toxic mold illness, and most of all, we need to make accessible all the information available to warn the public at large about the dangers toxic mold exposure presents to their health and their lives.

We must not let those with vested interests argue that our finest scientists are wrong about mold illness when the resulting health problems are so obvious. We must not let politicians or powerful contributors stand in our way, denying us the right to safe workplaces, homes and schools.

Dr. Shoemaker's book is an important step in raising public awareness of the health problems linked to mold exposure. We must support him and other physicians who seek therapies to treat those exposed to toxic mold. We must stop the politically inspired arguments that prevent those who are ill from being treated fairly and from being returned to health. We must support HR 1268, sponsored by Congressman John Conyers, D-MI-14 which would bring hope and help to countless mold patients; this legislation it's waiting to become law. Please let your representatives in Congress know you want to see this legislation enacted.

Reading Dr. Shoemaker's work is your first step to understand what I know: mold makes us ill. Dr. Shoemaker has proven it with his latest research, and *Mold Warriors* tells the true stories of how a pioneering rural family physician came to reveal one of the 21st Century's most complex medical mysteries.

Join me in learning more about mold illness with these brave *Mold Warriors*.

Bianca Jagger

Bianca Jagger is a Council of Europe Goodwill Ambassador, a member of the Executive Director's Leadership Council for Amnesty International USA, and a member of the Advisory Committee of Human Rights Watch America. Ms. Jagger also serves on the Advisory Board of the Coalition for International Justice; is a member of the Twentieth Century Task Force to Apprehend War Criminals; and a Board member of People for the American Way and the Creative Coalition.

For her long-standing commitment to human rights issues, social justice and environmental protection, Ms. Jagger has received among others the Right Livelihood Award, known as the "Alternative Nobel Prize"; the World Achievement Award from Mikhail Gorbachev; the United Nations Earth Day International Award; the Green Globe Award by the Rainforest Alliance; Amnesty International/USA's Media Spotlight Award for Leadership; and The American Civil Liberties Union Award.

PREFACE

In the past, viruses and bacteria were entirely unknown, yet had massive effects on health. In a similar way, today many illnesses are due to exposure to biotoxins, especially those made by molds growing inside our buildings, yet many aren't yet aware of it.

They soon will be, though: biotoxin-induced illness will become one of the most important medical topics in the 21st Century.

As you read this, you're surrounded by biotoxins. So are your loved ones. In fact, an aggressive explosion of unique indoor molds that carry chemical toxins on their spores has caused an explosion of illnesses associated with the inflammation these toxins cause. Biotoxins have made many of you ill, but your physicians have told you instead that you're sick from fibromyalgia, stress, depression, Chronic Fatigue Syndrome or some other vague and unsupported diagnosis.

Yet paradoxically, if they knew what to look for, biotoxin illness is easily diagnosed and, if caught in time, readily treated. Hundreds of my patients, once debilitated, now lead normal lives.

What I'm saying is new; this book isn't just another health book. It's the first major book that explains how biotoxins cause those illnesses, and what you can do to make sure you or family doesn't get them. If you're already ill, it reveals a new treatment protocol that has already helped hundreds get well. And it explains how to go about educating your doctor so that he can adequately diagnose and treat biotoxin illness.

Whether or not biotoxins cause you or your loved ones harm largely depends on your genes; your previous, present and future toxic exposures; and what you decide to do with the information in your hand.

Whether the source of mold is your child's flat-roofed school, the office building without proper ventilation or the family home with its moldy walkout basement, the illness is no different. The first step to healing begins with learning what *Mold Warriors* can teach you. This book opens a New World for your physician, your health care provider and most importantly—**you**.

Don't expect everyone in the medical community to quickly embrace the facts in this book. Major changes in Medicine come only after fierce battles against limited, narrow thinking. As you read this book, you'll see the fierce battles many have fought just to remain healthy or stay alive.

These battles, which I've called the "Mold Wars," continue.

The Mold Wars are a classic case of right vs. wrong, with quite a bit of David-and-Goliath thrown in to keep it really interesting. Sounds dramatic? Overstated? Hardly. Here's a sample of what insightful patients have said and the traditional responses they've received:

Mold makes me chronically fatigued and takes away my ability to think.
The CDC and the Institute of Medicine says that's not proven.

I'm sick because of mold.
No, you have fibromyalgia or you're depressed.

Mold causes many physiologic problems that my doctor can measure.

No, I've never heard of those tests.

Mold illness has a genetic basis.
No, we don't know anything about HLA and its relationship to biotoxin illness.

Mold illness causes scars on the brain that look like MS.
No, those scars are non-specific.

Mold illness is chronic unless treatment is begun.
No, mold illness is an allergy.

Mold illness causes learning disability in children and unusual neurological problems in adults, including seizures and Parkinson's disease.
No, the government says that isn't so.

Mold disrupts immune responses, turning on antibody attacks against our own bodies.
No, where are your data proving that?

You don't really have any knowledge about what I'm saying about mold illness, do you?
No, and besides if I ever admitted that I agreed with you, it might cost my employers millions of dollars as well as cause them a lot of embarrassment.

The good news is that the truth is now out.

In *Mold Warriors*, the truth wins against overwhelming forces.

Mold Warriors will show you how mold makes us sick. You'll meet real people with real health problems, who will tell you what you must know to survive the molds in our homes, businesses, schools and public facilities.

The magnitude of the mold problem in the U.S. will stun you.

Fortunately, there is treatment for mold illness, if the illness is caught in time. And I offer multiple areas of research that hold hope for those who get a diagnosis too late.

Ultimately, *Mold Warriors* is a book about people. You'll meet them in their own words as they tell their own stories.

The Mold Wars involve physicians who don't truly and fully listen to their patients. It's about making sloppy diagnoses, like, "You must have a bad case of fibromyalgia, Mr. Mold Victim. Here's a prescription for some antidepressants. Have a nice life."

But the Mold Wars also involve physicians who want to help, but who don't know how; they can do the wrong things, things that don't help and that may hurt.

The Mold Wars attract attorneys on both sides of the battle, each with their entourage of experts and aides. You'll meet some unattractive people when we tell you about the legal cases. And you'll meet the Mold Warriors, people who win justice despite being opposed by the huge financial reserves and often the political power of their adversaries.

So *Mold Warriors* is about political issues too. In September 2004 Rep. John Conyers of Michigan introduced House Resolution 1268, a bill that has the potential to help many people injured by mold.

When huge corporations, many fueled by investment money from insurance companies, are threatened by mold laws and mold litigation, what do you think they'll do? Will corporate America use its vast resources to lobby Congress to support the rights of its citizens to live, work and learn in a toxic mold-free environment? Doing so might mean that those insurance companies would have to pay out many millions—perhaps billions—of dollars in loss claims.

I don't hold out too much hope, and neither should you.

Toxic molds grow indoors wherever we give them what they need: water, food and cover. Rapidly growing molds make spores. Mold spores spread colonies wherever air, your clothes, furniture or papers take them. Spores carry hundreds of toxin molecules that make those of us with a certain genetic make-up, called HLA DR, sick. And *25 percent of us* carry those susceptible HLA genes!

Schools with flat roofs and reduced air circulation, possibly with defects in plumbing and leaky roofs or walls that collect condensation, are a threat to our children. Wherever moisture indoors lets molds grow, remediation is needed, but what school system has all of the money they need for proper maintenance and repairs?

Mold Warriors are no different from you, but they've learned that in their cases, countless health problems—fatigue, obesity, memory problems, chronic respiratory symptoms, unusual seizure-like activity, chronic joint pains, chronic abdominal pains and many more—are *not* caused by allergy, stress, depression or fibromyalgia.

They've learned about new uses for established blood tests that prove in clearly measurable ways what exposure to toxin-forming molds does to our bodies. They know that mold illness is chronic and can be progressive.

But there's good news too. Mold illness, once detected, is treatable for nearly all sufferers. Sadly, some victims don't respond to proper treatment; then the illness is disabling, robbing sufferers of vitality without bringing rapid, merciful death. It's for those small groups of patients that Mold Warriors fight the ultimate battle of life and death.

If mold illness can be diagnosed and treated early and effectively, why do Mold Warriors need to fight? Why aren't mold illnesses as controversial as the common cold? After all, we've known mold makes people sick since Biblical times. Are there forces at work trying to suppress the truth about mold? Who and why?

The list of enemies of truth about mold illness is long because mold illness is acquired from flaws in the indoor environment. If there is a flaw, then someone has caused the flaw. And that means liability. With liability, someone might have to pay to correct the flaw and *the illness*. Given the number of people in buildings with too much moisture, there is a staggering number of dollars potentially involved.

Do you think powerful insurance companies or construction companies, building owners or even governments would always volunteer to pay up? Would they admit responsibility for illness and property damage because that would be the morally and ethically correct thing to do?

Not in the USA.

Considering the influence of massive corporations with a vested interest in the status quo, how can Mold Warriors get past Congressional lobbyists who want to bar funding for scientific research on mold illness? Do powerful medical groups turn their backs on mold illness because it's in their economic interest to do so?

Where is the public demand for effective screening of children for mold illness? Screening only takes five minutes!

Where is the demand for mold health screening for adults in "annual physicals?" It only takes a few minutes and a few tubes of blood. Where is the media in all this? AWOL.

When will owners of stale-air buildings wake up to establish baseline health measures and serial monitoring as a risk management tool? It takes less time than their weekly waiting time for elevators.

One of the biggest barriers to truth in mold illness comes from "corporate medicine" groups.

As you will see, there are physicians—opinion leaders in influential medical societies—who are part of an entrenched "damage control" consortium. They see themselves as conservators of proven truths, whereas in fact they are truth suppressors. We have and will meet the same physicians repeatedly in court cases and consensus panels. They come, equipped with academic appointments and long résumés, to deny the health effects of mold, creating smokescreens of abstraction and diversion. A few of those physicians are also the willing pawns of vested interests, essentially selling their opinions for large sums, while at the same time receiving *public* funding. They are in the "expert-for-hire" business.

These physician enemies of the Mold Warriors never see patients in any meaningful way, spending much of their professional time rendering "expert opinions" on mold when they have yet to lay their "caring" hands on their *first* mold patient. That these physicians form an unholy alliance with defense attorneys will not be a surprise to mold patients.

One such esteemed professor has made his career testifying for the tobacco industry, for corporations accused of poisoning the Love Canal in New York, for the asbestos industry, for the manganese industry, for fungicide manufacturers and, now for insurance carriers in mold cases. His opinion never varies: he says that the toxic chemicals in these cases have never been proven to harm anyone. And this man sounds so authoritative, some people even believe him!

When the defense needs a phalanx of locked-step medical opinion—and those lawyers are well funded and resourceful—there's a stable of medical school graduates volunteering to chime in.

Armed with facts propelled by slingshots of truth, Mold Warriors cannot afford to back down against the massive financial resources of their adversaries. For us, it's about quality of life and in some cases, it's about death. As you learn the truth about mold, the cynical opponents who would have us ignore mold illness will fall like statues of Saddam, pulled down by the strength of the raised voices of millions of people demanding that mold illness and mold patients receive the national attention they deserve.

Already, Congress has heard the outcry from victims of toxic molds. U.S. Rep. John Conyers introduced legislation several years ago that would provide benefits to people injured by exposure to toxigenic fungi. His original bill, called the Melina Bill, has been re-introduced into the House and Senate in 2004 as House Resolution 1268. As part of the run-up to the Resolution's introduction, Congressman Conyers held a press conference and a Congressional briefing Sept. 22, 2004. Along with mold activists from multiple groups and human rights activist Bianca

Jagger, I was asked to provide testimony regarding the human health effects caused by toxigenic fungi. My testimony serves as an introduction to *Mold Warriors* as well.

• • •

"Honorable Representatives, good morning.

"I appreciate the invitation from Congressman John Conyers to speak to you. I would like to talk to you about the 'hidden' health threats posed by exposure to toxin-forming fungi, such as *Stachybotrys, Aspergillus, Chaetomium, Acremonium, Actinomycetes* and other organisms growing inside water-damaged buildings that are our homes and workplaces.

"My comments are supported by my daily medical practice and the results of studies conducted by our research group, The Center for Research on Chronic Biotoxin Associated Illness (CRBAI). My experience with national health policy regarding a different health threat from a toxin-forming organism, *Pfiesteria*, gives me a perspective on human illness and public health that most other physicians don't have.

"You might have heard from authoritative groups such as the Centers for Disease Control and Prevention (CDC) and Institute of Medicine (IOM) that 'exposure to mold illness has not been shown to cause chronic health effects beyond respiratory illness.' They are completely wrong. It is my opinion that we must change public policy to bring change in the treatment of chronic illness caused by exposure to mold. This change would provide much needed help to hundreds of thousands of patients suffering from a treatable, chronic illness. Rep. Conyers' bill, HR 1268, is an important step.

"For those patients made ill from mold exposure for which treatment is not too late, we can make them whole again. For those children with mold illness that impairs their ability to learn, we can restore their ability to progress academically. For those unfortunate patients whose illness has caused irreversible health effects, including chronic fatigue, neurologic damage and rheumatologic injury, in addition to respiratory problems, we must find new approaches to improving their health.

"Many people with mold illness are currently being diagnosed incorrectly with common quick and easy labels: fibromyalgia, stress, depression, chronic fatigue, deconditioning or other unsupported illnesses. Those diagnoses are wrong.

"We offer effective, inexpensive screening with a solid case definition of mold illness. Moreover, our diagnostic protocols give patients a chance to see what is truly wrong with them printed on a lab report. We have identified measurable, reproducible diagnostic markers for mold illness that also help us monitor therapy.

"Fortunately, we now have information on any patient's genetic vulnerability to mold illness. Now we also have validated biochemical markers that document what is wrong physiologically with patients sickened by contaminated indoor air. There is no longer any rational basis for mold patients to be misdiagnosed by the medical profession.

"How can I make such statements? I am a Family Practice physician from Pocomoke, Maryland. Since 1997, I've diagnosed and treated patients made sick following exposure to water-damaged buildings. We found that their illnesses were due to amplified growth of indoor resident, toxin-forming organisms, including a small number of mold species. Patients from all over the United States, as well as international patients, have been referred to me for treatment. To date, I have diagnosed and treated over 2000 patients with mold illness, including 250 children.

"Why do so many people come from so far to see me for treatment of their mold illness? In part, they come because there is no logical Federal health policy, based on sound science that establishes a mechanism for detection and treatment of mold illness.

"How can I claim to know more about mold illness than the IOM and the CDC? I personally examine and interview mold patients; I treat their illness. I see what happens when mold illness strikes and when patients relapse with re-exposure to a school or other contaminated building when just days before they were fully healed by treatment.

"I am able to tell you the biochemical changes that develop after just four hours of exposure, 24 hours of exposure, 48 and then 72 hours as well. Every week, I lower astronomically elevated levels of compounds that cause inflammation and then correct the abnormal immune functioning found in these illnesses. I can tell you what genes make a person susceptible to mold illness. Specifically, those genes are part of the immune group called HLA DR. Practically, those mold illness-genes are found in 24% of the population.

"Members of research organizations who themselves don't treat patients will never have such first-hand experience. Their work will always lag behind clinical researchers who actually provide patient care.

"On Sept. 10, 2003, following my presentation at the 5th International Conference on Bioaerosols, Fungi, Bacteria, Mycotoxins and Human Health, held in Saratoga Springs, NY, Dr. Stephen Redd of the CDC kindly said, 'There is more data here on mold and human health per square inch,' than he had ever seen before. Dr. Redd addressed you in July 2002, as I am doing today, telling Congress that there were no data confirming human illness risk from mold exposure, but that he felt it prudent for individuals to not be exposed to indoor air environments contaminated with mold.

"What is the reason to avoid exposure to indoor mold if illness didn't follow such exposure?

"I asked Dr. Redd if he would permit me to present my data to the CDC in Atlanta. He told me he would wait for publication of the IOM report. I am still waiting for my invitation from the CDC. Today, I ask you, what effort is being made by authoritative groups to link to the clear data rooted in the work of physicians who actually treat patients and who know what microbes growing in moldy buildings do to people?

"I have written several papers on the so-called Sick Building Syndrome (SBS) in collaboration with my research colleagues, statistician Dennis House and U.S. EPA neurotoxicologist, H. Kenneth Hudnell, PhD. Two of those papers are available to you from Rep. Conyers' office today. We have the hard scientific data on causation of illness, showing a direct connection between exposure to biotoxic mold in buildings and subsequent acquisition of illness. I believe you now have the direct science and rock solid evidence needed to change public policy.

"I invite you to look at our studies. I also invite you to meet some real people with real illness. Let them tell you that indoor mold can make people sick.

"Bianca Jagger is here to tell you her story of profound health impairments solely due to mold exposure. I have evaluated her and speak with her permission. She looks normal when you review her standard blood tests. When you look at the labs that actually show what cytokine response to mold toxins does, however, she is a physiologic disaster. From absence of important regulatory hormones, like MSH and astronomical elevation of inflammatory elements such as MMP9 and C3a, to no control over cytokine responses, we can show why she is exhausted with minimal activity. While she may appear to be 'normal,' she is a hollow shell of what she could be, with multiple symptoms from multiple body systems that hang on her like lead choke chains.

"I invite you to go with me to nearby Prince Georges County, Maryland. Let us talk with the 55 police officers made ill by exposure to the fungal contamination in Oxon Hill Station or Clinton Station. Lorrie Taylor will have them all ready for your interviews.

"Go with me to Eastern Correctional Institution, a state prison in Somerset County, Maryland. Let us talk with Sue Donahue and the cohort of sickened workers who were forced to fight for their Worker's Compensation benefits. While we are there, we'll talk with Judge R. Patrick Hayman, who single-handedly forced the State of Maryland to close down the Somerset County District Courthouse because of the massive indoor growth of mold and sickness caused by *Aspergillus, Chaetomium* and *Stachybotrys* fungi, even as the toxin illness was literally blinding his right eye.

"Perhaps you would rather go to Hampton Bays, New York, to meet Pat Romanosky and her fellow teachers at the elementary school with 1000 students. I think you would get an earful from Pat or mothers of children injured by exposure to molds in the school. Be prepared to hear about the absurd barriers against the truth about mold illness erected by the administrators of the school. Some teachers, however, might not want to come forward out of fear. The Hampton Bays case is an incredible example of just how vicious the battles over mold can be. It is also an example showing the arrogance of power that can be brought against those physicians and teachers who say mold is a health problem.

"Consider listening to Melissa MacDonald, former counsel on intellectual property to the House Judiciary Committee who lost her job this year because of illness caused by fungi growing in this very Rayburn Building. If the anthrax and ricin attacks were scary enough to grab the attention of you and the media, consider that biotoxic mold in the HVAC, maybe even here, also maims innocent bystanders.

"Maybe someone you love has an illness with multiple symptoms that affects so many body systems. Perhaps your loved ones have been given one of those meaningless medical labels. Now that you know that mold illness is real, do you wonder if mold might have a role in their illness? Maybe you know someone who works in a moldy building but who is tired all the time, can't concentrate and for whom work is a painful task that leaves them exhausted by lunchtime.

"And if you have a child in a flat-roofed school with water leaks and moldy books in the library, you already know that there are sick teachers and students in that school.

"If these anecdotes still don't create a desire for action, come with me to meet Carol Anderson. She has been so terribly sickened by her condominium at the Ritz-Carlton Residences, located just a few blocks away. I'm sure she would not mind if you watched her become ill again after re-exposure to her condominium. She can't afford any more health hits from mold. Her illness has already taken her vitality and creativity. And no, if she is exposed again, waiting for her illness to appear won't take much of your time. The illness will come on so quickly in her now it will stun you. Why? She is primed for mold illness; we can show diagnostic changes in routinely performed lab tests in just a few hours after she is re-exposed for a few minutes.

"Carol will probably would tell you she is so sick now, without much hope of improvement, that little difference would be made in her 100% disability by the addition of a few more mold toxins even though she would be even less able to care for herself.

"When you finally fund the research on the value of replacement of alpha melanocyte stimulating hormone (MSH) we need so desperately, perhaps then there will be hope for Carol and those others chronically injured from exposure to indoor toxigenic molds.

"Honor the sacrifice made by those who proved that mold makes people sick by talking with Martha Knight of the Baltimore Washington Conference of the United Methodist Church in Columbia, Maryland. She'll tell you first hand all about the mold illness of her fellow workers in one office building. They were treated, only to fall victim to the toxin-bearing spores retained in the documents they took into their second office building.

"You already know the common denominators that dominate these real cases: money and health. Look at the costs in dollars and ruined lives that result from mold. Environmental illnesses, like the *Pfiesteria* cases, force a conflict between those made ill by a disturbed environment and those responsible for disturbing the balance of Nature. Who is responsible for mold illness?

"Mold illness won't go away just because corporate groups deny its existence.

"How about seeing proof with your own eyes? Invite me to your district. We'll make it a point to visit buildings where there are insufficient dollars spent for maintenance on public schools, county offices, hospitals and courthouses. Let me show you how easy it is to screen patients. Remember, the IOM told us in May 2004 that there is a dearth of evidence of illness caused by exposure to mold. Nonsense! Give me the time of you and your staff for a week and I will show you how to identify hundreds of cases of mold illness that are currently overlooked. And better yet, let me screen you and your loved ones. I'll do the testing off the record if you prefer.

"You will see what I see everyday in my mold illness medical practice, if you only open your eyes. As we say in Pocomoke, Maryland, 'The illness is as obvious as a steamroller running over your foot.'

"Thank you for your attention. I hope I will have a chance to meet with you individually to answer your specific questions. If there are questions now, however, I would be happy to take them."

• • •

In *Mold Warriors*, you will find true stories of courageous individuals who would not give in to lack of knowledge from their doctors or economic pressures from employers or insurance companies.

In Chapter Four, you will find the complex physiology of mold illnesses, translated from "Shoemakerese." While this chapter has lots of technical terms, learning the language of mold illness is the first step to understanding why mold illness is so predictably diverse. Armed with this new knowledge of mold illness, you can help your doctor become fluent in mold physiology as well.

In the end, the Mold Wars will be over when primary care physicians open their eyes to new insights about illnesses they have not recognized before. Every time a patient with an unsupported diagnosis of chronic fatigue, fibromyalgia, irritable bowel syndrome, depression, stress or chronic pain syndrome is returned to a normal life by the proper diagnosis and treatment of mold illness, a previously unseeing, but truly caring primary care physician will add his or her voice to those who are demanding societal change regarding mold.

We can no longer allow school administrators to ignore the health and learning threats posed by flat-roofed schools in need of maintenance. We can't permit employers to deny work benefits for occupational illness caused by moldy, water-damaged work-sites. We must demand that construction techniques be upgraded to prevent water entry and trapped condensation and that the deliberate denial of responsibility for construction errors stop now.

Mold illness is so common, yet under diagnosed. It is so treatable, yet therapy is not provided. It is so easily recognized that the failure of our esteemed medical organizations to act to protect the rights and health of so many of us, is a national disgrace.

When it comes to the truth about mold illness, there is no barrier too great for Mold Warriors to overcome. Mold Warriors never give up when they are right. And we are right.

Welcome to the world of *Mold Warriors*. Welcome to 21st Century Medicine.

The Biotoxin

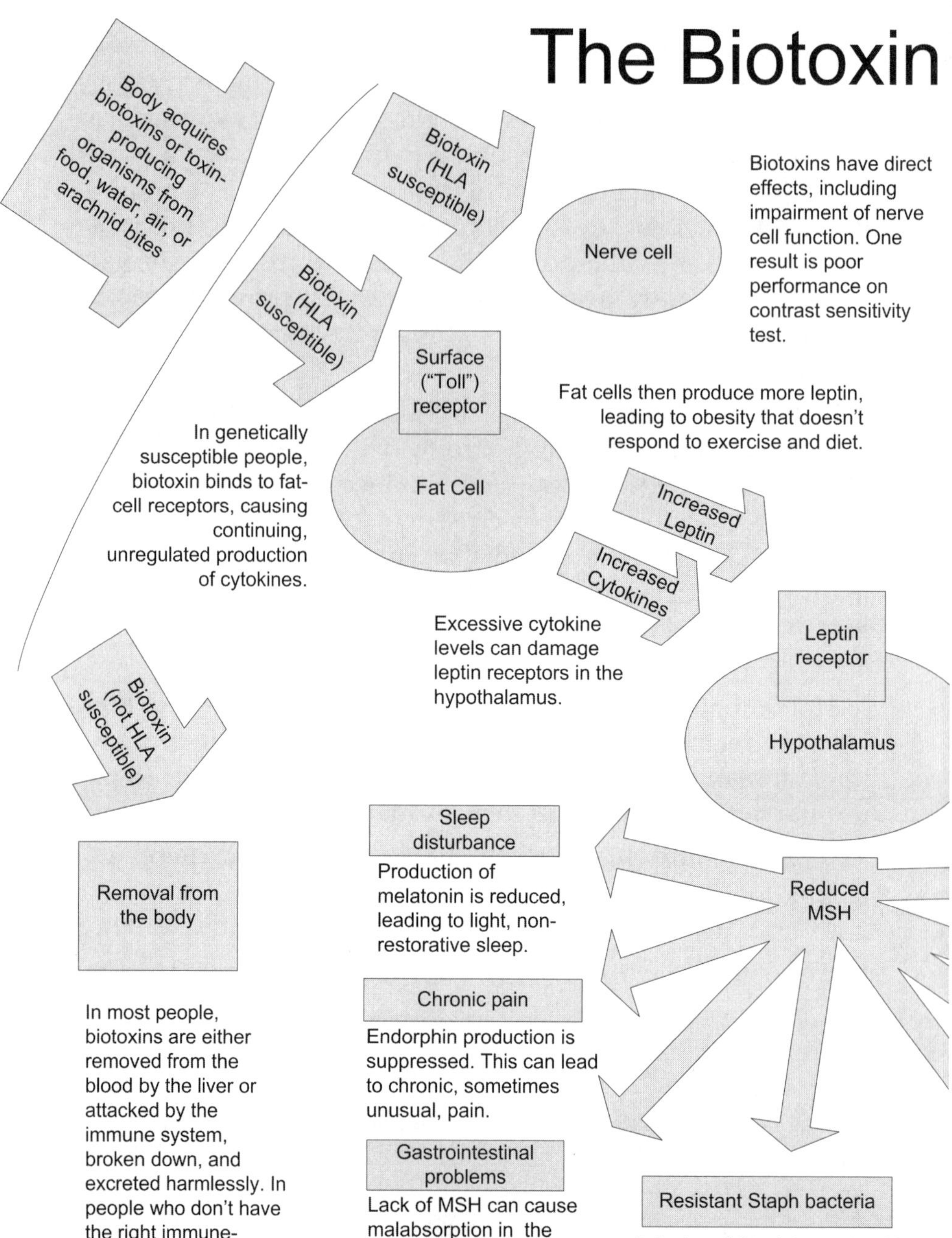

In most people, biotoxins are either removed from the blood by the liver or attacked by the immune system, broken down, and excreted harmlessly. In people who don't have the right immune-system genes, however, biotoxins can remain in the body indefinitely.

Pathway

Ritchie C. Shoemaker, MD
www.chronicneurotoxins.com

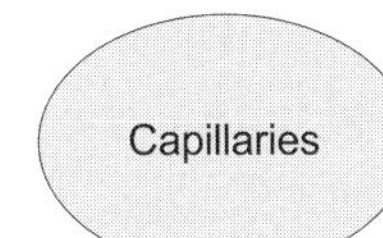

High cytokine levels in the capillaries attract white blood cells, leading to restricted blood flow, and lower oxygen levels. Reduced VEGF leads to fatigue, muscle cramps, and shortness of breath (may be over-ridden by replacement with erythropoietin).

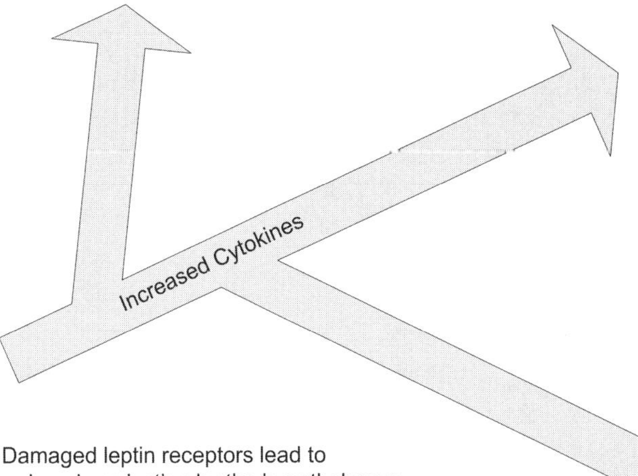

Immune system symptoms

Patients with certain HLA genotypes (immunity-related genes) may develop inappropriate immunity. Most common are antibodies to:
-- Myelin basic protein (often from fungal biotoxins; affects nervous-system functions)
-- Gliadin (affects digestion)
-- Cardiolipins (affects blood clotting)
The “complement” alternative immune pathway may be triggered (detectable as an increase in levels of the protein C3a).

Damaged leptin receptors lead to reduced production by the hypothalamus of MSH, a hormone with many functions.

Cytokine-related symptoms

High levels of cytokines produce flu-like symptoms: Headaches, muscle aches, fatigue, unstable temperature, difficulty concentrating. High levels of cytokines also result in increased levels of several other immune-response related substances, including TNF, MMP-9, IL-1B, and PAI-1. MMP-9 delivers inflammatory elements from blood to brain, nerve, muscle, lungs, and joints. It combines with PAI-1 in increasing clot formation and artierial blockage.

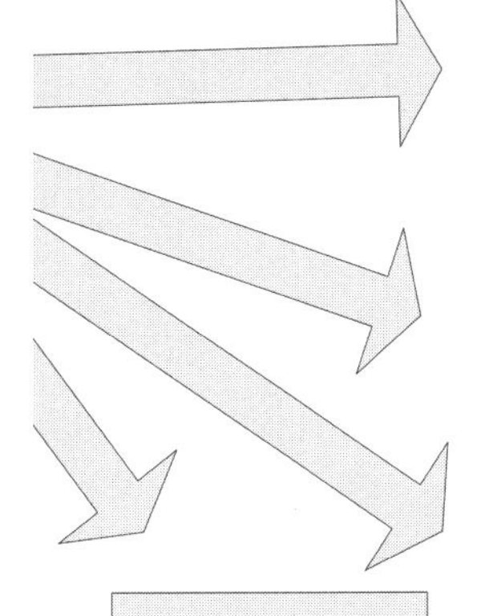

Reduced ADH

Reduced MSH can cause the pituitary to produce lower levels of anti-diuretic hormone (ADH), leading to thirst, frequent urination, and susceptibility to shocks from static electricity.

Reduced sex hormones

Reduced MSH can cause the pituitary to lower its production of sex hormones.

Prolonged illness

White blood cells lose regulation of cytokine response, so that recovery from other illnesses, including infectious diseases, may be slowed.

Changes in cortisol and ACTH levels

The pituitary may produce elevated levels of cortisol and ACTH in early stages of illness, then drop to excessively low levels later. (Patients should avoid steroids such as prednisone, which can lower levels of ACTH.)

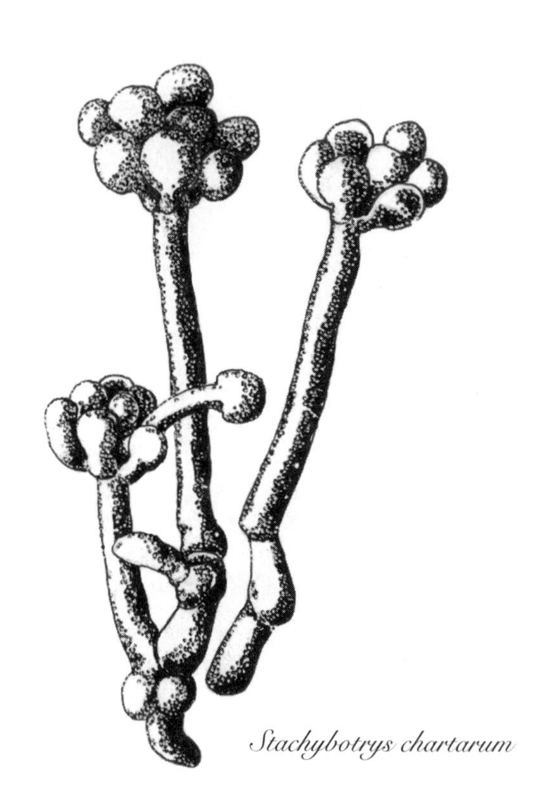

Stachybotrys chartarum

CHAPTER 1

Birth of the Mold Warriors

I'm far from home, walking through isolated woods, when I find a small hole in the ground. I'm curious, and so I look into the rabbit hole to see what's there.

Rather than finding a rabbit, I see that the hole leads to a door. Hesitantly but unafraid, I open it.

I find myself in a whole new world; and being here teaches me simple lessons, brings to my mind images and ideas I've never before seen or thought, and offers wondrous answers to help those back home. This land uses an innovative language completely alien to the old world on the other side; how will I tell everyone about the things I see, when there are no words to describe them?

And what I see is magnificent: waterfalls, each leading to spectacular sights downstream; hills and valleys; and people climbing uplifted mountains distant on the horizon. I realize that things that are impossible at home are possible here.

I know *many people* who have a sickness and wounds that festered forever on the other side of the door. But here, the sense of hopelessness and loss I felt from being unable to help them suddenly lifts, like fog dissolving in a warm dawn. The untreatable suffering responds to balms from novel plants in the new world. Excited, and filled with new hope, I rush home to describe my new findings.

To my surprise, only skepticism and derision greet me.

"A new world full of healing and meaning? Right! Are you some kind of kook?"

• • •

That land on the other side of the door came to me in a dream I had during some of the worst battles I fought for my patients with neurotoxic illness. I'm a family practice physician who found a new cause for some of the most entrenched illnesses of modern medicine—sicknesses given accepted names such as fibromyalgia, chronic fatigue syndrome, Post or chronic Lyme disease, irritable bowel syndrome, multiple chemical sensitivity, deconditioning, hypochondriasis, stress, depression, somatoform disorder, medically uncertain symptoms, as well as many others.

Whatever their diagnosis, my patients describe unremitting fatigue, weakness, muscle aches, chronic pain, cognitive and memory problems, headaches, joint pain, morning stiffness, sensitivity to bright light, blurred vision, mood swings, sweats, unexplained thirst, shortness of breath, mysterious inflammation and weight gain that never goes away regardless of what they eat. Not all patients have every symptom, but none of these symptoms—and there are so many more—ever go away for long.

On a day with few symptoms, and there are usually a few here and there, my patients often try to do something special, like mow the lawn or go food shopping, but they pay for it afterward with several days of exhaustion. Treatments help only temporarily. Antidepressants, asthma medicines, tranquilizers and sleep apnea machines might briefly help, too, but all eventually fail.

Maybe you or a loved one suffer from one of these modern mystery illnesses. Modern physicians don't speak your language, and *their* solutions are impotent. Probably they've ignored your complaints or they've used words like "hypochondriac, whiner," and terms like "malingerer," and other insulting labels. It's even worse when the victim is a child; it's so difficult to watch a mother seek help for her child over and over again.

I found radical new ways to detect and heal these chronic illnesses, which are caused by biotoxins manufactured by invisible microorganisms in our environment. You're exposed to these organisms, perhaps in your home, work, school or local recreational area. When some people think of "environmental exposures" they think of kids sneezing or churchgoers eating contaminated mayonnaise at a Sunday social function.

But the exposures we'll discuss are subtler and not so benign. They routinely destroy health, degrade function at home and work, and split once-tender marriages. Some of these exposures eventually can kill, but only after a protracted illness brings a merciful end to long-term suffering from an undiagnosed cause.

You or a loved one may pick up one of these illnesses by chance; it might be from spores carried throughout your home by a faulty air conditioner; it might be from a pond, lake or estuary near your house. Or maybe a tick will bite you.

Biotoxins are the cause of so many different illnesses; I've come to believe that unveiling their role in medicine is as revolutionary as the discovery of bacteria and viruses. Their massive effects—and their victims—cannot be ignored any longer.

Many diverse organisms make biotoxins, including some algae, spirochetes (like those that cause Lyme disease), dinoflagellates that contaminate seafood, recluse spiders and special types of mold, among others. The molds thrive inside our homes, schools and businesses, and aren't merely limited to the media-hyped "black molds." Many anonymous species are just as dangerous. It is those anonymous species that help make mold the leading source of biotoxin illnesses.

Regardless of the source or the biotoxin's name, toxins can't be measured by routine blood tests. Many types of diseases—not just toxin-induced illnesses—can not be found immediately by simply looking at a few blood samples. But ignoring biotoxin illnesses due to a lack of basic lab tests is both foolish and dangerous.

As future research unveils the layers of complexity of these illnesses, perhaps the mechanisms I'll describe within these pages will seem primitive. Why would that surprise us? History repeats itself in medicine, too, after all.

Don't forget, it took massive resources to identify the HIV virus. Research continues to expand the basic science, making the HIV discussions of the early 1980's seem primitive. And even the physicians who've been willing to learn the new language of biotoxins have struggled to keep pace with the ever-increasing amount of biotoxin-formers in our everyday environment.

Despite its inability to be measured easily, the damage to our bodies caused by biotoxins is easily shown by indirect methods. Like the nuclear physicist who "sees" a neutrino by the path it leaves behind, experienced toxin-treating physicians can see the distinctive fingerprint the biotoxin leaves behind; just like a fingerprint, it's not like the trail left by anything else.

Some people escape getting toxin-induced illness because they have genes that remove the toxins before they can do any damage. But unrelated friends living or working in the same place, that don't have the protection of those genes, may become unable to function or merely tired after exposure.

Regardless of whether you have a genetic vulnerability, this book will teach you to identify the clear signs of biotoxin-induced illness in you and your loved ones. You may be stunned to find that many suffer from biotoxin-induced illness. This book will also teach you how you can eradicate those toxins.

• • •

Let's look at what we're fighting as Mold Warriors—specific types of mold growing indoors in buildings with too much moisture. Molds appear to be primitive, single-celled organisms feeding on rotten oranges and week-old bread that "aren't important" in the overall scheme of life and death on Planet Earth. But without molds, yeast, mushrooms and fungi—and they're all related—we wouldn't have much life. And it wouldn't be just because we wouldn't have beer and cheese!

Molds might ruin a loaf of bread or cause a brief bout with allergies, but usually molds growing on food or in the outdoors environment are harmless. The few mold species that can cause infections in a few people aren't our concern here, either. The small numbers of mold species that live and breed indoors and routinely make toxins are our real concern.

Molds are as complex as any other animal. A mycologist, a scientist who studies molds and fungi, might say they're even more complex. They reproduce by making seeds (spores). They send out enzymes from within their sturdy outer walls to digest the food around them. Smaller pieces are easily used by the fungi, not big pieces. In other words, they eat backwards, bringing in pre-digested nutrients. As they digest their landscape, they break down and recycle huge amounts of organic material, supplying a tremendous amount of available food that other organisms use for fuel. It's no surprise that bacteria and a host of other micro-critters like to live around molds. And it's no wonder, too, that molds make antibiotics, including the familiar penicillin and cephalosporins, to keep marauding bacteria at bay.

Like any animal, molds need food, moisture and protection, which is why our homes' indoor temperature is ideal for mold growth. Cellulose, found in the paper and wood products used to build most homes, is their favorite food source; they can unlock all the sugars trapped inside. We wouldn't think of living out in the open, and neither do fungi. They like it nice and dark.

In any given house, just because you don't see molds, doesn't mean they aren't there, munching on drywall, insulation and plywood. When mold is happy, it's ready to breed. It sends its offspring adrift as spores. Pity the poor spores; how can they survive the aggressive predators who love to eat them? But put a toxin —call it a mycotoxin or biotoxin, if you prefer—on the spore. Now send it away with its self-contained army of biological weapons on board. The spore now has a fighting chance.

Primitive? Perhaps, but attaching a toxin to spores is an age-old, successful strategy for survival.

You'll see that the health problems posed by molds in our buildings are best corrected by information, preventive action, reasonable building techniques and prompt cleanup of water damage. Repairing the damage—called remediation—is expensive only *after* the building is damaged. Preventing something bad from happening is always cheaper than remediating it.

Mold Warriors will tell you how mold causes chronic illnesses; you'll meet a few people who were very ill, most of whom have regained their health. You'll learn about the special testing and the treatments that helped them recover. You'll give up the naïve notion that mold illness is rare and you'll see that the main battles about mold are about money. No one wants to accept responsibility for someone else's mold illness, so you'll watch some legal games, some more despicable than you can imagine, as they're played out in court. You'll also learn that the new world of medicine provides some answers to the many questions we still have about mold illness.

Mold illness is real, not rare.

The story of the Mold Warriors—patients, physicians, indoor air experts and lawyers—begins in rural Pocomoke, a town with three stoplights along Route 13 on the Eastern Shore of Maryland. My family practice began there in 1980. Few of my classmates at Duke Medical School in 1977 wanted a primary care practice; for me, it was an idyllic life. The wetlands and swamps drained by the blackwater Pocomoke River were my ecology classroom. I loved small-town life, where patients said hello as friends in the grocery store and where a handshake was better than a signed contract.

Beginning in late 1996, though, a change in my practice and my life arrived with new blooms of a toxin-forming dinoflagellate, *Pfiesteria*. This algae-like creature had killed millions of fish in the estuaries of North Carolina and now it was active in our tributary of the Chesapeake Bay. We heard about some research scientists in North Carolina getting sick from tanks of *Pfiesteria* growing in their labs. So when some of my long-time patients started telling me about their memory problems, fatigue, coughs, chronic pains, diarrhea and sensitivity to bright light, it made sense to wonder about human illness from that toxin-former.

Even as the illness reports were picked up by the *Washington Post* and the *Baltimore Sun*, but loudly denied by the States of Maryland, Virginia and North Carolina, luck was on my side. One of my first *Pfiesteria* patients had terrible diarrhea along with her headache, memory problems and cough. Nothing had helped her—antibiotics, Pepto Bismol©, Imodium©. I gave her some cholestyramine, a cholesterol lowering powder that was one of my favorite diarrhea treatments because it was such a terrific bowel plug. Sure enough, when she called a few days later, her diarrhea was gone, as expected.

But she said her memory was better, her cough stopped and her headache was gone, too. Hhhhhhmmmm, I remember thinking; what was really going on here?

Cholestyramine isn't absorbed into the body; it simply binds cholesterol and bile salts in the intestine, preventing reabsorption further down in the bowel. Did cholestyramine bind something made by *Pfiesteria*? A toxin, perhaps?

One after another, patients from all over the Chesapeake Bay came to Pocomoke for evaluation and treatment of the new mystery illness. I didn't have any markers to use in diagnosis; nothing was ever abnormal in the lab tests. I only had symptoms, exposure, absence of other explanations for the illness and response to cholestyramine to support my claims of a new biotoxin-induced illness.

There wasn't a surge of endorsement of my work among the academics from Johns Hopkins or the University of Maryland, even after I published papers on diagnosis and then treatment of the *Pfiesteria* human illness syndrome. The Centers for Disease Control and Prevention (CDC) got involved in 1998, calling my treatment protocols "premature," and convening a panel of health officials from a variety of East Coast states to try to define "my" illness.

Despite the fact that none of the expert panel members had ever diagnosed or treated any *Pfiesteria* patients, the expert panel came up with guidelines for the new illness, calling it "Possible Estuarine Associated Syndrome (PEAS)." It didn't matter that the panel failed to recognize the complexity of the illness, ignoring many of the symptoms reported in my paper. What was important that the new illness had a name that organized medicine had to accept.

My work wasn't popular with politicians or the tourist industry, either. Who wants to eat *Pfiesteria*-laden fish or seafood from a sick Chesapeake Bay? Who is that kook doctor, anyway? Speaking for the State of Maryland, one PhD, suggested I needed to "learn more about science," even as a state-appointed group of physicians confirmed my findings.

Patients continued to come to Pocomoke for treatment. As the numbers of successfully treated *Pfiesteria* patients piled up, so did the criticism from the State-employed scientists. Why did I have so many patients when no one else did, they wondered?

A quick look at my waiting room would have given them the simple answer. One *Pfiesteria* patient after another went to one physician after another seeking help, only to be told their illness was all in their head—maybe they were just stressed out or depressed. Some were told they had fibromyalgia or sleep apnea.

No, they didn't; they had PEAS. No wonder they came all the way to the Eastern Shore. They wanted to feel better! And patients were feeling better on cholestyramine (CSM).

Maryland responded to continued criticism of their failure to act to protect human health by starting a study that followed both patients exposed to estuaries where *Pfiesteria* might have been active and other patients, who were not exposed. No one was treated. As exposed, sickened study patients who had received no treatment talked to others with the same illness who'd been easily treated; one by one they came to Pocomoke to enroll in my protocols.

I didn't add to my list of admirers when I commented at a 2000 CDC conference on *Pfiesteria* that failure to treat patients by a state—any state, not just Maryland—was no different than failure to treat syphilis in the infamous Tuskegee, Alabama study.

But what I said was true.

Help in the *Pfiesteria* Wars came from Ken Hudnell, PhD, a neurotoxicologist at the U.S. Environmental Protection Agency (EPA) in Research Triangle Park, N.C. Ken had long been interested in assessing the neurological functions of vision. In fact, some of his work on visual contrast sensitivity (VCS) testing from the early 1990s showed it was a powerful tool in demonstrating the effects of toxins.

Contrast is a brain function; essentially, your eyes and brain work as a team to distinguish black from white and dark grey from light grey. By using a special card that only measured contrast, measuring the ability to distinguish a light grey line from a darker grey line on a light grey background, Ken could test the contrast part of vision without involving all our common eye limitations. In other words, his special card removed near, far, static, motion, peripheral and color vision problems so he could analyze the effects of toxins on vision. Basically, Ken's VCS test turned vision into a neurologic yes/no test that showed distinctive deficits in toxin-exposed patients, but not in unexposed people.

When Ken found VCS deficits in *Pfiesteria* patients, he and I quickly collaborated on reporting on a series of patients as a Grand Rounds in Environmental Medicine for the prestigious journal of the National Institute of Environmental Health Sciences (NIEHS), Environmental Health Perspectives (EHP) in May 2001. We followed that paper up with another, which also appeared in EHP, using VCS and symptoms to *screen* patients potentially exposed to *Pfiesteria* and compare them to controls. We had a powerful tool and a compelling statement: we could use VCS as a marker for the illness, show improvement after treatment with cholestyramine, show recurrence of the deficit with re-exposure, and show repeat recovery of patients with re-treatment.

It wasn't too long (1998) before I started treating Florida patients who were exposed to blooms of a newly invasive species of blue green algae, *Cylindrospermopsis* that was overrunning Central Florida lakes. When you added up their exposure to toxins from the lake, symptoms and VCS deficits, without confounders and their response to cholestyramine, it was a crystal clear marker for the illness caused by those toxin-forming algae.

The State of Florida didn't want to acknowledge my work, even though I reported it to the CDC and to various state groups. Curiously, when the state was awarded a $3 million grant from the CDC in November 2003 to train physicians in diagnosing and treating patients exposed to algae, no one from the state called me to ask for input, even though by that time I had the largest case series of diagnosis and treatment of blue-green algae patients anywhere. I'd shown those data to the Florida Department of Health experts and had been invited to give talks on the subject by State agencies and people living close to the algae-covered lakes in July 2003.

Later in 1998, a few patients with symptoms and VCS deficits following exposure to buildings with lots of mold growth indoors came in for evaluation. Their symptoms were no different from those with illness from dinoflagellates and blue-green algae. Did they have a biotoxin illness as well?

They sure did.

If I thought the *Pfiesteria* Wars in Maryland were rough, or the Blue Green Algae Non-Wars in Florida were a tragic failure of a state to do the right thing, they served an important purpose: preparation for the intensity and viciousness of the Mold Wars to follow.

The stimulus for the next step in understanding biotoxin illnesses came from Lyme disease. In 1999, researchers Dr. Sam Donta and Dr. Mark Cartwright found a biotoxin made by *Borrelia burgdorferi*, the causative agent of Lyme disease.

Lyme disease was first recognized as a distinct entity in 1975 even though there were many reports of it earlier. Like other biotoxin illnesses, it creates multiple symptoms and organ effects, as did biotoxin illnesses caused by dinoflagellates, blue-green algae and fungi. Many Lyme patients took antibiotics for the infection and were cured. Many others took antibiotics, but didn't improve at all. Still others afflicted with multiple symptoms were diagnosed with Lyme disease and treated with long-term antibiotics; some improved, but most did not. Could Lyme disease involve a toxin that could be treated with cholestyramine?

Beginning in May 1999, I treated 112 patients with symptoms refractory to antibiotics following a tick bite or exposure to areas where others had tick bites. In a paper presented to a regional meeting of the American Society for Microbiology in April 2000, *my data clearly showed that Lyme disease also involved a biotoxin*. As opposed to the benign course that followed treatment of mold patients with cholestyramine, however, over half of the Lyme patients initially felt much worse.

Now what was going on?

The answer was that biotoxins have the ability to disrupt the mechanisms our body uses to produce inflammation. I learned that health and recovery from illness required a close look at pro-inflammatory cytokines (called cytokines for simplicity here), matrix metalloproteinase-9 (MMP9), and tumor necrosis factor alpha, (TNF; more on these compounds later). The worsening of the Lyme patients after taking cholestyramine, complete with changes in blood levels of inflammatory chemicals, allowed me to identify critical features of the abnormal immune response in biotoxin illness. Clearly, biotoxins interact with and disrupt *both* our inflammation system and our immune response causing them to act without control.

Just like the mystical wonderland world beyond the unopened door, biotoxin knowledge is new. It unveils 21st century illnesses —fibromyalgia, chronic fatigue syndrome, chronic Lyme disease, autoimmune disorders, treatment resistant anxiety and depression, leaky gut, "medically uncertain symptoms," and so many other health challenges as biotoxin illnesses. The inflammatory cascades that underlie many other modern illnesses, including atherosclerosis, diabetes, obesity and multiple sclerosis, are off-roads of the main Biotoxin Pathway.

The Mold War battles that followed the discovery of the mechanisms of biotoxin illness follow the same themes as the *Pfiesteria* Wars. We'll find money and vested power opposing the truth about environmental illness and what caused it. In *Mold Warriors*, we'll find deliberate deception, professional jealousy, hired guns masquerading as professional witnesses, governmental intrigue and stories of valiant struggles, as one David after another fights against a horde of Goliaths.

In short, you'll find in *Mold Warriors* the themes repeated throughout the history of the struggle for a new truth as it emerges from the cauldron of skepticism.

• • •

As I left the wonderland and shut the door, I took notes about what I had found, putting special emphasis on material about inflammation, cytokines, hormones and antibodies. I clutched them close to my heart, since many lives hung in the balance.

I returned to the old world, intent on sharing my discoveries. I'd gathered materials on curing unique illnesses, and recorded the highly specialized lab results that would reveal the cause of many uncertain illnesses. I began to use these findings and saw clinical improvement and cure in many. So I shared with friends and peers the stories about hopeless people with challenging illnesses who had recovered.

As I shared findings from my trip to the previously unseen land with medical experts and specialists in toxicology, I expected everyone to be amazed and awed.

Instead, they told me, "These fungi aren't new, they're everywhere. They don't make us sick. We've never heard of your biotoxin words or your lab tests. Why should we take time to learn new ways to heal?"

Why don't they see that some species of fungi have developed the ability to release toxins to survive? How is it the naysayers reject new ways of healing? Do they believe they learned every truth in medical school and there's nothing new to learn?

Undeterred by those who won't believe, I continued to provide evidence supporting my claims. One by one, I showed others the entry door and the beauty of the new world. Soon, they see what I see; in time, we all learn that the new world has its roots in the old world medicine.

But when I began to fight for recognition from the old world, its many protectors went on the attack. Perhaps they feared the powerful insights provided by the new world would render their "old" knowledge useless and their methods out-of-date.

The battles for truth began in earnest then, as more people recognized the lessons from the new world were valid. Gradually, as patients, physicians, researchers, government officials and attorneys slowly came to see the power of the new medicine in fighting biotoxins, many joined forces to give birth to a new approach to Nature and Healing.

And so the Mold Warriors came to be.

In the end, truth wins out. Molds or fungi, make us sick. As more people see that, the numbers of Mold Warriors grow.

This war isn't a fantasy or a make-believe computer game. The mold pioneers have opened a new window on medicine, one in which fungi flourishing indoors make biotoxins that cause chronic illness. The stories in this book are true and the battles are real. Many still continue.

Proving that mold causes common illnesses that are rarely recognized by physicians has been a challenge. Swimming upstream in the white water rapids of current medical opinion has not been easy. Proving that water-damaged buildings are harmful to some people, against the massed forces of building owners, insurance companies and their lawyers, each with a cadre of highly-paid "experts," has taken six years. Along the way, withstanding unfounded criticism, unsubstantiated accusations and unwarranted personal attacks from the old world has become a way of life for Mold Warriors. But that life has brought victories in courtrooms and in recovered health of patients for whom hope had been a lost four-letter word.

Indoor growth of toxin-forming molds makes people sick in ways most physicians don't yet fully understand. The causes of the illnesses include domino-like immune responses to biotoxins and all of the later effects those additional responses create. Learning about mold illnesses (or Sick Building Syndrome) will first involve learning some new words and about some exciting new findings in this area.

It's time for you to come take a walk through the gardens and walkways on the other side of the rabbit hole.

Learning about biotoxins may change your life, or start a new life of healing for your child, spouse, sibling, parent or friend. A few ideas and words will initially seem new, but understanding them will give you power. Once you know about biotoxins, current answers to complex health questions will be exposed as useless. You'll understand biotoxins and how our chemically altered world is their source. You'll become fluent in what biotoxins are and how they hurt so many of us. If you're sick from biotoxins, you'll learn what you need to know to remain healthy or to regain your health.

And if you have one of the illnesses I mentioned earlier—chronic fatigue syndrome, fibromyalgia, irritable bowel disease, unusual anxiety and depression—read carefully; you may have been misdiagnosed. You may actually have a biotoxin-induced illness. Many patients who've been mistakenly diagnosed with these mystery illnesses have come to see me and have proven to have treatable, biotoxin-induced illnesses like Lyme disease. Many have improved on the protocol I've developed.

While there's still much to be done, the story of the Mold Warriors is finally ready to be told.

Aspergillus

Mold Invader

By Pam Carson

8/23/04

Body alarm sounds
  sending helmet headaches
  as the nightmare begins.
Eyes plagued by visual disturbances
  and light sensitivity.
Vanishing memory, skill gaps come and go.

Paralyzing depression
  unrealistically framing life as hopeless.
Speech slurs and slip-ups
Disobedient words pass lips
Sent forth on a mission
  never reach their destination
  fall on ears as distorted clones

Disaster descends to weeping nose scabs,
  a throat on fire, infected sinuses
  to congested lungs.
First soprano solo voice has been taken away,
  laryngitis its shadow.

Chills replace drenching sweats.
It's not menopause.

Dizziness; weakened, numb, tingling limbs
  crippled by airborne assailants,
  lose their flexibility
  guarantee stairs an agony.

Joints that ache and burn
The limp—the harbinger
   of legs that eventually won't work at all.
Legs at night
Unwillingly house traveling electric maggots.

Skin that hurts to touch
   becomes reddened and raw
   sports rashes that ooze sticky brown

Recipient of patronizing explanations
   for phantom symptoms evading
   diagnosis and treatment.

Ordinary tasks become Olympic events
   as gripping lids, brushes and pencils
   becomes a formidable mountain.

Puzzling indifference of coworkers and friends
   who bear same symptoms,

Crushing emotional pain to hear:
   "There is nothing wrong with you!"
   "You don't look sick."

Failed VCS test is the ticket to homecoming
in Pocomoke City
   as Dr. Shoemaker welcomes me:

"You've come to the right place."

CHAPTER 2

Proving Mold Makes People Sick

The Battle Begins

Hampton Bays, Long Island: November 2002

Perched behind the wheel of her 1997 Olds Cutlass wagon, Pat Romanosky stared at her clattering windshield wipers and then cursed out loud.

"This rain ... I'm so sick of it!"

It was a cold November. Dark silver clouds and chilly Atlantic Ocean winds blew past her car. *Another grinding winter in beautiful, downtown Hampton Bays,* she thought.

Waiting at another traffic light on the short ride to work, the 51-year-old Romanosky slapped the dashboard in utter frustration. What if she threw her head back and yelled at the other frazzled drivers around her, "Look at us—we're nothing. We're nothing more than work dogs ... sheep!"

Glaring at the rain-slicked pavement, she tried to take a deep breath—and immediately began coughing again. They were deep, wracking spasms, and with each one, her tortured lungs felt like they were on fire. It used to be just a mild raspy throat and muscle aching, but now her lungs were killing her. When would this coughing and wheezing stop? Just a few stairs produced shortness of breath, watery eyes, muscle aches, insomnia and now memory problems. For the past six weeks, the veteran Hampton Bays Elementary School phys ed teacher had been struggling with what she had described as, "The Mother of all Flu-Bugs."

Nasty? You bet. And she wasn't alone.

Pat was on her way to teach at one of Long Island's largest elementary schools, with more than 1000 students in grades K through 6. Ever since it got cold, about 50 of the 200-plus teachers and staff had been coughing, aching and fighting off endless attacks of maddening diarrhea. Like Pat, more than 25 percent had blurred vision and bouts of mental confusion. For Pat and her students, merely concentrating on a blackboard today was a monumental strain.

Were the kids getting the same germ as Pat? In the past 31 years of her career, Pat had never seen kids so limp and fatigued —even during play. Something was going on with them; something was wrong.

The allergy medications and expensive antibiotics Pat kept taking all failed eventually. Her co-worker, Lisa, had been in her doctor's office so many times that the doctor now simply called in another prescription, no office visit necessary. Pat and Lisa knew something was very wrong.

"*Am I over-reacting*?" Pat wondered. Flu-attacks were a common occurrence in November and December in students and teachers alike.

Maybe.

But this school year was different. For one thing, the student absentee rates had been recently soaring, double that of past years. And so many teachers had been calling in sick lately that some days "subs" outnumbered the fulltime teachers.

Frowning now—she absolutely *hated* to be late—Pat reached for another tissue box while the traffic crawled. Wearily, she watched the red taillights winking through the rainy haze.

She'd had 12 hours sleep, and yet, as she drove to school, she was utterly tired, woozy and out of sorts. Everything hurt, and she occasionally felt strange stabbing pains, like ice picks from out of the blue.

"What was that?" she thought.

Maybe it was some new type of flu. Maybe Hampton Bays had been attacked by some unusually powerful strain of influenza—the words "Hong Kong Flu" drifted slowly through her dozing brain. Maybe some new bug had yet to make it to the local news.

She put her foot on the accelerator and eased the Cutlass forward a few feet. Another day of feeling tired, foggy and just not right. She just had to get through another day ...

"If I only had some answers," she thought. But Pat was too tired to search for them.

Tracking a New Kind of Disease: Toxic Mold

While Pat Romanosky struggled through her day at Hampton Bays Elementary, I sat in front of a microscope in Maryland and asked myself a disturbing question: "*What the hell is going on here?*"

Perplexed, I leaned forward and took another look at a series of "tape lift" specimens of mold from a courthouse that I'd been studying for the past several days. As I slowly focused the instrument, the clump of mold on the tape, magnified 400 times, gradually came into view. Fascinated, I noted the delicate, almost alien structure of the "fruiting bodies" and the appendages that formed the key anatomical features of this increasingly common toxic mold. I couldn't get around the images from old sci-fi movies, featuring creatures from distant planets forming "flowers" that then sprayed lethal gases and chemicals at helpless Earthlings. The fungi were like something from the Little Shop of Horrors.

"Stachy," (*Stachybotrys*) was an especially nasty fungus—a moisture-loving, poison-manufacturing organism that thrived in indoor places where leaky roofs, trapped condensation or porous basements allowed water to seep into a building's structure, permitting mold to grow. It damaged drywall, insulation and wood. If a building part had cellulose, fungi ate it. But molds could thrive almost equally well on metal air ducts, too. It made sense: fungi like to live at the same temperatures we do. All they needed to grow was protection from predators, plenty of moisture and food.

We commonly give molds all three. We grow them with water leaks in any part of the home or in a commercial building. We then spread our crop of indoor fungi around by using heat or air conditioning (HVAC) ducts to carry spores to other waiting fields in homes and workplaces. After being carried throughout your entire home and workplace by the ductwork, they at first grow like invisible weeds on carpets, papers, sofas, file cabinets, photographs, drywall, and ceiling tiles—everywhere.

With a sheltered indoor environment, the spore seeds lack major competitors. Some of these spores are from a very few species of biotoxin-producing fungi. The indoors environment kills other mold species so that the toxin-formers own their turf like a gang. Without natural competition found in wild woods or even your home compost pile, the indoor fungi grow explosively. They spew billions of spores every week! So many of our buildings with wet basements, leaky roofs, in-wall condensation or plumbing leaks are ripe for hostile takeover by fungi, it's no wonder we're always hearing about "Sick Buildings."

Beneath my microscope, this toxic invader looked like the "twist-up" balloons that delight children at fairs and carnivals. Imagine a foot-long, rubber sausage with one end twisted into floppy ears to form a grinning dachshund, and you'll have a reasonably accurate mental image of Stachy-under-the-microscope.

But these microscopic sausage-forms weren't cute, and they certainly weren't going to be delighting any children. Packaged on the outside of the reproducing spores were biotoxins—biological warfare defenses the fungi send forth to guard its young.

Even if the spore died, the biotoxin stayed, ready to unleash its poisonous effects on whatever—or whoever—came in contact with it. Biotoxins were forever.

Along with a handful of others like it (including *Aspergillus*, *Penicillium* and *Chaetomium*), the fungus on my tape lift was a serious poisoner: a fungal invader armed with a suite of powerful biotoxins that could make human beings very sick, indeed. Although most of the medical community didn't realize it yet, these formidable fungi were the culprits in a growing epidemic of biotoxin-related illness, of which "Sick Building Syndrome" or "SBS," was but one example.

I couldn't stop asking, "If there were *thousands* of kinds of fungi growing outside, why were there only a *few* species that were *always* found in buildings where I also found sick people?" Where was the biological diversity of the indoor fungi? Did the biotoxin-makers have some kind of monopoly on water-damaged indoor environments?

But answers were slow to come. While Pat Romanosky struggled to get to work each morning in Long Island, fighting an impossible fatigue just to get out of bed, I was already treating scores of Maryland patients who'd complained of symptoms linked to indoor mold. Only a few days before, in fact, a frightened mother had dropped by my office seeking help for her 12-year-old son, Bobby.

The worried mom explained how her son had been experiencing "a whole lot of trouble at school."

"They told me he's got attention deficit disorder, Dr. Shoemaker, but that just doesn't make any sense," she said. "Bobby's always been a terrific student, and he usually brings home A's. But about two months ago, he started getting sick all the time. And then he couldn't seem to concentrate. He has no energy, can't learn or remember things. His math and reading comprehension have fallen off sharply. He seems unable to learn. Nobody can figure it out. Our family doctor wants to put him on that Ritalin, but I think he has some kind of disease we haven't figured out yet."

Bobby had the same symptoms as dozens of other patients I'd treated during the past few years. By the fall of 2001, I'd already treated more than 500 such patients. They reported the same headaches, muscle cramps, blurred vision, mental confusion and inability to concentrate that troubled Bobby. Just a few months before, I told Dr. Ken Hudnell, my neurotoxicologist research ally from the Environmental Protection Agency, that the diagnosis of Sick Building Syndrome was an easy one to make—it was as obvious as a steamroller running over your foot, once you starting asking the simple questions that defined the illness.

You just had to ask specific, basic questions. Using my patient computer system, a clear pattern soon emerged.

Was Bobby another SBS victim? He had toxic mold exposure and the many typical symptoms—fatigue, headache, cough, confusion, memory problems, unusual belly pains and muscle cramps. Growing pains? No growing kid should have all these problems.

Sure enough, when I asked Bobby's Mom about mold in their home, she quickly described a "black, fuzzy powder" that had been growing along the edges of the poorly assembled doorway of her pre-fab home. After a period of recent drenching rains, the mold was now climbing up the side of at least one living room wall.

Within a matter of hours, I found myself standing beside Bobby's mother in her musty living room, carefully lifting mold-samples from the wall with adhesive tape. His Mom was right; Bobby had an undiagnosed illness. The silent, invisible disease-causing mold was growing without any restrictions—and it was releasing toxins.

After initial bewilderment about what I was seeing in so many samples like the mold from Bobby's home, the secret was revealed. My recorded roster of more than 100 patients from over

43 buildings served us all, by leading me to mold samples in many other buildings, all over Maryland and Virginia. These mold-filled buildings made me realize again that biotoxins can come from many sources and from *entirely different kinds of organisms*. Mold has bloom conditions, just like *Pfiesteria*, my first biotoxins teacher, which wiped out massive numbers of fish and made many people very ill. Mold and moisture means mycotoxins; the sick person is proof.

And yet to public health bureaucrats, problems in patients like Pat and Bobby didn't exist. Ever feel like your doctor doesn't get it? Today's modern medical professional often isn't able to be a patient advocate. Large organizations such as the Centers for Disease Control and Prevention (CDC) and the National Institutes of Health (NIH) don't always react to new information with open arms because their funding depends on carrying out only those tasks assigned by government. That's why getting bureaucracies to agree with something new and important is like pushing a parked school bus uphill. Remember the first research reports into the cancer-causing chemicals in tobacco smoke? People like Pat and Bobby just don't fit with "established" medicine.

The CDC and NIH opinions dictate medical policy for the U.S. government. Most of their experts continue to insist that the experiences of countless patients like Pat and Bobby were not acceptable as evidence that mold causes human illness. According to these organizations, there was no absolute confirmation of cause and effect. From their Ivory towers, they would explain that the research studies suggesting that there was proof that fungi made people sick were "flawed" in one way or another. Unfortunately, when I read those studies, I had to agree—the initial studies were weak. The proof wasn't there. Sure, there were studies showing moldy buildings had lots of sick people, but good scientific studies, *strong enough to convince the skeptics*, were glaringly missing. But as a Family Physician, I still had to treat all those patients who came to me. I didn't have the luxury of ignoring them, not even for a week. I was a treating physician, not just a bystander expert with a degree on the wall.

But during the past several months, I'd been sending the bureaucrats at both agencies dozens of case studies that show how mold was making people sick and what could be done to relieve the chronic symptoms. What was the CDC response? Nil. While *Stachybotrys* and *Aspergillus* and their nasty cousins flourished in the dark—and while an ever-increasing number of offices grew debilitating toxins—was the Medical Establishment asleep?

Sitting in my work-cluttered lab that morning in November 2001, I knew I faced an uphill task that would challenge Hercules. How did mold-attacks hurt people? What caused the spread of biotoxins to the human victims, and how did the toxins operate, once inside a person? How could we measure the illness in our offices?

Sure, the cholestyramine (CSM) protocols I'd perfected in the *Pfiesteria* epidemics worked well as the first step in making these patients better, but where were the laboratory tests that showed the mark of the illness? And why was it that not everyone exposed to mold toxins got sick? And why did some exposed people escape chronic toxin mold illness after a few days of being sick? And why did the illness look so different in different people? Some people had severe pneumonia-like symptoms; while others looked like they had fibromyalgia and Chronic Fatigue Syndrome or some seemed to have rheumatoid arthritis.

And then there were the many eccentric biotoxin symptoms or "diseases" that surprised everyone, including me. These mold-induced ailments baffled the neurologists and many other sincere physicians. Some patients had Parkinson-like movements, unusual tremors or bizarre muscle jerking. Others included kids with unusual "seizures," or people with "multiple sclerosis," complete with neurologic defects and scars on the brain, until they were treated. The brain lesions stayed, but the symptoms disappeared. Among my treatment "successes" were people that had once been labeled with all these diagnoses.

So many questions and inexplicable patient symptoms—enough for a career!

• • •

I was putting my slides back in the cabinet when the telephone rang.

"Hello, this is Dr. Shoemaker. How can I help you?"

"You don't know me, doctor," said the voice on the telephone, "but a friend in my office at the Somerset County Library recommended that I give you a call."

The library again? So far, I'd treated several Princess Anne County library patrons for SBS-like symptoms: blurred vision, muscle cramps, headaches, confusion, short-term memory loss and unexplained weight gain. Before treatment, they each averaged more than 15 health symptoms. Only one thing connected these patients and the illness—the library. As I found out later, all of them frequented the same reading room in the downtown library building.

When they were treated with cholestyramine, they'd start feeling fine. But if they went back into the library without the cholestyramine shield, they'd be sick again within days. The repeated cycles of illness and recovery, illness and recovery that paralleled exposure and re-exposure was clear proof that the library was making them sick.

"I know you're busy, Dr. Shoemaker, but how soon can I get an appointment?" she asked.

"Can you get down to my office this afternoon?" I asked.

"Today? You mean it? Just name the time," the woman said.

"I'll see you at four o'clock," I replied.

"Help Me Make It to the End of the Day!"

After 31 years of teaching at Hampton Bays Elementary, Pat Romanosky felt like it was her first day. Pat now had to learn how to concentrate and pace herself so she could last through the long school day. *Just take it one class at a time, and don't hurry,* she told herself. *Three o'clock will come soon enough, and then you can catch up on your paperwork and head home ... Maybe Joe would cook again tonight if I can't get off the couch again.*

As soon as she walked into the ancient gymnasium where she taught most of her classes, Pat's heart sank. Originally constructed in 1923, the decrepit gym was a creaking wreck—a rundown dinosaur of a building in which clanking steam radiators hung from discolored walls and some of the grimy windows hadn't been opened in decades.

Who knows when the gym began to make so many people sick? The roof had leaked forever, but it didn't do any good to complain about it. And once again today, the temperature in the musty gym soared past 90 degrees, due to the steam-heat that remained trapped inside. She looked at the black marks on the walls, like side-growing stalactites, and thought, "I'm working in a damn mold-growing greenhouse."

With a weary sigh and another sore throat coming on, the teacher walked around the gym, flipping on one giant-sized fan after another to cool the room enough to make it bearable for students. Without the fans, the temperature could easily soar close to 100 degrees, making the basketball court and exercise rooms unusable.

But the perpetual heat wave wasn't the only problem with the Hampton Bays gym.

For the past several years, students and teachers who used the facility had noticed moisture creeping down the plaster walls of the basketball court, obviously the result of a leak in the roof. Sometimes the gray-painted masonry simply appeared to be damp, but on other occasions—especially after heavy rainstorms

—bright beads of water glittered in the gym lights as the moisture inched onto ceiling tiles and down the walls. More recently, Pat and two other instructors had also noticed a growing patch of greenish-black mold that covered several square feet of wall in the locker room below.

Water follows gravity; mold follows water.

At least three times during the past year, the gym teachers had requested help from the Grounds and Buildings unit at the 1,000-student school. Finally, after the teachers begged for help, on one occasion the janitors had been dispatched after much ado and delay to the gym, armed with rags, buckets and scrapers. They'd spent maybe an hour scrubbing away at the stubbornly resistant organisms clinging to the walls of the locker room, to no avail. Nothing could touch it.

After activating all of the fans along the gym floor, the weary instructor grabbed the key ring on her belt and began to unlock the Equipment Room in order to gather up the supply of basketballs and volleyballs required for today's classes. As she slid her key into the door-lock, a voice called out behind her.

"Hi, Pat! How's it going?"

She turned to greet Paul, one of the other two instructors on her team. "Hey, Paul. Do you want the truth, or do you want to hear the usual cheery crap?"

He stared at her. "What do you mean?"

She shook her head and sighed. "The truth is, I feel terrible. I'm so sick and weak, I don't know how I'm going to make it through the day."

Paul nodded. "Yes, I know what you're talking about. I'm not feeling well, myself. I think I must have come down with ..." He trailed off, and then burst into a violent fit of coughing. Pat watched him hacking away for at least 20 seconds.

"Geez, Paul that sounds awful. Have you been to see a doctor?"

Eyes red and wet, the gasping teacher finally blurted: "I've been *twice*. Told him about the coughing, headaches and the muscle aches, along with the blurred vision. In the middle of the afternoon I fall into some kind of … it's like I'm dozing, mentally. It's like being in some kind of helpless trance."

Pat nodded. "Me too."

"The doctor says he doesn't know what's wrong with me, that I should see somebody about stress and maybe start on an anti-depressant. He's given me every test in the book. It's not flu, and it's not a persistent cold. For a while, he thought it might be some sort of food allergy, but he gave up on that diagnosis two weeks ago. Last week, he told me flat-out: I'm stumped! So, since he can't help me, he says I need a shrink."

Pat Romanosky's eyes grew wider. "He sounds exactly like *my* doctor, Paul. The sicker I get, the madder he gets—like it's somehow my fault. He sends me to an ear, nose and throat specialist just about every year, but last year was the worst, after it got so bad that I lost my voice. Remember my 'laryngitis semester?' He told me it might be 'vocal cord abuse,' from so many years of shouting at the kids on the basketball court. I ended up having vocal cord surgery, but even that didn't help. Bottom line: I ended up with a prescription for nasal steroids and a pat on the back. It didn't matter. Every time I'd go in that gym, I'd get sick again and couldn't talk right. I had to go out and get a *bullhorn* and take it to school every day."

Pat's voice had been creeping up, and now she was angry.

"And what about all the *other* staff who've gotten sick during the past few months?" Pat added loudly. "Everywhere around here, you see tired teachers with red eyes and headaches coughing into handkerchiefs. How many of them would you say have got some version of what *we've* got?"

Paul reflected. "Well, out of a staff of 200 or so, maybe 60?" he guessed.

"Paul, that's 30 percent of the faculty," said Pat.

Her colleague shrugged his wide shoulders. "I don't think it's an exaggeration. Some days, this is more like a hospital ward than a school!"

Still coughing, Paul headed off to grab his *third* big cup of coffee of the morning, while Pat wandered into the Equipment Room and began to load the gear she needed for her classes onto a wheeled cart. And then it happened again. Reaching for a yellow volleyball, she felt a streak of pain go flashing through her left elbow. For the past several months, she'd been experiencing these "pain flashes"—moments of stabbing discomfort in which the tendons in her arms and legs felt as if they were being subjected to vibrating electric shocks.

Weird, she thought. Was it tendonitis? Arthritis? Or was this the first sign of the condition she most feared: menopause? At the age of 51, Pat knew she was within shouting distance of the "change of life," and she often wondered if all her recent health problems were linked to a "menopausal syndrome" that included mood swings: Husband Joe could tell you something about them! Might the sudden night sweats, muscle cramps, headaches, blurred vision and short-term memory loss be the start of menopause?

Maybe. But there was also a major problem with *that* explanation: What about the endless coughing and red eyes? Menopause symptoms didn't include red eyes, sore throats and raspy voices. What in the world was the matter with her? Should she see a neurologist—maybe even a psychiatrist? And what about her friend who went to that Fifth Avenue internist in Manhattan two years ago who'd charged $150 for a half-hour consultation—before concluding that her *real* problem was "fibromyalgia." Well, what was fibromyalgia anyway—some sort of vague problem in which your muscles hurt for no reason, you couldn't concentrate or sleep right, and you soon found yourself taking huge amounts of super-expensive drugs to combat the inevitable "fibromyalgia-related depression."

No thanks, none of that crap for me, she thought.

Groaning inwardly, she pushed the equipment cart into the center of the basketball court, and then jumped nervously when a loud bell clanged in the distance. Already, she could hear the sound of approaching, sneakered feet. *Welcome to Phys Ed 101.* She prayed inwardly for a moment, reciting a prayer that had become quite familiar in recent months:

"Dear Lord, help me to get to three o'clock!"

Showdown at the Library: Stachy in the Stacks!

Located on a quiet downtown street in sleepy Princess Anne, Md. (population: 2,313), the Somerset County Library was a throwback to another era—the slower-paced and simpler world of 19th Century America, in which the green-grassed and stately "courthouse square" was the center of public life. Walking along the tastefully landscaped entranceway to this graceful, brick-and-granite library was like walking into a museum of America's past, and I half-expected to find a liveried doorman in white gloves awaiting my arrival: "Dr. Shoemaker, welcome. We've been expecting you; right this way, please!"

The reality turned out to be somewhat different, however.

After determining that my latest patient from Princess Anne—the middle-aged clerk who'd phoned me—was indeed suffering from Sick Building Syndrome, I'd journeyed to Princess Anne for a consultation with the building manager. He agreed to let me sample for mold and do the special visual contrast testing (VCS) on each of the staffers who were at work that day.

Visual contrast testing involved no needles, big noisy machines or radiation. The results were reproducible, no matter who did the test, from an internist in Savannah, Ga. to an optometrist in Seattle, Wash. And it allowed us to test unique nerves involved in vision that run from the eye to the back of the head. The nerves cross and fan out on their way to the back of the brain. Why is this so important? They allow us to see if toxins are disrupting, even faintly, part of the brain's function. *The test is so sensitive that just about anyone with normal visual acuity will show up positive if exposed to various biotoxins, and so specific that a VCS deficit wasn't some kind of false clue.* A few patients with biotoxin-associated illness might pass the VCS test but healthy people don't fail.

Unlike many physicians in 2002, I understood that a brand-new disease threat was emerging in America—a cluster of "biotoxin-associated" illnesses had been hatched in our rapidly changing environment, where 70,000 different industrial and agricultural chemicals have combined in recent years to create new "environmental niches." These were creating many debilitating ailments like Chronic Fatigue Syndrome or Chronic Lyme Disease, not to mention the empty, non-diagnosis of fibromyalgia.

The previous summer I'd written *Desperation Medicine*. I warned of fungi that were flourishing inside climate-controlled buildings across the country producing a strange new illness: "It's an alarming fact, according to the latest data from the U.S. Occupational Health and Safety Administration (OSHA), more than 20 million American workers—nearly 15 percent of Uncle Sam's work force, in fact—may now be affected by toxins from SBS fungi." Think about it; are you one of the 15 percent? How about your family, friends or neighbors? These illnesses from mold are incredibly common, yet aren't being diagnosed or treated.

"Add in the 10 million U.S. school kids who are also exposed daily to building-related fungal toxins, and it's easy to see why the word 'epidemic' isn't an exaggeration, when it comes to describing the recent surge in SBS," OSHA had said.

I wasn't alone in my assessment of how serious the threat was, either. Describing the growing problem in the Boston area, the director of the Pediatric Environmental Health Center at Children's Hospital, Michael W. Shannon, M.D., had only a few months earlier warned the nation that SBS was a growing threat to the health of children who attend the nation's 110,000 public and private schools.

"These kids are coming to us from multiple school districts with a variety of complaints caused by their school buildings—everything from chronic headaches, shortness of breath, burning eyes, nose irritation and difficulty concentrating. I think it's now clear that SBS is an increasing health problem for America's schoolchildren, as air quality continues to deteriorate in many of the newer, 'sealed' school buildings around the country.

"In recent years, we have identified many sick schools, and in some cases we've found schools where literally hundreds of students and teachers are symptomatic," Dr. Shannon explained. "This is too complex for most pediatricians. In most cases, it takes three or four hours to properly evaluate a child with suspected 'Sick-School Syndrome,' because our evaluation includes examination of the school environment and then close interaction with public health authorities. Treating kids with this condition is so demanding, in fact, that I usually tell pediatricians to simply refer their SBS cases to the nearest environmental health center."

Manuel Lopez, M.D., the chair of the Environmental and Occupational Disorders Interest Section of the American Academy of Allergy, Asthma and Immunology, had also been telling reporters in the summer of 2000: "I think people are starting to realize now that a big effort must be made to really start studying the indoor environment. And that's especially true for schools. I do think there's growing evidence that building-related illnesses are increasing in these new 'tight' buildings that have been sealed to save energy."

Ruth Etzel, M.D., editor of the Handbook of Pediatric Environmental Health, had noted: "As pediatricians, we're now taking the threat of SBS much more seriously than we would have 20 years ago." Then she'd gone on to describe the key symptoms of SBS: headache, fatigue, shortness of breath, irritation of eyes, nose and throat, lethargy and inability to concentrate.

Unlike these physicians, however, researchers at the federal health centers and university laboratories were lagging behind the new science when it came to understanding how America's "changing indoor environment" had triggered a new epidemic of biotoxin-linked diseases that were caused by fast-growing mold in offices, factories, schools and homes.

As for the politicians and bureaucrats who controlled America's public health sector: forget it. As summer turned to fall in the first year of the brand-new millennium, both the state and federal health establishments were in full denial of the new outbreaks of SBS. In state after state, the same mantra was being repeated with mind-numbing frequency: "We don't have a problem with mold in this building—and nobody has gotten sick. All we're seeing is a severe outbreak of the flu!"

So I wasn't too surprised that when I reviewed my findings with the library building boss, he wasn't too concerned that 9 of his 10 employees had deficits in visual contrast testing, the bedside test helpful in diagnosing biotoxin illness.

"Thank you very much for dropping by," said the manager on that rainy November day, "but we don't think we have a medical problem, and we aren't concerned about the possible presence of mold on these premises. We'll replace the ceiling tiles that are discolored because they are unsightly, but not for any health reasons."

No illness threat? The ceiling tile from the break room had started making me sick within 20 minutes. It was certainly growing some mold that made toxins.

When the report from P and K Microbiology came back two weeks later, saying that the tile had "massive contamination with *Stachybotrys*," I sent a copy to the manager, but heard nothing more from him.

How would you like to work for a defensive man like him? Perhaps you or a loved one already does.

Hazmat Suits and Respirators

It wouldn't be until March 2003 before Pat Romanosky found her way to my clinic in Maryland.

While consulting with an expert on Lyme disease—she was willing to try anything at that point—the veteran Long Island Phys Ed teacher had mentioned the leaking roof and the spreading mold in her rattletrap gymnasium at Hampton Bays Elementary.

The specialist, Dr. Joseph J. Burrascano, Jr., did a double take. "Are you telling me that you've got mold growing in the gym?" he asked.

"I am, doc. I've also got more than 50 teachers who are more or less permanently ill—and all of them are struggling with the same symptoms I've been telling you about today," she said,

Dr. Burrascano was peering at her. "Have you considered the possibility that you might have a buildup of biotoxins in your body from the mold?"

Pat shook her head. "It never occurred to me."

The doctor sighed. "Pat, I'm getting more and more patients like you all the time. I treat them with antibiotics if they have a bacterial infection from Lyme disease that's making them sick. But I can't kill a mold toxin with antibiotics. These fungi have somehow found a way to create toxins that *remain* in the human body, almost indefinitely."

The teacher watched the physician reach for his notepad. "I'm going to send you to a physician down in Maryland. He's been specializing in chronic diseases like Sick Building and Post-Lyme Syndrome for some time now. And he seems to have a very high rate of success with patients like you—most of them lose their symptoms within a few weeks."

Within two weeks, Pat was sitting in my office. She told me her story, like so many others I hear every week. Chronic water intrusion, mold growth, the administration doesn't do anything and laughs off the possibility of illness.

"It's just a little mold, Pat," they say. "Mold never hurt anyone."

No investigation is done, but Pat now hears that she's considered to be a troublemaker.

The administrative position became a little more threatening when Pat began to hear rumors about her upcoming retirement. Though no one actually said to her, "You don't want to cause a lot of trouble when you're so close to getting your pension, do you, Pat?" the implication was clear: *Keep quiet. Don't ask questions we don't want to answer.*

Badly frightened still, the patient blurted out, "I've been infected with toxic mold, haven't I? It's proven now. What if I can't get rid of it? Am I going to have to live with these miserable symptoms for the rest of my life?"

"No, Pat," I said. "I suspect that the worst part of your ordeal is over. I've been treating patients like you for several years now, and I know how to make you better. But there are probably more people involved than just you. I'd guess that at least 25 percent of your co-workers are ill. There's a way to go about your evaluation that would also help many others. How many kids are in the school? A thousand? And how many kids graduated during the last 15 years?"

I knew that her political battle was likely to get pretty rough.

"The School Board won't be real happy to think about burning down the only elementary school in the district to get rid of the mold. Sounds like the contamination there is so bad, it probably would be cheaper to build a new school than to try and fix the mold castle," I said. "Any kind of remediation would mean evacuating the building, mandatory use of Hazmat suits and high-grade, small-pore size respirators. Once they open up the walls of that school, the spores will swirl out of the wall cavities like something out of an Arabian Nights horror movie. There won't be any way to contain the toxins. What a nightmare!"

I sighed. "And the voters won't be real happy with you, if and when they're facing a school construction bond. They'll attack you, attack me and do whatever they can to avoid taking any responsibility for hurting students' ability to learn and the staff's right to work in a safe environment. I see it everywhere, every day. And they will be wrong."

She didn't hesitate.

"I'll do it, Dr. Shoemaker, provided that you get rid of these headaches, red eyes, muscle cramps and memory lapses that have been driving me crazy," she said. "I'm ready, doc! When do we start?"

CHAPTER 3

Somerset County District Court

First Victory in the Fungal War

It Is a Universal Theme in Biotoxins: You Can't Kill Them with Antibiotics, Heat, Cold, Acid or Anything. Biotoxins Live Forever.

Any building can grow fungi that produce toxins—simply add water. Buildings become sick when growing fungi make spores that contain toxins, and then the spores are dispersed throughout a building, commonly via the HVAC system. Toxins are protective shields for spores—they help them survive attacks from all the other microbes living there. But the serious consequence of this "mold-against-the-world battle" is that once the toxins are distributed with the spores, these chemicals stay active, even if the spore itself dies. Fungi can make other damaging compounds too, but toxins are the worst.

For mold to flourish indoors, and then to erupt with toxin-laden spores, mold needs indoor temperatures, cover, moisture and food sources. Paper, insulation and plywood lead the long list of mold's favorite foods, but there isn't too much inside a building that fungi can't use as nourishment. Indoors, where the climate is just right and the food service is excellent, fungi also enjoy safety, as competing predatory microorganisms simply don't thrive indoors: they require more than the basics that fungi need—and get—inside a typical building.

It doesn't really matter how water gets in and stays inside, but water will make any healthy building sick.

Water can come in through the roof—valleys can leak, flashings can fail (especially if they aren't installed correctly) and edges alongside chimneys can act like sieves. Water can come in through defective siding, improperly installed outer walls, shabbily installed windows, loose plumbing joints or loose fitting doors. Condensation in cavities without a vapor barrier is another source of water.

Put a building on a concrete slab at the bottom of a slope and watch the water migrate upwards from the slab into carpets and walls through the porous concrete. Dig a basement without solving the problem of ground water pressure and watch the water run in, following the path of least resistance. A thin film of waterproof paint on the interior of a basement won't stop water. That walkout basement, complete with rec room, laundry room and storage area under the stairs, usually becomes a walk-in home for mold.

Judge Hayman's Mold Greenhouse

The Somerset County District Court Building in Princess Anne, Maryland, provides a good example of the most common problems I see in typical sick building site investigations. This is a true story; it's been the subject of numerous newspaper articles, TV shows and radio programs ever since November of 2001.

It was Judge R. Patrick Hayman who called me.

"Ritchie, I think I have a sick building here," he offered. "The entrance foyer smells like a four-day-old sock. You can see the black mold on the ceiling tiles. A lot of us are sick. What can you do for us?"

It seemed ironic; the debate about whether or not a building makes people sick often ends up in court, or is settled before trial. This one was already in court!

I agreed to do a site visit and test everyone who worked there using visual contrast sensitivity (VCS) testing. Contrast testing allows a rapid measure of biotoxins' effects on the brain. VCS is such an elegant diagnostic device; portable, non-invasive, fast and inexpensive, with results that are easily reproduced. VCS shows an obvious deficit in patients hurt by biotoxins. The perception of subtle differences between white to grey to black is lost in sick patients but not in healthy people. All you need to do VCS testing is the right light source, eyesight (acuity) better than 20/50 and about 18 inches of desktop space. And documenting a patient's neurotoxin history takes about as much time as a good cardiac history, once you know how. I was ready to go.

"Don't forget your cholestyramine, Ritch," I could almost hear my wife, JoAnn, say. No problem there—during the past four years, I've sampled molds and tested patients in so many sick buildings, I can usually tell within 20 minutes of entering a site whether or not mycotoxins (fungal toxins) are present. Cholestyramine (CSM) is the medication used to eliminate fungal toxins.

There's nothing else that gives me that distinctive hot taste on the sides of my tongue, together with a queasy stomach, headache and sensitivity to fluorescent lights. Once you're sensitized to mold, like I am now, chemical sensitivity is a big deal. After being in so many of the worst buildings, I'm extremely sensitive to smells, even from innocuous sources like a computer. Visitors to my office wonder why a fan near my desk blows upwards next to the computer—it's because I need to get those fumes from the heated elements in the machine away from me! Fortunately, cholestyramine, taken as a preventive measure, blocks the initial symptoms of the illness caused by toxin-forming fungi.

The Saga Begins

Armed with my standardized light source, symptoms lists, VCS equipment and assessment sheets, and prescriptions for CSM,

I'm off to the jewel of our judicial system in rural Somerset County, Md. The single-level, colonial style building is about five years old.

Built on a concrete slab and surrounded by a paved parking lot, with marginal drainage at best, the modern colonial style one story building has several angles in the roof that invite leaks. The easy to install, low maintenance, moisture retaining carpet inside covers every bit of the concrete floor. Even if the roof didn't leak, the run-in of water from the pavement alone could provide all the moisture needed for the unchecked growth of toxic fungi. The drywall, wood and composition ceiling tiles provide a welcome source of nutrients for the fungi, too. As water enters the structure, through the roof or from ground seepage, the mold can grow freely, hidden from view on the underside of the cellulose construction materials like sheetrock and plywood. A casual inspection won't show anything abnormal. But when you look in the hidden areas where mold grows, don't be surprised at how much mold you find. To be a fungus finder, one has to look where the sun doesn't shine!

The courthouse was designed to satisfy building codes. And when it comes to building codes, I've come to believe in a concept I call the "Nuremberg Trial" argument. Basically, the codes are enforced by building inspectors who just *follow orders.* We know the orders are wrong, and the inspectors likely know as well, but who will defy the regulations? One of the culprits in the mushrooming number of sick buildings is a basic lack of understanding of fungal habitats by the bureaucrats who write the building codes.

Of course, it really isn't anyone's "fault" that the sewer pipes backed up shortly after the building opened—the flood from the malfunctioning plumbing exhaust could have easily started the fungus ball rolling. And we will never know if that first water intrusion created the sanctuary that later nourished staggering numbers of *Stachybotrys, Aspergillus* and *Penicillium*. Likewise, we can't dump all the blame on the plumbers because the overly elaborate front entry lets in water when the wind comes from the West, as it usually does. And we can't entirely blame the framing carpenters who are responsible for moisture seals around the doorway. They are off the hook, too. After all, they can't be blamed just because the concrete slab—poured as a foundation to save money—was the guaranteed destination for water run-off from the parking lot. Still, why didn't *someone* recognize that concrete slabs are portals of entry for moisture, especially in low-lying, high water table areas?

Most sick buildings and homes don't have the multiple sources of water intrusion like the Courthouse. In a lawsuit, blame for the water intrusion is often arbitrarily apportioned. Who's likely to be sued? The self-employed subcontractor or the big developer? Lawyers involved in mold lawsuits pick a "responsible party," usually one with deep insurance pockets, as a target of negligence suits. The lawsuit is just one part of the battle. Your attorney might win that part. And your protests might win you an expensive retrofit mold treatment. And how about the buildings burned because it's cheaper to burn them than fix them?

But all the litigation, settled claims, expensive retrofits and abandoned buildings won't do what our treatment protocols do: help to restore the health of affected patients.

Hopefully, we'll identify the susceptible patients in time, before the "fallout" from toxins made by fungi, bacteria and other organisms causes the diverse array of damage to critical hormone systems or normal defense mechanisms. Once the exposure has gone on too long or too intensively, we have to do a lot more to help mold victims regain energy, think more clearly and enjoy a better quality of life than they would without correct treatment.

As we'll see, mold exposure initiates a series of illness generators, hurting immune system responses and altering blood flow in many small blood vessels. A mold victim has a "multi-system, multi-symptom" illness; they're among the most difficult to treat, but it can be done. Sadly some genetically vulnerable patients exposed too long and too intensively never make a full recovery; they are primed for more toxin attacks and consequent damage for the rest of their lives.

They're the patients I find the most challenging and love to help.

• • •

Just imagine, though, if the sick building is a school. These are the patients I find most painful to see.

All too often, schools are built with flat roofs: invitations for leaks we call them. In these days of limited school funding, maintenance budgets are slashed before the art teacher and music teacher positions are cut. When I see kids labeled as being "problem learners," or having learning disability when the problem is simply mold growth in the school, I want the responsible parties to stand in front of Judge Hayman.

How will we know if the school has been filled with mold toxins for years? Low test scores? An increase in learning disorders? An increase in diagnoses of attention deficit disorders, followed by a rise in Ritalin prescriptions? The Ritalin prescriptions are a symptom of learning impairment, not a cure. Don't forget that all those attention deficit diagnoses are diagnoses of exclusion. That means treatable conditions that look like ADD and ADHD, such as mold exposure, must be recognized and treated first. If Ritalin were a cure, then it makes sense that American children—who now take more than $1 billion worth of Ritalin annually—would have higher test scores than the pre-Ritalin generations.

They don't. How could they, when Ritalin does nothing for learning disability caused by mold?

When Little Johnny can't learn, because his brain has become fouled by mycotoxins, then it's too late. Send him to a sick school building every day and Little Johnny will be lucky to remember his teacher's name.

• • •

So who was negligent in constructing the courthouse? Who should compensate the victims? Which water intrusion allowed the mold to grow? Who should pay the dollars for cleanup and remediation? The insurance companies surely are interested in the answers to these questions. If everyone blames someone else, who lets the buck stop on his or her desk?

As we get deeper into mold, keep one caution in mind: indoors isn't outdoors. Molds outside are attacked by an incredible diversity of enemies. Not so inside.

When patients ask me why we see neither hundreds of species of mold nor "monocultures" of fungi in sick buildings, the answer is simple. A sick building simply lets fungi survive, reproduce and release mycotoxins and other compounds into the air the inhabitants breathe. With many organisms sustained by food, water, cover, and a chance to reproduce (the courthouse building amply provided all of those), it's a fact of biology that the fittest survive—multiple species will compete for small niches but only a few will win the battle for survival.

Every time the front courthouse door opened, for example, or when the Judge came in on a Saturday to do some paperwork, wearing his freshly fertilized gardening shoes, a new opportunistic invader could be introduced. But the reality is that biological structures in established ecosystems have their own controls. You won't see incredible dominance of one kind of mold or another in soil or in water outdoors, because there are too many organisms that already live there to allow one kind of critter to dominate. We sometimes see blooms of algae or massive red tides of dinoflagellates, but these explosions of single-organism populations are usually self-limited events in Nature. Biological control is re-established, bringing in diversity of species over time.

In contrast, the indoor air environment in sick buildings isn't an ecological system in balance. A select few species of fungi rule the scene.

I wasn't surprised to find the same symptoms I saw in workers from countless other sick buildings presented by the employees in the courthouse. Everybody had many symptoms—they were wide ranging and included fatigue, weakness, aches, muscle cramps, unusual stabbing pains, sensitivity to bright light, tearing, blurred vision, headaches, sinus congestion, coughing, shortness of breath, abdominal cramping, rashes, skin sensitivity, memory impairment, confusion, difficulty in concentration, trouble finding words in conversation, funny metallic taste, numbness, tingling, spinning sensations and mood changes. All were found as a distinctive grouping; this wasn't depression or mass hysteria. This was a classic presentation of chronic, biotoxin-associated illness.

Not all patients had all of these symptoms, and the symptoms changed from day to day. None of the affected employees had no days without symptoms. If the person had a good day and felt like doing something extra, like clearing out a closet or raking a pile of leaves, he or she wouldn't just bounce back the next day —they'd feel bad for a couple of days, with a delayed recovery from normal activity.

The second step in screening patients exposed to the building involves doing the visual contrast sensitivity test. It only takes five minutes to obtain results you can believe. Do we see the distinctive marker of brain damage from biotoxins, impairing the ability to detect contrast, or not? VCS scores in the District Court workers were predictably abnormal. They all showed the typical fingerprint results of the biotoxin, a VCS deficit. While the special fingerprint is easily recognized, it does *not* tell us if the biotoxin comes from mold, Lyme disease, unusual algae or a host of other biotoxin makers. With this simple visual test, however, we're able to discern the presence of toxin by assessing optic nerve function from the eye to the back of the brain. The test results can tell an experienced investigator that the presence of health problems is clearly linked to a biotoxin's effects.

Eliminating confounding exposures and diagnoses—a key part of the SBS case definition—is essential to complete the neurotoxin case histories of those who worked in the Somerset County courthouse. In the end, we had six sick workers, each of whom had prolonged, close exposure to the greatest concentrations of mold. Four others were well. After another four workers on "the other side" of the partitioned offices also agreed to be tested, we identified two more cases.

Blood tests showed the typical normal values for all the standard tests we use in medicine. These normal results are one reason why biotoxin illness is missed in routine medical care. The exception to "normal blood tests," are the results of the "unusual tests" I use routinely. Why do I order such "unusual tests," ones that some doctors don't know about? Because they show the illness! Not surprisingly, in the District Court workers we found elevated levels of immune proteins called "inflammatory cytokines." At the cellular level they function in a manner similar to the "command and control centers" we heard so much about when the U.S. was attacking Iraq. Cytokines alter cell functions. If the world of cytokines seems complex, just remember that individual cytokines are responsible for multiple biological activities. Individual cytokines interact with each other. When illnesses involve disruption of cytokine effects, like mold illnesses do, look for symptoms involving the entire body.

I didn't test for antibodies to the fungal species: a positive antibody test doesn't tell me the specifics of the time, location or duration of exposure, nor about exposure to mycotoxins. Other physicians routinely employ antibody tests to help in their work with mold-exposed patients, but I don't.

Remember, I'm not talking about a fungus growing in deep body tissues like the brain. The patients with those life-threatening illnesses are usually immune suppressed. They must have strong fungus-killing antibiotics, or else they may die. I know that some physicians treat mold illness patients with the same antibiotics, but I rarely see much benefit in those cases. I treat patients who have *toxin*-associated illnesses which are much more common, and are due to exposure to areas where fungi and other microbes thrive indoors.

The attitudes of the courthouse employees regarding my findings were mixed. Some were angry with the building owner, who apparently had cleaned up the plumbing mess, installed new ceiling tiles when the roof leaked and fixed the carpet after the front door flooded, saying, "Everything was fixed." Some were scared; if they spoke out and said they were sick, they might lose their jobs. Some were downright devious, pretending to have no symptoms when they had obvious ones. One red-eyed employee said she never noticed any redness of her conjunctival membranes. A manager said his ability to assimilate new knowledge was fine, but it took four separate explanations for him to learn how to do the very simple VCS test. Still, despite the transparent attempt to sabotage the investigation, we didn't have just one sick patient (a complainer); or two (a conspiracy); we had at least three, and three makes a cohort. And that means a real group with a real illness that we could track for improvement with CSM and relapse with re-exposure without protective use of cholestyramine.

The Judge was going blind in one eye in addition to having a typical mold illness. Several other workers were very ill. They started on my treatment protocol right away, with prompt improvement in VCS scores and abatement of symptoms. When the medication was stopped and they continued to work in the building, relapse began within 36 hours. When they were retreated, they improved again. Fortunately, the site investigation coincided with the time when I had the amazing Heidelberg Retinal Flowmeter in my office, on loan from the manufacturer, Heidelberg Engineering.

Using Heidelberg's dual laser Doppler to measure blood flow to the eye, we could look at three tiny areas—the retina, the nerve tissue around the optic nerve (neural rim) and the deeper part of the optic nerve where the separation between the body and the brain is basically just one cell thick, also known as the lamina cribosa. Examining these sites reveals the effect of cytokines binding to their receptors in the capillaries: reduced blood flow, the marker of a biotoxin patient. If VCS testing appeared to be low-tech, the Heidelberg was about as high tech as one could imagine.

I was able to show a clear deficit in capillary blood flow to the neural rim of the optic nerve and the retina in the sick people, but healthy persons had normal capillary flow rates. With treatment, flow rates increased, beginning in 12–24 hours, making the sick people look like the healthy ones. With re-exposure without protective medication, flow rates fell within 12 hours.

So we had biomarkers to spare. The VCS scores matched the cytokine levels and both matched the Heidelberg flow rates, and all matched changes in symptoms, confirming both a decline in health following exposure and improvement in health with treatment. Since the Heidelberg wasn't portable and the VCS was, and the results agreed, we could use VCS with confidence. To secure the diagnosis of Sick Building Syndrome beyond any doubt, I still had to show that there were no possible variables of alternative (confounding) exposures—nothing else that might cause a neurotoxic syndrome in these patients.

So far we had showed both the presence of neurotoxic effects and confirmed exposure to toxin-forming mold species. However those findings did not rule out Lyme disease or chronic fatigue syndrome, for example, or exposure to indoor mold from another, remote source. as the cause of the illness. Some patients might have multiple sources of fungal exposures or multiple contacts with biotoxins. Could they have *Pfiesteria* or the new blue-green terror algae making them ill, as well?

We take the patient as we find him: If he had sinus congestion unrelated to mold exposure, he will still have the congestion following treatment for mold. But having sinus congestion from another source doesn't rule out the significance of the mycotoxins as the cause of his chronic biotoxic illness. We treat this patient with CSM.

For final proof that will silence any *logical* objections to a biotoxin diagnosis, we perform repetitive exposure trials. (We can't use this protocol in the few patients whose illnesses have gone on for too long, producing irreversible damage.) These trials use the sick patients as identifiers to report the exposure that makes them sick. After the patients improve with CSM, used exactly as prescribed, they go back to work in the same building. Call them blank slates now, if you will.

Now, take the patient out of the contaminated building, stop CSM and expose them to the "ubiquitous fungi of the world." We document the result. If the contaminated building is the illness source, nothing bad happens. The "ubiquitous" fungi surrounding the patient do *not* re-create the illness. Now the patients can be studied "prospectively." Practically, that means if the suspect building is safe, they won't get sick again when they go back to work.

Only this time, they don't have the protection of CSM. We already know about any possible confounding exposures, including multiple tick bites, or sickness from eating red snapper (ciguatera), or occupational exposure to solvents (the fumes give a positive VCS) or a visit to relatives who live in an estuary of the Chesapeake Bay (*Pfiesteria*). Exposure to the building is the only thing new. We look in on the patient in just *three* days. If the building is harboring biotoxins, the previously sickened patient will get sick *again* right away.

Most patients from the courthouse reported that their symptoms returned within *2–3 hours*. The CSM protocol was re-instituted, with the same significant rate of recovery. Theirs was an illness that followed exposure to only one source of biotoxins: the building. Mind you, this elegant protocol won't tell you which biotoxin made the person sick, just if the toxins that cause the illness are in the building and nowhere else.

The next step was intriguing. If we could treat the illness, could we *prevent* it as well? It turns out that our prevention protocols worked in some patients to prevent re-acquisition. The molds will keep reproducing, increasing levels of spores and toxins, making exposure to the building a greater risk to all. Of course, it's far better to remove the patient from the contaminated environment than to prevent the illness with CSM. CSM works by removing toxins *after* they've entered the body—it's better not to have them in us to begin with. Even better would be for the building owner to remediate the building correctly, eliminating all those toxin-formers.

In my experience, that almost never happens.

CSM isn't an easy medication to take. Bloating, reflux and constipation are predictable side effects. Sure, we can find clever ways to get around most of the gastrointestinal complaints and we know some pharmacists who make up special CSM preparations that don't contain sugar, aspartame or unwanted additives. Human nature being what it is, drinking a mix of CSM four times a day on an empty stomach, 30 minutes before eating or taking any other medications, just won't happen for long unless the patient is truly motivated. We use less frequent dosing of CSM and some additional medications for maintenance, but even then, some patients stop their preventive doses.

"Prophylaxis" means eliminating mold toxins *after* they're already inside us. As you'll see in later chapters, for some patients, the damage is irreversible. Prophylaxis won't help them. The problem for those who aren't irreversibly injured by toxins is that between initial entry and subsequent elimination in bile, is a long series of potential targets for biotoxin damage.

Better to clean up the building, or evacuate it if clean-up fails.

Why the Courthouse Was Dangerous

Repeated and/or prolonged cytokine responses take their toll on our body organs and us. Some patients don't recover all their mental capacity after the fungal illness is treated. Some don't have all their joint pains go away. Some acquire new organisms, growing happily in freshly altered "econiches" after cytokine responses change mucus membrane defenses. These "squatters," in the nose, for example, can create their own harm by causing additional cytokine responses. Even worse, some patients suffer damage to delicate pathways in the hypothalamus, affecting the nerve impulses charged with regulating hormone levels, resulting in out-of-control immune responses.

These patients usually develop intractable pain; chronic, nonrestorative sleep, unexplained weight gain and lessened mental capacity. They also usually develop increased amounts of compounds that *deliver* cholesterol out of the blood stream and into the areas under the blood vessel walls, creating the foundation for clots and blockages. The problem for mold patients isn't the total amount of any kind of cholesterol (LDL, HDL, triglycerides) floating along *inside* blood vessels; it is the cholesterol types that cause blockages that are deposited *outside* blood vessel walls. Immune responses to mold elevate the efficiency of the cholesterol delivery process to a standard that would make FedEx be proud.

So if biotoxins can cause immune mechanisms to put cholesterol in the blockages developing outside blood vessel walls, might this mean that heart attacks are made more likely following exposure to toxic fungi? There's no question that *clot-promoting* compounds increase with exposure to mold and fall with treatment. Just when you thought that the explosion of heart disease, diabetes and obesity was due to sedentary lifestyle and increased fat consumption, now you have to factor in a huge variable: inflammatory cytokines from biotoxin exposure. (You may be interested in reading Chapter 14 of my previously published book, *Lose the*

Weight You Hate, entitled, "Environmental Acquisition of Diabetes and Obesity.")

Many SBS patients also begin to notice that they become more sensitive to fumes, smells and chemicals. With repeated exposures, the sensitivity for some becomes more pronounced. In the full-blown sensitive patient, someone with Multiple Chemical Sensitivity (MCS), just a few seconds of smelling fumes is overwhelming. Mere seconds of "off-gassing" coming from computers and phones, new paint, new carpet, freshly printed reading material, or even just a ream of copy paper, can make patients sick for weeks. Our treatment protocols for "Multiple Chemical Sensitivity" may bring order to this difficult-to-confirm diagnosis if the illness is caught quickly after it appears. To date, having seen over 500 MCS patients, I have yet to find one who wasn't made ill early in the illness by exposure to water-damaged buildings. I continue to look for sources of the origin of MCS other than mold exposure—so far without success.

One particular patient nearly died from repeated exposures to the courthouse, when those exposures caused acute bile stoppage and near liver failure. The liver is a site often damaged by cytokines. Normally, the liver uses a wonderful "detox" system in which cells that line the tiny bile ducts use a transport system to preferentially secrete negatively charged organic biotoxins, like mycotoxins, into the bile. From there, the toxins go into the upper intestine. CSM grabs the toxins, preventing reabsorption, and escorts them into the commode.

But the transport system can be overwhelmed by cytokines and stop working. Moreover, bile flow can stop. Sure enough, the patient who almost died had liver tests that showed severe damage. Fortunately, we can use a medication that revs up the toxin transport system and defeats the cytokine effects; in this case, it saved the patient's liver.

If you're wondering why didn't everybody in the building get the same illness? Most of them did, but some were worse than others. The answer likely lies within our genetic make-up. When the *Pfiesteria* problem rocketed my life into a different career, one question troubled me for years. In a group of 10 people who played water volleyball, went water skiing and overall had a great time one week before the big *Pfiesteria* bloom and fish kill in the Pocomoke River in 1997, only three became ill. How come?

Over the years, the role of antibody production, controlled by the immune response genes, HLA DR, has become a big deal in immunology. Antibody production is dependent on recognizing foreign invaders, like parasites, viruses, bacteria and biotoxins. Our HLA DR genes control recognition. Was HLA DR associated with biotoxin illness?

Sure enough, individual HLA DR genotypes nearly defined 100 percent of those who would get sick after biotoxin exposure. So it was with *Pfiesteria*, mold, Lyme and all the rest. No wonder the illnesses caused by biotoxins all seemed to have the same symptoms—they all followed the same basic mechanism.

So far though, patients with the worst problems after treatment for mold illness all have two particular HLA subtypes, unusual types found in less than two percent of the entire population, and not found in patients who don't have symptoms after appropriate treatment. Is it possible that the reason some patients develop chronic tiredness, and many other symptoms is due to genes controlling chronic fatigue? Certainly is possible.

While I haven't screened a large number of school kids with learning disability to see if they have a particular genetic make-up, the idea that mold illness—causing difficulties with concentration, recent memory, word finding and decreased assimilation of new knowledge—might be underlying learning disability, isn't far-fetched. We are already seeing that in my school age patients with mold exposure from school and HLA susceptibility, there is a tremendous increase in learning disability.

Verdict and Sentencing

Anyone who has ever heard Judge Hayman speak knows why he sits on the bench. Never one to trip over a word or to run from an opinion he could support, Judge Hayman makes the spoken word a razor-edged, golden sword. He looked at the evidence concerning his Courthouse and its link to human health, including his own: water intrusion; massive amounts of toxin-forming mold; proof of illness in his staff; their positive response to treatment; absence of relapse elsewhere off CSM; near immediate relapse with unprotected re-exposure to the building. He spoke for all to hear:

Verdict: Guilty.

Sentence: Abandon the building, NOW.

At first, the State of Maryland raised an objection: they wanted a second opinion from their own physician, Dr. Cheung, the State Medical Examiner. We will meet him later, after he is no longer working for the State, when he's hired to provide a second opinion on Pat Romanosky's illness, among others.

Dr. Cheung didn't have much hands-on experience with mold illness back then. No one did. Judge Hayman was not impressed with the cursory exam the doctor performed for the second opinion and told me he thought Dr. Cheung just went through the motions in his exam.

Chief Justice Martha Raisin had the final say on Judge Hayman's demand that the building be vacated immediately. Remediation costs were estimated to be more than $1 million. What choice could the Chief Justice make but to leave the building even before the lease was completed?

Still, she was not sure what to do. After all, Dr. Cheung didn't agree with my opinion. Judge Hayman asked to read the report submitted by Dr. Cheung to Justice Raisin. He announced his opinion, in classic Hayman style:

"This report ignores every plain fact."

• • •

That did it. Vacate the building. The precedent was now established.

This case has a happy ending; most do not. The State of Maryland decided to move the District Court out of the building, even in the face of the expense and inconvenience involved. Such a rational response to SBS is as unusual as it is encouraging.

• • •

We now have multiple requests to do "prevalence studies" in local buildings, where the employees think the building makes them sick. As you might expect, those responsible for maintaining a safe work place in several of these buildings aren't too happy with the prospect of health screening on the premises. I simply respond that if the building has no contamination with toxin-forming fungi, they have nothing to worry about. Like the *Pfiesteria* example of several years ago where water-quality tests were called normal, as if that had anything to do with the growing number of sick patients, in sick buildings, air quality tests aren't important in the face of finding sick people. If environmental reports provide confirmation of the presence of toxin-forming fungi, that is evidence that strengthens the case against water damage and microbial growth as the factor hurting people. However, the presence of *human illness* is the single most important indicator to determine whether it's a "sick building" or not.

If workers in a building show a typical biotoxin illness, the correct response is simple. Those responsible must figure out where the problem is, stop the water ingress, mitigate the building, fix the health problems and get on with the day-to-day operation of the building. The problem arises, however, when someone decides to deny responsibility for maintaining a safe work place. You might read Chapter 10 of *Desperation Medicine*, "The Appearance of

Good Science," to see how this typical scenario plays out. I see it every day.

Maryland took the lead at first, in dealing with the problem of SBS by acting in the patients' best interest at the Somerset County District Court. I hope that other states will follow Maryland's lead, and soon, because the fungal species that we have good reason to fear are growing in thousands of buildings. We need an organized, patient-friendly approach to help the many actual and potential victims of SBS.

Chaetomium fumigatus

CHAPTER 4

The Biotoxin Pathway

"Ritchie, I've read your paper on the Biotoxin Pathway. I've been over the diagram you drew for me last week, I've looked at the charts and the graphic (after preface) you gave me, and I'm almost there. I still have some questions and I want to go over the language and concepts one more time."

It was Jim Frasca, medical researcher for New Dominion Pictures, calling. He'd worked with me on a Discovery Health Channel TV show about the health threats of *Pfiesteria*, entitled, "*Dangerous Catch*," in June 2003. He'd recently acquired a biotoxin-induced illness and had come to my office in Pocomoke for evaluation and treatment.

Jim admitted that until *he* began having the very same symptoms that patients described on his TV show, he had doubts about my theory that biotoxins cause human illness.

"But the stories your patients told were all so consistent. I must have talked to 25 or 30 people," he said. "They all were sick as hell and got better after you treated them. Everyone I talked to said the State of Maryland did nothing for them because they didn't need to be treated. But you corrected the illness with cholestyramine (CSM). No one can persuade that many people to tell the same 'made up' story."

Jim's innate skepticism about *Pfiesteria* and illness was also heightened when some of the academics he talked to on the show criticized the new theory and the doctor proposing it, but when he asked them if they'd ever treated a *Pfiesteria* patient, they all said no, they hadn't.

"*How could they* ***know*** *much about an illness,*" Jim reflected, "*when they'd never changed the clinical course of a patient who had that illness?*"

Pfiesteria had been the public's first example of a biotoxin-induced illness. It also showed how criticism is never in short supply when someone makes a revolutionary breakthrough, even when it's based on sound science. All those *Pfiesteria* docs who'd been so critical of my theory and me were proven wrong. Now, these new "medical mold experts," physicians who didn't have experience treating mold patients and who didn't know much about the physiology of mold illnesses, weren't reticent in their criticism, either. Déjà vu all over again.

In the end, science prevailed in both situations. I was fortunate to find many excellent scientists who helped shape my clinical thinking and research design. Whenever my ideas about biotoxin illnesses needed unbiased critical review, help was usually just a phone call or an email away.

"*What's that you say, Ritchie?*" asked Jim. "'*Always challenge today's hypothesis tomorrow?*'"

I nodded. "Something like that, Jim. But you're right—if the observed data don't fit the model, change the model. When what we see isn't what we predicted, it's the basis for the prediction that's wrong and not the data."

"*It seems like ages ago I was telling you about the time I was two blocks from my house, yet I had no idea where I was. Incredible, really scary.*" *Jim recalled.* "*I suddenly felt terrible. Although I finally understood what the Pfiesteria patients felt like, I didn't have Pfiesteria. It was mold. I still don't believe that it only took a short time to get sick. I'm glad it only took a short time to get better. And thanks for the cholestyramine."*

"Jim, did you ever send off a sample of the mold that made you sick?" I asked.

"Well, no," admitted Jim. *"I was moving anyway, and my new place is dry and mold-free. At first, I thought I might have gotten sick from my girl friend, Tami. She's sick, too, you know, and I need to talk to you about her. Oh, I'm sure I've got a mold illness—no tick bites, estuaries or ciguatera. But I want to fully understand the Biotoxin Pathway, so I can explain it to Tami."*

"Are you well enough to remember what I say?" I asked gently.

"I'm OK," insisted Jim. *"My brain fog is gone. And those fancy words you use for cognitive problems, 'decreased assimilation of new knowledge' and 'difficulty with word-finding,' are all gone. My memory is fine. I just need to be able to translate your jargon into words people —especially Tami—will understand."*

• • •

My "new" language, and the concepts it describes, is foreign to most people. Unlike most new patients, Jim had previously heard me use the unusual words that describe the players and the processes that make up the Biotoxin Pathway.

"Jim, the words I use might not make sense to you at first, but they're the language of *your* illness and Tami's, too," I told him. "You don't have much choice: you have to learn them so you'll understand how to get better. The concepts of how your illness causes your symptoms will become clear, in time."

• • •

Genetic Susceptibility & HLA DR

Explaining the Biotoxin Pathway is a complex undertaking that takes some time—I can't explain it in a few paragraphs or a few minutes. For most patients, it takes several conversations and a few question-and-answer sessions. I usually begin by explaining HLA-DR, the genetics behind why some people get biotoxin-induced illness and others don't.

"Jim, let's take a look at the Biotoxin Pathway," I said. "I'm going to explain as you follow along with the diagram and the drawing. Use them to help you picture what my words are describing and stop me when you aren't clear."

The illnesses I treat are caused by exposure to biologically produced neurotoxins, also known as biotoxins. Genetics decides who will develop a chronic illness following exposure and who won't. People are exposed to biotoxins all the time, and mold toxins are the most common. Most people won't get sick from mold because their immune system's response genes make antibodies to the toxins, but there are those—the genetically susceptible—who lack this antibody-making capability, and that's where the trouble begins.

"Take a look at the table in the back to find your genotype in the mold column. And you'll find Tami's under the column, 'chronic fatigue,'" I told Jim.

"That's the HLA DR, right?" asked Jim. "I've got the mold-susceptible genotype of HLA and Tami has the 'dreaded genotype,' the one you see over-represented in the group of people who often won't get better once ill. I can expect a big improvement based on my genotype, but Tami probably won't do as well. It's incredible that you can tell so much from a simple blood test."

Some people who aren't genetically susceptible to chronic effects from biotoxins may suffer a short-term illness following a biotoxin exposure, such as a sore throat, red eyes and cough. But because those individuals have immune response genes that produce an antibody to the biotoxin and quickly eliminate it from the body, the illness is short-lived. Biotoxins can cause many symptoms, but if the body can eliminate the toxin with an appropriate immune mechanism, the symptoms disappear.

"Your parents determine your genotype. You get two sets of linked HLA genes, one from Mom and one from Dad. So how come my older brother doesn't have my illness?" asked Jim. "We have the same parents, so we should have the same genes."

"He isn't guaranteed to get *your* genes," I pointed out. "Jim, remember Gregor Mendel and all his experiments with peas you learned about in high school? And how he figured out what the genes of the offspring and parents were doing?"

"High school? I remember Suzie and her mini-skirt," laughed Jim. "OK, I remember a little bit of genetics from biology class. So because my parents each have two copies of the HLA genes, and I only got one from each, maybe I got one kind and my brother got the other kind from each parent. If we were twins, we would have the same sets of genes, right? OK, so maybe my brother isn't genetically mold-susceptible or maybe, if he is, he just hasn't been exposed to heavy growths of toxin-producing indoor mold yet."

A simple example of susceptibility comes from observing people who work in a moldy sick building, one with amplified growth of toxin-producing fungi due to water damage. With exposure, non-susceptible patients may feel ill after several days inside the building, but when they leave for the weekend or go away on vacation, self-healing occurs. If they stay away from the mold, they stay well. Some people can literally swim in mold toxins and not suffer any chronic effects.

But that's not so for others who have the genetic predisposition for biotoxin-related illness. The genes responsible for the steps involved in antibody formation, known collectively as HLA DR, are found on chromosome 6 and are associated with failure to clear the illness following exposure. Because the toxin isn't recognized as something foreign, as an antigen should be, it's "allowed" to persist in the body. Because these toxins have a unique structure—we call them "ionophores" in chemistry class —that prevent them from being metabolized or excreted, these patients stay ill, even following removal from exposure. For these people, even short-term exposure can be dangerous, especially if they've been primed by prior exposure.

All we need to determine susceptibility to particular toxins is a simple blood test most labs do every day (*see sample lab order sheet in appendix*). We have a registry of which HLA DR genotypes occur in people and at what frequency. Certain genotypes are consistently identified in those who become ill after exposure to biotoxins made by one kind of organism. The susceptible genotypes aren't the same for different biotoxins. I use the words, "mold susceptible," or "Lyme susceptible," for example, to indicate the genotype. Conversely, those unaffected by exposure have another type of genotype, one capable of mounting a sound immune response defense against such toxins.

"Wait a minute," said Jim. "Whether you get sick from Lyme isn't determined by genes, is it? Aren't we all at risk if we go where there are lots of deer, mice and ticks?"

"Right you are, Jim. The problem I'm talking about isn't the acute illness from the infection," I explained. "It's the chronic illness due to toxins that continues *after* antibiotics are finished. Antibiotics can clear the infection, but they'll never eliminate toxins. It's the toxins left behind that make Lyme a chronic illness for some patients."

"Are you telling me that all those Lyme patients who stay ill after antibiotics are suffering from toxins? That would mean that the 'fibromyalgia' my Aunt Nellie has after her Lyme isn't fibromyalgia at all. It's toxins! Ritchie, this is incredible."

We've learned a lot about the "defective immune response" from the attention that anthrax has received in the last few years. We know that the foreign antigen has to be packaged with HLA to result in antibody formation. The dendrite cell, which is a specialized white blood cell, "presents" foreign antigens to an unprogrammed (called "naïve") T lymphocyte for more processing, after it coats the antigen with HLA. Then the T lymphocyte, now programmed to respond to the antigen/HLA complex, presents the combo to the immune lymphocytes (B cells) that actually manufacture the antibodies. Now, understand this: one of the toxins anthrax makes, called "lethal factor," actually *blocks* HLA- attaching activity in the dendrite cell. So, if lethal factor—a toxin —is present to stop the immune antibody response before it starts, then no toxin-specific and toxin-clearing antibodies are made. The B cells might have been able to make an antibody before, but not after toxins stop the process.

"Geez, this immunology is complicated," said Jim. "The dendrite cell acts like a scavenger, constantly monitoring blood for invasion by some foreign thing. It sucks up anything foreign, puts a special molecule—that's what the HLA genes code for—called the HLA molecule, on the antigen. Then it shows the tasty package to another class of special white blood cells, the T lymphocytes. The T lymphocytes grab the tasty package and take it in. Is the antigen like some kind of hot potato, one that white cells keep tossing around? The T-lymphocytes in turn program the antibody lymphocytes, the B guys, to work on making antibodies to the package. Right?"

"Yes," I said, glad he understood this so far. "Just look at all the kinds of ways the immune response could malfunction. When germs sabotage the immune response, we call it immune evasion. It's amazing we can fight off anything! And a lot of people think that immune evasion is the problem in immune failure in illness."

"Does that immune evasion idea apply to all biotoxins?" he asked. "It seems like the whole antigen-presence-to-antibody production pathway can get fouled up in lots of steps and the foul-up occurs after something happens, like an illness. That makes sense. That would explain why I wasn't sick from mold until I was exposed to amplified growth of indoor mold, even though I was born with the mold-susceptible genes."

Jim paused for a moment, and I could tell he was thinking.

"Have you done the basic research to prove the HLA 'toxin theory'? It seems like once the HLA is expressed, then that's when susceptibility to toxins gets started," he asked.

"Not yet," I replied. "The problem is, we don't have any good way to identify many of these toxins in blood. In time we will, but for now we have to rely on 'observational science'—what we see in the registries of sick people. We know what frequency these genes have in the 'normal population,' and we can see what the frequencies are in sick people. Then we compare the two frequencies to come up with a 'relative risk.' If the ratio of sick-to-normal in gene frequency is greater than 2:1, then that's statistically significant. And that's why I worry about Tami. Based on the populations of people with her genes I've studied, we know her genotype won't tolerate too much biotoxin exposure."

• • •

Complement and C3a

The body has a second way to clear out offending antigens, one that's *complementary* to the antibody system. This second system, cleverly called "complement," is comprised of more than 30 proteins in the blood. It's the third most common group of proteins in blood, behind albumin and the antibodies.

Later on, in the chapter on 21st Century Medicine, we'll talk about the innate immune response and the acquired immune response. "Acquired" is basically all the antibodies our immune system learns to make; "innate" centers on complement.

Complement works like the guards who patrol the perimeter of a military camp. The guards' job is to detect a foreign invader, engage the invader and most importantly, to alert the rest of the soldiers, "Help is needed *now!*" When the bulk of the forces arrive on the scene, the foreign guy runs away, is captured or is killed. Complement consists of a few guard proteins already primed and ready for action. As soon as a foreign antigen dares breach the body's perimeter, the complement proteins cascade into action, bringing in reinforcements, interacting with other elements of the immune response and making us well again.

That system works well, until a few parts of the complement cascade start to activate *spontaneously*. That would be like if the guards went nuts and starting spraying gunfire everywhere inside and outside of the camp. In biotoxin illnesses, as you'll read, that happens a lot. Over activation or self-activation of complement, however, doesn't cause symptoms: it's like the termites in the foundation beams of a house. Quietly, unnoticed, the support of the immune response is undermined, as complement *turns on* all sorts of pathways that hurt us. We would never know complement is hurting us if we didn't look. Complement is a fabulous first defender against foreign antigens, but it mustn't stay aggressively protective all the time.

Complement helps the antibody system protect us from foreign invaders in two ways. The first way, called the classical pathway, requires an invading antigen and the antibody made to it to be fused in a combined group (antigen-antibody complex) before the pathway can be activated. Complement will coat this combination, adding to the inflammation already going on, making the invader easily taken in and destroyed by special attacking white blood cells. The second mechanism is called the alternative pathway, one in which there are no antibodies made and the antigens only (in this case, the toxins) turn on the inflammatory process. We now know that once the alternative pathway is turned on, look out, it can *stay turned on forever.*

"So that's why I had to have that follow-up blood test, what was it? C3a, or something like that? Was that the one that was so high in those Pfiesteria patients that did poorly over the long term, even after you'd treated them and they'd thought they were well? Do you think the dreaded genotypes are ones that let C3a stay up too long?"

Hold on, Jim. Let me finish one idea before you go on to the next one. The one protein made when the third component of complement is activated, we call it C3a or "anaphylatoxin," is really important in biotoxin illnesses. If you want to know if you have a chronic biotoxin illness, ask your doctor to measure C3a, because C3a levels go way up within *hours* of biotoxin exposure if the person can't make an antibody to the toxin (the genetically susceptible patient). *Persistent* C3a elevation is often seen in patients who develop an acute biotoxin exposure who then go on to develop the symptoms of Chronic Fatigue Syndrome.

If my HLA hypothesis—the one that says there's no antibody response to toxins—is correct, which pathway in complement will be involved? If there isn't an antibody response, then it *has to be* the alternative pathway. Sure enough, that's exactly what we see. Antibodies to biotoxins don't play a role in the alternative pathway or in biotoxin illnesses.

"HLA susceptibility is based on my genes," said Jim, summing things up. "If I get exposed to a toxin invader, it won't be cleared by antibodies formed by the HLA instructions. The instructions don't arrive. Then the toxin activates complement, which you can measure by increased C3a, and that protein in turn shoots up immediately following exposure. Then you can see if the toxin is going away by measuring C3a as you're treated. But some people don't clear the extra C3a even with treatment, and they're the one who develop chronic fatigue."

"Slow down!" I laughed. "We'll come back to Chronic Fatigue Syndrome later. All we know is that some genotypes, including Tami's, are tremendously over-represented in those who have Chronic Fatigue Syndrome and that many of those patients have elevated C3a. We have to do the tests to see if there is cause and effect."

"If the toxins aren't cleared by antibodies grabbing them, how do they stay in the body? Won't the liver, the main detox organ, break them down, or won't the kidneys excrete them in urine?" Jim asked.

Neither occurs. Without antibodies directly clearing toxins, patients with genetic susceptibility stay ill. The only mechanism they have to eliminate the offending toxins is found in the cells that line small bile ducts (canaliculi) in the liver. There, a series of proteins, called the organic anion transport system, removes biotoxins from blood and preferentially secretes them into bile.

Bile is like the liver's sewer system; the liver devotes a lot of energy to throwing away what it doesn't want. The bile eventually flows into the upper intestine, following a short stay in the gallbladder if you have one. But unfortunately, toxins don't *stay* in the intestine to be dumped into *real* sewer systems later on; they get reabsorbed before they reach the colon, so they remain in the body. Toxins will remain forever in genetically susceptible patients unless treatment removes them.

"OK, let me make sure I've got it," said Jim. "For an individual with a given HLA-DR, a defective toxin-clearer, if you will, the toxins aren't recognized as foreign invaders and the liver doesn't order a 'detoxifying' response. In those individuals, when exposed to mold, Lyme leftovers or particular algae, the immune system doesn't recognize the toxins as a threat, and so it doesn't produce antibodies to attack and destroy them.

"Unchecked, the toxins hit my body, causing illness symptoms anywhere and everywhere, since biotoxins are able to pass easily through all blood barriers and membranes, due to their special structure. And once a person hosts a toxin, it doesn't leave on its own. And that's why some people stay sick forever after they're exposed. They don't self-heal. But that's where you come in. You bind the toxins with CSM and the primary illness dissolves."

Jim stopped for a moment to gather his thoughts. *"Tell me more about treatment. I think I'm in love with Tami, but she's always been sick. You've got to get her better so she can have a life again. And I hope it is with me!"*

"OK, Jim here goes. Part of the strategy I use to attack and remove toxins begins by trying to understand the biology of those that cause chronic human illness. Don't forget that when I use the word 'biotoxin,' I'm not talking about every toxic substance made by living creatures. I'm talking about toxins that are ionophores, have a similar structure and weight, and that activate particular receptors. Got all that?" I asked.

Toxins have molecular structures that consist of an inner, water-loving (hydrophilic) system surrounded by a fat-loving (hydrophobic) group of molecules. Biotoxins are a lot like the emulsions we use in cooking, with a mixture of active agents divided by their affinity for water.

Curiously, the size of the toxin's innermost part is identical to the size of a water molecule. What an extraordinary finding in evolution! It seems as though the structure of the toxins made by fungi, blue-green algae, bacteria, dinoflagellates, apicomplexans, spirochetes and recluse spiders (and what else?) all evolved based on the *structure of water*. I haven't seen the same structural similarity in toxins made by plants.

This dangerous, water-like structure allows the molecules to diffuse easily across cell membranes, causing damage. It's this property of toxins that makes them part of a larger group of compounds called ionophores. Not all ionophores are toxins, but to date, the biotoxins in my work have all been ionophores.

Ionophores can move quickly across membranes and go from cell to cell, distributing themselves throughout the body. Amazingly, they do not need blood to travel, but can jump from one cell the next. So a biotoxin in the nasal passages can end up in your skin, heart, liver, lungs, muscles and brain.

Fat Cells and Toxins

Much worse, however, is the ability of these highly mobile biotoxins to bind to receptors, called "Toll receptors," on the outer membrane of *fat cells*. The biotoxin receptors apparently aren't the same kind of receptors used by some bacterial toxins. The bacterial toxins end up being taken in by the cell. Modern science knows that biotoxins simply send a message from the world outside the cell, and that message is translated from the outside surface of the cell membrane into a signal that eventually homes in on the nucleus of the cell.

If you're like most Americans with extra pounds, you'll find many topics in this book that relate to weight, rarely appreciated in weight loss books. Contrary to what most people believe, fat cells are extraordinarily active metabolically and aren't just unwelcome globs of fatty acids. Many experts feel that the fat cell is probably more important in the *combination* of *endocrine* function and *immune* function than any other type of human cell.

"So," said Jim. "The receptor idea is like an electrical plug that requires a special outlet. Or a reserved parking space. This toxin parks here, not over there. If the toxins temporarily bind to fat cell receptors, and the cell doesn't engulf them, then there's going to be a time when they ***leave*** *the receptor, travel around and then eventually, they're dumped into bile and then the upper intestine. And that's when you can bind them with cholestyramine (CSM) because CSM has a positive charge with the same size as the negative charge of the toxin, and opposites attract. The toxin is stuck on CSM by electrical glue, so down into the commode it goes."*

"You're so right," I congratulated him. "You *are* feeling better, aren't you?"

"Jim, I want to go over how the toxin *outside* the cell can turn on genes *inside* the cell. It's the key to the chronic illness that toxins cause. If all toxins did were to float around endlessly, there would be some direct acute effects and that would be all there was to the illness. But toxins wage *gene warfare* against us. Think about it: It's the inappropriate activation of genes that leads us to the additional layers of illness in the Biotoxin Pathway, as you'll read. Here's the reason these biotoxin illnesses end up being so complex. They manipulate both the endocrine and immune mechanisms of the fat cell and turn that machinery against us. And the attacks never stop. Ready?"

Once biotoxins bind to a fat cell receptor, they release a second messenger, a compound that moves from the membrane, floating into the cell. There it binds to and activates release of a "cloaked" molecule, a third messenger called nuclear factor (NFkB), that swiftly moves to the nucleus. It's called nuclear factor because it turns on gene transcription of a segment of our DNA that in turn codes for cytokine production, which leads to inflammation. When NFkB acts on genes when it shouldn't, due to biotoxins, you'll have chronic inflammation.

"So NFkB becomes a target of toxins," Jim said. "It sounds like we use NFkB to live when it causes inflammation at the right time, but if we're over-using it, then we get sick from all the cytokines the fat cell makes. All these steps are linked, like dominos falling in many directions."

"Right, and that's the reason that biology isn't physics," I explained. "In physics, one action will cause one reaction. In biology, one action, like a toxin sending a message to DNA, causes a *cascade* of multiple effects. The excess cytokines can do many other things on their own, Jim. But let's slow down a minute."

"No, keep going. I'm with you. Ping-pong isn't pool. In ping-pong, you hit the ball and one thing happens—hopefully, it goes across the net. In pool, you hit the cue ball and you can make 16 balls scatter, each hitting different balls and different sides all at once."

The nuclear factor then enters the nucleus of the fat cell, where it binds onto a special site in the DNA called the cytokine nuclear receptor. Here's the source of so many chronic symptoms—fatigue, pain, brain fog and weight gain—in mold patients. This receptor is like a switch that controls a small group of genes that make cytokines. Block NFkB or block the cytokine receptor and illness symptoms are reduced. When NFkB turns on the nuclear receptors, they signal the cytokine genes to crank out the cytokines. This is great if you're ill and need cytokines to help you, but it's terrible if toxins are turning on the cytokine-generating machinery at the wrong time. This selective DNA regulation is a critical factor that determines how an organism responds to signals from its environment.

Think of specific genes linked as separate strings of Christmas lights, each controlled by a light switch. Turn on the switch and all the lights go on. The switch is the same as the receptor. Every time the *receptor* is turned on, *all the genes* under its control are activated as well, but *only* those genes in that particular string are turned on, not every Christmas light in the neighborhood. When the receptor is no longer activated, the genes are turned off.

You'll see that this is very important to how you feel. In general, healing your illness involves turning on some genes and turning off others, much like going through your house and turning on some lights in rooms that need it and turning off lights when no one is using the room.

When we talk later about preventing intensification reactions, manipulating the cytokine genes is critical. If we're trying to lower matrix metalloproteinase-9 (MMP9), leptin and plasminogen activator inhibitor-1 (PAI-1) in mold-exposed patients (you'll see how important that is!), we'll be talking about using this kind of targeted *gene therapy*. Soon you'll meet medications (call them TZD) that can soothe those patients made ill by excessive cytokine responses to biotoxins. They control the cytokine genes, finally calming the revved up cytokine producing biotoxin system and bringing relief.

Cytokines

The genes turned on by NFkB are necessary for life: they regulate fat cell production of cytokines. Cytokines are special proteins responsible for healthy immune and inflammatory processes. They're also made by the white blood cells that are killers of infectious agents and cancer cells. The cells that line your blood vessels also make cytokines. Cytokines regulate the cells of the immune response.

"Slow down, Ritchie," said Jim, momentarily confused. "Cytokines sound really important, but what are they?"

There are two main kinds of cytokines, I explained: the pro-inflammatory cytokines and their counterparts, the anti-inflammatory cytokines. Both are involved in the immune response. We need each to live, and neither exists without the other's contribution.

The pro-inflammatory cytokines are the focus of this discussion so I'll call them 'cytokines' to ease the distinctions between pro- and anti-inflammatory cytokines and make the language a little more tolerable. Cytokines *recognize* foreign invaders like viruses, bacteria and out-of-body proteins, and set off an alarm that *recruits* additional cytokines for war. In addition to these warrior cytokine proteins, the immune response recruits additional defense cells, including white blood cells like macrophages and lymphocytes, multiplying the defenses. The immune cells make even more cytokines, including the big boys, the *killer* cytokines, like tumor necrosis factor and interleukin-1-beta. In the end, as the immune response recognizes and eliminates the foreign antigens, it's the heavy hitters, tumor necrosis factor alpha (TNF) and interleukin-1-beta (IL-1B) that finish the job. When the cytokine system recognizes that the foreign antigens are gone, the recognition cytokines jump in and says, "All is clear," the cytokine cascade turns into a trickle and then turns off.

"But that wasn't happening to me until I took CSM," said Jim. "That got rid of the toxins and the heavy cytokine bombardment the toxins caused. And it hasn't stopped for Tami has it? Her cytokines are working overtime, so she stays feeling bad, like she has the permanent flu. She's exhausted and everything hurts, too. And no kidding, her earlier doctors tried to hide when she came into their clinic. What doctor wants to listen to 30 health complaints from one patient? All those symptoms were from cytokines? How is my flu different than her daily flu feelings?"

The symptoms you had from your flu last November were due to a *cytokine response* to the virus, but not to the virus itself. Your body sent out cytokine armies to fight off the flu antigens when you felt tired, achy and had a headache. You covered up in blankets when your cytokine-caused chills came on and threw them off when the cytokine-caused sweats followed. Maybe you felt a little confused. More cytokines, but this time in the brain. We all know when the cytokine response is over; we begin to feel better, the aching stops and the fever breaks. Grandma could tell there was a cytokine conflict coming on by looking at a child's eyes—that glazed, sick look is from the cytokine response. She was right, cytokine effects are readily recognized, even if you don't use fancy immunology lingo.

But what would happen if there were no controlling, "cease-fire" signal from the recognition cytokines that stopped the cytokine response cascade? What would happen if biotoxins continuously turned on the fat cell cytokine nuclear receptor? The receptor would just keep cranking out its gene products. We would have too much of one or more of the gene products, including TNF, MMP9, IL-1B, leptin, PAI-1. Imagine these products endlessly manufactured and released inappropriately, making the biotoxin illness chronic. This loss of control of cytokine response is one of the features of a biotoxin pathway.

"It's as if we didn't have a thermostat to turn down the heat in our homes in winter or if we rarely took our foot off the gas pedal," Jim said. "I get it: too much cytokine response going on when it should be shut off makes us sick. The excess cytokine production is the second devastating layer of the illness."

Chronic biotoxin illness involves a series of failed control mechanisms that predictably follow exposure to and retention of biotoxins in the body. These products of the cytokine nuclear receptors are necessary for normal life, but they *must be regulated.* The over-production at the wrong place (in fat cells) and at the wrong time (constantly) is like a fire in the wood stove in your home in winter. The heat from the stove keeps everyone's body warm and watching the fire in the stove warms your soul. But if embers from the wood stove aren't regulated and are instead glowing in the middle of the living room, burning up the rug and the sofa, you have a real problem.

Uncontrolled, ongoing production of cytokines made by fat cells provides a second mechanism for producing symptoms like fatigue, pain or cognitive dysfunction. For the biotoxin-sickened patient, there are no days without cytokine-related symptoms. Fat cells stimulated by biotoxins no longer respond to control mechanisms, and a series of predictable abnormalities follow. We can look for those abnormalities as *markers* for biotoxin-illness in the blood. When we see the markers improve with treatment of toxins and worsen with re-exposure to toxins in multiple studies of patients, we have a model to help us understand what's going on inside an individual patient

MMP9

Recently our understanding of illnesses characterized by chronic cytokine excesses has changed dramatically. Now we're able to measure matrix metalloproteinase-9 (MMP9) levels with a simple blood test. Cytokines cause white blood cells to release MMP9, so we know that high MMP9 means high cytokines. People can be tested quickly to determine if cytokines and toxins are contributing to the illness by looking at MMP9 levels. If you've been exposed to biotoxins, and the MMP9 is too high, it means you have a high toxin load. Some patients with biotoxin illness *won't* have a high MMP9, however, so you have to look carefully at *all* aspects of an illness to understand where MMP9 fits in.

One reason why patients with biotoxin-associated illnesses always feel so bad is high MMP9. The MMP9 won't tell you which toxin is the source of their illness, and there are some illnesses not caused by toxins that also have a high MMP9. MMP9 won't make the diagnosis, but measuring MMP9 is part of the diagnostic process. One of the consequences that result from excessive cytokine production is increased MMP9 production. MMP9 is a protein-digesting enzyme that dissolves complicated molecules in the matrix, the medical name for the tissue beneath the cell membranes in blood vessel walls.

"My MMP9 was more than 600 and it fell to 250 as I felt better. I remember you said normal is less than 300, right Ritchie?" asked Jim.

I explained that MMP9 is the "delivery enzyme" that lets inflammatory compounds get out of the bloodstream and into target tissues like nerve, muscle, brain, joint and lung, where they do major damage. Show me a person with high MMP9, and I'll show you someone who feels ill due to a cytokine response. We'll come back to this point repeatedly.

In order for any compound to get out of your blood vessels and into the underlying solid organ tissue, those compounds must be delivered across a series of barriers. The first two barriers are the two membranes of each cell that line the blood vessel. Next, there's the thick gristle, we call it the basement membrane, which provides the underlying scaffolding for endothelial cells. Finally there's the thick, sticky ooze that is the matrix itself. MMP9 makes the tunnel, "eating its way" through the dense matrix, acting as the "deliveryman."

Later when we talk about cholesterol, keep this picture in your mind: the bloodstream doesn't give up its molecules easily. You need a delivery mechanism to do the work.

Elevated MMP9 levels are commonly seen in biotoxin illness patients. It's MMP9 that delivers very powerful inflammatory elements from the blood in many serious illnesses. In the brain, MMP9 delivers the inflammatory perpetrators that potentially cause the plaque formation that looks like multiple sclerosis (MS); into joints, where it causes inflammatory arthritis; into muscles, where it adds to the delayed recovery from normal activity that's so common; and into the lungs, where it produces coughing and contributes to shortness of breath.

We measure MMP9 repeatedly, since it's one of the best markers for disease progression available. MMP9 levels can change rapidly, as illustrated by our work with patients who suffer from immediate exposure to fungal toxins, recluse spider bites and blue-green algae. Levels can go from a comfortable 200 to an agonizing 800 in 48 hours. Lyme patients know what rocketing MMP9 levels feel like when they take antibiotics and the bacteria start to die, causing a Herxheimer reaction (AKA "die-off reaction"). This incredible worsening in symptoms is simply the result of effects from rapidly rising MMP9 and tumor necrosis factor (TNF). What a combination! High levels of either TNF or MMP9 make us feel awful; in the Herxheimer, both attack at once.

Why do we rely on MMP9 as a marker for cytokine activity? Couldn't we just order a blood panel that would give us accurate readings of all the cytokines? I'd like that! But we don't have good assays available from the standard labs yet. Even worse, the blood readings might not reflect what's actually going on. Why? Cytokines can bind to their own receptors along blood vessel walls, so a blood test would report only the *unbound* portion, which wouldn't necessarily reflect their true levels. Fortunately, MMP9 levels are excellent markers to show activation of a number of cytokines.

If the attending physician hadn't measured MMP9, how would she be sure that patients really were feeling bad? When the physician doesn't know about MMP9, and then tries to say the problem is psychological or psychosomatic, patients are being mistreated. I remember when one patient was so relieved to find a sky-high MMP9. She told me, "At least I now know it's not all in my head."

"Ritchie, isn't this like the patient with high blood pressure?" asks Jim. "If the blood pressure isn't measured, how do you know if it's part of the problem? When they have a stroke or heart attack that came from untreated hypertension, are physicians going to say the illness is stress or depression? Come to think of it, isn't that what happened to Franklin Delano Roosevelt when his blood pressure was so high? They said it was ***stress*** *and sent him to some resort in Arkansas to relax, if I remember correctly. Goodness, the mold-naysayer physicians are repeating the same mistakes of history! A biotoxin turning on all those cytokines, without any feedback control sounds bad enough, but I know there's a lot more to these illnesses. Go ahead, I'm with you so far."*

• • •

VEGF, Hypoperfusion and Energy

We're not done with all the bad things excessive, uncontrolled cytokine release can do to us.

Cytokines result from the *effect* of the initial biotoxin illness, and once unleashed, they are the *cause* of so many additional problems. Cytokines cause complications by disrupting *other* chemical pathways. Let's go back to the basic mechanisms of how excessive cytokines harm us, in order to understand the *third layer* of attack on biotoxin patients. There's another term you'll need to understand—"hypoperfusion"—the medical term for reduced blood flow in capillaries. Follow along with me to see how cytokines first knock out control of production of vascular endothelial growth factor (VEGF) and then watch what happens when uncontrolled VEGF knocks out normal perfusion.

Capillary beds are tiny hair-like blood vessels that are the final road traveled by blood to get to the tissues. These microscopic-sized blood vessel walls are lined with endothelial cells. Endothelial cells can make cytokines and they also have receptors for them. Having both local production and binding capability for cytokines means that blood vessels can react to cytokines made anywhere in the body, as well as to locally produced cytokines. In turn, that means that there's an intricate interaction of endothelial cells and cytokines that enables specific targets in blood vessels to receive both general and local signals. That specificity for cytokines and for targeting in turn guarantees that there's coordination of immune response from cytokines throughout *all* the beds of capillaries in the body. Any attack on blood vessels by a foreign invader will be greeted with a coordinated, body-wide defensive response.

Once activated by cytokines binding to their receptor, the endothelial cells release their "glues" (adhesins and integrins), which "grab and hold" circulating white blood cells in place at the receptor site. The aggregation of white cells, all flocking to the receptor like Japanese beetles following a pheromone in a trap, and now held in place by the glues, creates a logjam in the small capillary, functionally narrowing it.

With the acquired obstruction to blood flow in place, little blood gets through to the other side. This reduced flow of blood in capillaries, called capillary hypoperfusion, reduces the total amount of *oxygen* and other vital nutrients that are normally delivered to the rest of the capillary network on the other side of the blockage.

The concentration of the oxygen in the blood isn't changed, but the hypoperfusion starves the downstream cells for nutrients and oxygen. The reduced oxygen, called hypoxia, is sensed by regulatory cells, which in turn react by producing a gene controller called hypoxia inducible factor (HIF). HIF produces vascular endothelial growth factor (VEGF), the next big player in biotoxin illnesses. Later on, when we talk about erythropoietin, don't forget that HIF turns on production of that "anti-cytokine" too.

"Ritchie, I'm boggled by all these dominoes. This series of events at first sounds complicated, but now it sounds like what happens is easy —one, two three. You start with the toxin, add the cytokines, bind the cytokines, plug up the capillaries and guess what? You've got the oxygen boys reacting," said Jim.

Jim was getting it! I continued.

VEGF stimulates the blood vessel growth, and in healthy people, under normal circumstances (without an excess of cytokines), it corrects the hypoxia by *dilating* blood vessels, increasing oxygen. If the low-oxygen problem persists, VEGF amazingly creates new blood vessels, bringing in more oxygen and nutrients. VEGF is getting a lot of research attention these days, because too much VEGF might feed cancer cells—give them more fuel to grow faster. Low levels of VEGF, like what we see in biotoxin illnesses, are rarely discussed in today's science. What a mistake!

Unfortunately, the high cytokine levels found in biotoxin illnesses *suppress* VEGF production. The effect of this suppression is profound. If VEGF isn't turned on, then mere modest exercise can be overwhelming. If VEGF isn't turned on, there's no way to increase blood flow in capillaries in your tired tissues during exercise. Persistent capillary hypoperfusion—in other words, not enough blood flow—means persistent fatigue, cognitive problems, aching and all the rest of the systemic symptoms associated with biotoxin illnesses.

Once again, we see a failure to regulate normal physiologic events. VEGF is just one more example to add to the list.

Linking chronic symptoms to low VEGF levels is easy. We can measure what low VEGF does to oxygen delivery in capillaries by giving patients a metabolic stress test.

Done in a hospital pulmonary lab, this 10-minute test measures the metabolic response to exercise. Low-VEGF patients won't be able to generate much oxygen delivery while they're being monitored riding a bicycle.

The maximum available oxygen (VO2 max) throughout the body in a low VEGF patient will be quite low. In fact, their low oxygen level in capillaries is so low, it is identical to what we see in patients with advanced cardiac disease!

Moreover, the "*anaerobic threshold*" becomes markedly reduced. Remember all those fake blondes in their Spandex on TV doing aerobics? They're pushing themselves to exercise maximally within their ability to use oxygen to burn calories.

Aerobic means exercise with oxygen; *anaerobic* means exercise without oxygen. Low VEGF people can't deliver the extra oxygen required to exercise. They can't exercise effectively to improve their fitness! The speed of onset of fatigue in these people is incredible, or stated medically, the threshold at which they can no longer deliver extra blood flow and therefore extra oxygen, to muscles is drastically reduced. That's the anaerobic threshold.

• • •

If you're low VEGF, you'll be low in delivery of oxygen to capillaries, too. And that deficiency is made tremendously worse with activity. The active cell, looking for oxygen, won't have the right amount available to burn glucose for energy. Normally, the cell gets two ATP (energy) molecules from the initial breakdown of sugar (glucose), generating breakdown-products (pyruvate and lactate). Those compounds are full of locked-up energy and can be processed further in mitochondria, but *only if oxygen is present.*

The mitochondria are called the "powerhouses" of the cell, because they generate so much extra ATP. Using oxygen, the two sugar breakdown fragments are eventually broken down into water and carbon dioxide, creating ***thirty six*** additional ATP molecules along the way. If there isn't enough oxygen available during exercise the cell acts like it doesn't have any mitochondria. The cell starts to be energy inefficient, burning a huge amount of sugar but giving the cell only two of the required ATP at a time, not two plus the thirty six.

Then the cell quickly begins to consume all the stored sugar (glycogen) from the cell's warehouse. But the amount of the storage glycogen is limited and must be replaced quickly—or else we die. So to restore glycogen levels, the cells reach for whatever alternative fuel sources it has on hand. Typically, that's your own body's protein, because protein is quickly broken down into amino acids. Two of those building blocks, alanine and glutamine, are rapidly turned from amino acids into sugar. Glycogen is replenished and we live! But the cost of burning lean body mass is enormously expensive for biotoxin patients. We can measure lean body mass too.

"So with this illness, Tami and I are burning protein, like a guy in some prison camp who is eating rice water and uncooked rat meat, we're storing fat when we shouldn't and we are wasting all our stored sugar. What a nice illness. Is that why I felt so tired after I tried to do a little extra, back when I was sick?" asked Jim. "It's like I had no way to get energy. Did that mean my mitochondria were just sitting around waiting for some oxygen so they could get some work done? And my muscles were getting 2 miles per gallon of glucose compared to 38 miles per gallon, like

they were supposed to? And that guy I saw in the Olympics running the marathon. It was like he was completely out of gas when he collapsed over the finish line."

"Correct, Jim," I said. "When you burned up the extra storage of glucose from your glycogen, the 'storage depot' of glucose for the cell, you burned protein inappropriately. Even worse, *since no one can replenish glycogen quickly, you feel bad for several days while the muscles patiently and steadily re-make the glycogen.* There are some fancy phrases the academic types use when they talk about this problem in Chronic Fatigue Syndrome patients, like 'post-exertional malaise' and 'delayed recovery from normal activity' that try to explain what happens when they exercise or do too much. What those empty phrases mean is that low VEGF creates chronic fatigue, due to an unregulated cytokine response to toxins."

"Now I get it," said Jim. "As my daily CSM removed toxins, the cytokine damage to VEGF production stopped, my VEGF went up and I was able to use glucose normally instead of wasting it. That's the reason why I actually gained weight when I started on CSM! I was replenishing the protein stores I had burned. And that amount was what I was programmed to have by my genes! I actually owed my body protein weight that had been consumed due to my illness. It makes sense. Ritchie, you know, this Biotoxin Pathway isn't so hard to understand, after all."

• • •

Autoimmunity

If only biotoxin illnesses involved just the three layers of illness discussed so far …

In reality, the effects of biotoxins are incredibly complex, involving many more mechanisms. The key is that excessive cytokine production can turn on autoimmunity. We've all heard of a couple of autoimmune diseases—systemic sclerosis and rheumatoid arthritis are some of the more common—and those illnesses involve out-of-control inflammation. Basically, there's an attack on our own organs by antibodies from our immune system. To use a military analogy like we used in explaining complement, these illnesses are like "friendly fire." Somebody should tell the immune response elements in the lupus patient to quit shooting at her own kidneys!

We know that autoimmune illnesses are inflammatory. The problem is that we don't know if inflammation turned on the autoimmune warfare first or whether the autoimmune genes were turned on first, causing the inflammation to then occur.

Whichever happens, autoimmune illness is really common in biotoxin patients. Maybe viruses alone are the culprits, and not biotoxins (*I doubt it*), setting off the series of events with inflammation and autoimmunity; or maybe intra-cellular bacteria, like *Chlamydia* or *Mycoplasma* are involved. We just don't know. One thing is certain: once the autoimmune genes are turned on, the inflammation process goes violently out of control. We use cytokine-binding drugs and immune-suppressant drugs to help lupus and RA patients. Too bad we just can't turn off the antibody genes; that would stop the attack on our own "self" for sure.

• • •

The same kinds of attacks happen in biotoxin illnesses, *especially* in those people who have long arms, or what we call "wingspan greater than height." I remember the first time I started looking for an association of long wingspan and autoimmune illness in biotoxin patients, I felt kind of sheepish. But sure enough, even though height is greater than wingspan in more than 90 percent of people, the *wingspan* group represented more than 80 percent of those with the autoantibodies. Associations between wingspan and autoantibodies, of course, don't necessarily imply cause and effect. The association is so incredibly strong, though; I'll keep recording the data.

There are three types of autoantibodies that show up a lot in biotoxin illnesses. Each of those antibodies, including those made against myelin basic protein, cardiolipins and gliadin, respectively, can cause or add to chronic illnesses. If you attack myelin, the covering of nerves in the body and brain, something goes wrong neurologically. If you attack cardiolipin, look out for abnormal clotting and abnormal circulation. Gliadin antibodies* are the most common in biotoxin illnesses: What havoc they cause with our diets!

This is a hot topic in research, so it's possible that the ideas you're reading about now may generate new breakthroughs and discoveries tomorrow.

Antigliadin Antibodies (AGA)

Gliadin is the 18-amino acid protein found in gluten that causes terrible gastrointestinal problems for people who make *autoantibodies* to gliadin. When the gliadin molecule is absorbed from the small intestine in these victims, which doesn't happen in normal patients, the immune cytokines see the gliadin for what it is: a foreign invader. The cytokine alarms go off and the antibodies made to counter the invasion, the AGA, are released. That antibody/antigen response turns on inflammation, including the classical complement pathway. Adding to the cytokine and inflammation burden hurts biotoxin patients. This is probably why so many biotoxin patients feel better when they eliminate gluten, the main dietary source of gliadin, from their diet.

Those same patients should also pass up all foodstuffs that contain the complex carbohydrate amylose. The no-amylose diet, summarized in the appendix (and covered in detail in one of my other books, *Lose the Weight You Hate)*, essentially eliminates

---

*AGA are a special category of "autoantibodies." They are combined with all the others for simplicity.

dietary gluten. These antigliadin antibodies are the same ones that we see in the genetically based gastrointestinal illness known as celiac disease, but AGA are found in a number of other rare conditions as well. Biotoxic patients are more common than celiac patients, so if you find AGA, look for a biotoxin illness! In contrast, AGA (we're not talking about celiac disease here) is almost never found in patients who don't have a biotoxin illness.

Anticardiolipin Antibodies (ACLA)

Antibodies to cardiolipin, ACLA, are common in the world of biotoxin-induced illnesses. You may know ACLA better by the name "lupus anticoagulant" or phospholipids. ACLA can reduce circulation in the small blood vessels. Whenever you see someone who has unusual problems with clotting and small blood vessel illness, think of ACLA attack initiated by biotoxin exposure. In people whose hands turn purple and then white with exposure to cold—the Raynaud's phenomenon—there's a subset of patients with serious biotoxin illness. Antibodies to cardiolipin aren't commonly found in healthy patients, or in any illness (not even in the textbook anticardiolipin ailments like lupus and scleroderma), but they *are* found in biotoxin patients with alarming frequency. Anticardiolipins can contribute to unusual clotting, from toes to brain; they're not a welcome addition to your immune response.

One frequent sad effect of anticardiolipins for biotoxic patients is fertility problems. Every obstetrician can tell you about the increased frequency of first trimester fetal loss seen in patients who carry anticardiolipin antibodies. When an environmental exposure makes you sick, it's bad enough, but the inability to carry a fetus to term resulting from heightened cytokine production is a tragedy.

Not too long ago, an Amish family came to my office for another opinion on treatment of Chronic Lyme disease. They'd been taking antibiotics for a long time without benefit. The mother had been pregnant 13 times, resulting in nine miscarriages and four living children. As it turned out, the problem wasn't Lyme disease but was actually due to mold exposure. When you hear that the whole family is sick, always ask about mold growth indoors. Her anticardiolipin antibodies had never been checked. If her mold illness had been diagnosed and treated properly, perhaps the tragedy of her failed pregnancies could have been avoided.

When ACLA occur with other biomarkers, like high C3a, many biotoxin patients need unusual medications, like heparin, to improve circulation in the smallest blood vessels. Heparin will lower C3a and lower the adverse effects of ACLA.

Myelin Basic Protein Antibodies (MBP)

The presence of antibodies to myelin basic protein (MBP) is another ominous development that we see in patients with exposure to toxigenic indoor fungi. Myelin is what covers your nerve fibers (axons); it's like the insulation for the wires of the nervous system. Without normal myelin function, all the nerves in your brain and most throughout the rest of your body are possibly compromised to immune attack. When one thinks of MBP, it's nearly automatic to think of multiple sclerosis (MS) and it's indeed the case that MBP antibodies are tightly linked to MS.

In a recent series of articles in the *New England Journal of Medicine*, antibodies to MBP at the onset of MS were significantly associated with persistent illness from MS at five-year follow-up. One of the hallmarks of MS is the presence of plaque, or scars, in the brains of patients, as seen on an MRI. These "unidentified bright objects (UBOs)," as the neuroradiologists might call them, aren't necessarily diagnostic of MS. In fact, they're sometimes found as a "non-specific finding" in normal patients, too

Because UBOs are seen with a much greater frequency in mold patients with elevated levels of MBP, UBOs shouldn't be discounted as being non-specific in those particular patients. The current work of our group, the Center for Research on Biotoxin Associated Illnesses (CRBAI), shows a significant correlation of these bright objects with MBP antibodies. If we show development of UBO with MMP9 and other markers for cytokine effects on nerve tissue, we might be able to prove that mold exposure *causes* plaque formation.

"Ritchie, you said that the autoantibodies are more common in long-armed people," said Jim thoughtfully. "I'm wondering, was the Amish mother long-armed?"

"Yes," I answered. "It might not make biological sense, and maybe it's just a curious finding, but by observation, wingspan greater than height is definitely associated with autoantibody production in patients with biotoxin illness. What we're looking at now is a series of wingspan-positive patients without biotoxin exposure and without autoantibodies. We'll follow them over time to see when and if they develop autoantibodies. If they do, we'll be on the lookout for biotoxin exposure."

"You're not ready to start telling people with long arms to stay out of moldy buildings are you, Ritchie?" Jim asked, only half jokingly.

"Well, I'd like to *tell everyone to stay out of moldy buildings*," I said. "But actually, simply because of its correlation, wingspan measurement is a baseline screening test I perform routinely, along with HLA DR, VEGF, MSH and some others."

"Don't tell me that you're saying that you can diagnose chronic fatigue by measuring wingspan," said Jim. "Some defense attorney will make fun of you, saying you're like those old time practitioners who said they could tell your future by the bumps on your head. They called them phrenologists, I think."

"Actually, one guy did that," I said, and I laughed, remembering, "Tried to make me look silly about wingspan, but it turned out that his fellow defense attorney wasn't laughing in the deposition—after all, he had a mold problem in his bathroom at home, often lost his train of thought, was always tired and his cough never went away. He always had to buy those extra long shirts so he didn't look like Ichabod Crane in court. After the case was over (they lost), he called the office to get a new patient packet."

"Ritchie, Tami has those bright spots on her MRI. The neurologist said, 'forget it,'" said Jim, a lot more quietly. The reality of the illness in his loved one was taking its toll now.

"Well, Jim, I would have said that, too, until I started working with neuroradiologists like Dr. John Rees a couple of years ago." (We'll meet Dr. Rees a bit later.)

The point is that there are many genetic linkages that might be there; we just don't have much data about them yet. Hopefully, now that we have the map of the human genome, researchers will become interested in how genes and environmental factors interact.

Our HLA DR registry is a huge potential library of information. Keep an open mind; don't let the critics bother you about wingspan. The genetics that underlie hundreds of illnesses are just now starting to be unveiled. Autoantibody production, once opened like Pandora's Box, creates a terrible possibility that environmental exposures to biotoxins are tied to our genes. The activation of those genes, interacting with cytokines, VEGF, MMP9 and all the rest, affects our diet, circulation and our nervous system.

• • •

Thiazolidinediones (TZD) and Gene Therapy

Fortunately, we have some weapons to help limit the biotoxins' ability to turn on the illness-causing genes. We can block the abnormal activation of the cytokine nuclear receptor with drugs, called thiazolidinediones (TZD) that were originally developed to treat diabetes.

Two members of the TZD family are now on the market, Actos and Avandia. Each has been shown to be safe for our livers, unlike the first TZD, Rezulin, which had to be pulled from the market several years ago.

The TZDs turn on genes, just like NFkB turns on the cytokine genes. The genes that TZD switch on aren't the same as the NFkB genes. In fact, the TZD genes directly *oppose* activation of the cytokine genes by NFkB. That means use of TZDs could lower TNF, leptin, MMP9 and PAI-1after biotoxins cause them to be overproduced! Even better, TZD can turn on the genes needed to increase VEGF production.

The nuts and bolts of TZD are that they do their job by activating a group of nuclear receptors called PPAR-gamma. PPAR in turn, controls production of two proteins that help move sugar more efficiently into cells—that's why we use TZD drugs to help treat diabetes. But the TZD benefit doesn't come from lowering blood sugar levels; instead TZD prevent sugar levels from rising after a meal. And there are so many other benefits from TZDs!

For example, PPAR genes control uncoupling protein, a compound that makes the fat cell burn fatty acids directly. They also control fatty acid uptake and enzymes that protect us from excessive digestion of HDL cholesterol. Don't forget, however, that the foods that contain the plant starch called amylose, a complex carbohydrate, are quickly broken down by saliva to make glucose. So when those foods are eaten, the blood sugar rises rapidly. Insulin will be released in tremendous amounts to counteract the rise in blood sugar. And insulin rising like that *prevents PPAR activation* by TZD.

When we prescribe TZD, making sure patients follow the no-amylose diet, we're actually prescribing gene therapy that not only helps prevent cytokine excess, but also changes how our body handles fats.

Oh, and it helps diabetics, too. Using these drugs in patients without diabetes is safe, provided the diet includes frequent small feedings and does not include amylose.

We've used the TZD drugs successfully in weight loss, reducing TNF (tumor necrosis factor) in Lyme patients, reducing levels of MMP9 and the accompanying MS-like symptoms that result from fungal toxins, and as an adjunct to lowering PAI-1 in atherosclerosis patients. A potential problem in using TZD comes from its use in patients with low leptin levels, as we'll discuss in the MSH section. If the patient is slender, with an inability to gain weight, avoid the TZD until increased fat in the diet has artificially raised the leptin.

• • •

Melanocyte Stimulating Hormone (MSH)

So far, I've talked about effects of cytokines, inflammation, vascular growth factors and autoimmunity in chronic fatiguing illnesses as well as some basics on how each of these elements function in the Biotoxin Pathway. I want to tell you a lot about MSH: I am amazed at its importance as the ultimate controlling hormone in so many pathways. Understanding this hormone has helped many patients heal. I think that MSH is similar to a "Boss" in a large corporation. There are so many individual departments in the corporation, but each eventually answers to the Boss. The body too has so many different systems, with hormones here, complement there, mucus membranes and cytokines each interacting as they do their own jobs. All eventually answer to MSH.

You can have problems with MSH and still appear somewhat normal. When MSH is out of whack, your ability to *control* each of the many systems under MSH control is markedly diminished.

But to understand the role of MSH in how you think and feel, we need to look at the brain first. Don't forget, the brain is separated from the rest of the body by the blood brain barrier (BBB). Substances in the bloodstream from environmental sources might damage the vitally important central nervous system, so it isn't surprising that the body tightly controls what gets past the BBB and into the brain. The BBB acts as a filter for many compounds, though some, like cytokines, cross it as if it wasn't there.

Do you know someone with Parkinson's disease? The characteristic syndrome, with its tremor and gait problems, is due to a deficiency of dopamine in one part of the brain. Why don't we just give dopamine to the Parkinson's patient? Simple: dopamine doesn't cross the BBB. How about the guy who is depressed and needs to correct chemical imbalances in the brain caused by reduced serotonin? Why not give him serotonin? Why do we have to give the patient a selective serotonin reuptake inhibitor (SSRI) to raise serotonin? Simple answer again: serotonin doesn't cross the BBB. There are many other examples.

Cytokines just cruise across the BBB, looking for a compatible receptor. They find one quickly, but unfortunately for the biotoxin patient, the cytokine receptor is in the hypothalamus, right in the middle of one of our biggest and most important neuroregulatory pathways. Called the POMC for short (proopiomelanocortin pathway is a jaw-breaker word), the pathway controls production of alpha melanocyte stimulating hormone (MSH).

This hormone—probably the most important one that most people haven't heard of—has been unwisely neglected by modern medicine despite being involved in many key functions. For example, MSH controls melatonin and endorphin production, as well as regulating cytokine pathways throughout the body. This hormone is incredibly active, regulating pituitary function as well as protective cytokine responses in skin (where it was found first), in the gastrointestinal tract and nasal mucus membranes, and possibly in the lungs, too. MSH is the boss regulator of many hormone pathways. By "patrolling the periphery," of skin, mucus membranes and gut (just like complement), MSH controls many of our *innate* immune responses.

The POMC pathway begins with leptin, a big molecule made by fat cells that also quickly crosses into the brain. There it contributes to the control of many hormones. Ideally, leptin should bind to its receptor (the same one cytokines attach to), in the hypothalamus, activating it and ultimately making beta-endorphins and MSH. The POMC system is tightly regulated, so that leptin feeds MSH production, but then MSH swings around to control leptin. Leptin levels are very often too high in most patients with biotoxin illnesses, making us wonder if there is a problem with leptin turning on its receptor.

In biotoxin patients, this is not trivial science. When the flood of *cytokines* made in the body blocks the leptin receptor, leptin's effects are blunted. No receptor activity, no activation by leptin and no MSH. The results of leptin resistance are catastrophic for the biotoxin patients: If leptin can't turn on its receptor, then MSH isn't made. When production of MSH, the source of all normal regulation of nerve, hormone and immune function is disrupted, many features of biotoxin illness appear.

When MSH levels are low, normal melatonin production is wiped out. Therefore, the patient gets no normal, restorative, restful sleep. Naturally, chronic fatigue follows.

Low MSH also means low endorphins. Without the "natural opiates" produced by the brain, the body has no way to shut down responses to pain stimuli. In low MSH, things that wouldn't hurt a normal MSH person, like a simple, loving caress, can cause agony. Arthritis in the neck that isn't a big deal normally becomes multiply magnified in the low MSH patient. When MSH is low, chronic pain follows.

If you wonder why biotoxin illness patients always hurt so much, and in so many unusual ways, just follow the straight line on the Biotoxin Pathway: exposure to toxins in genetically susceptible patients causes increased cytokine production, which causes reduced MSH levels, resulting in an endorphin deficiency that changes the patient's *perception* of pain.

Even worse, while leptin is being blocked at its receptor, fat cells continue to produce even more leptin, trying to overpower their way through the balky POMC receptor. When that happens, there's too much leptin in the blood. This failure of the leptin receptor to function normally is called *leptin resistance*. It's no different from the *insulin resistance* we see in Type II diabetes and in obesity, where there is too much insulin in blood. In both cases, the receptor for the hormone just doesn't work due to cytokine damage.

So what does all the extra leptin do? Leptin resistance causes wildly increased storage of fatty acids in fat cells. If you are high in leptin, you will gain fat and you won't lose it. No way.

As an example, the leptin-resistant patient will "*store*" the triple cheeseburger he eats better than the person who isn't leptin resistant. If the surge in obesity in the U.S. is associated with all the triple cheeseburgers made in fast food joints, how come it's only a select group who pay the price for their indulgence? High leptin levels mean, "*Forget any chance at weight loss using routine methods*."

Incredible symptoms in these biotoxin illnesses—they keep us fat, fatigued and constantly fighting pain. There's little hope to correct that "trio of misery" if disruption of MSH production continues. And to add insult to illness, genetically susceptible patients who are chronically fat, fatigued, and fighting pain are variously told they're gluttonous slobs. If instead the health care provider wants to be politically correct, he'll call the significantly overweight person, "*morbidly obese*." That term is both patronizing and insulting. Can you feel the pain of the biotoxin patient who goes looking for health help, only to be regarded as a hypochondriac, depressed, or instead, plainly ignored by their doctors. It's sickening to hear the same sad story from my patients over and over again, day in and day out. Their doctors just don't know what biotoxin illnesses are. And that is so sad.

"I remember reading in your weight loss book that some environmental causes of diabetes and obesity are due to cytokine effects from toxins on insulin receptors. Are you telling me that leptin and insulin both factor into the explosion of obesity and diabetes in this country?" Jim asked, a bit

incredulous. "I remember looking at you funny when you asked how much weight I'd gained when I got my mold illness. Now I see—you were asking when I developed cytokine damage to my hypothalamic leptin receptor."

I nodded. Jim was getting it.

MSH also controls cytokine production by our infection fighting, white blood cells. MSH deficiency damages the responses of these cells, so that infectious diseases often last longer, with more symptoms. When patients say, "My immune system is out of whack, I always have some flu or body aches, and that's why I have chronic fatigue," what they're usually saying is that they're MSH deficient. If they say their immune system is depressed, stop them right there: it's immune *over-activity* with too many cytokines *shutting off MSH* that's the problem, not under-activity.

The lack of regulatory control at multiple levels of body response is the signature of biotoxin illnesses.

MSH also controls hormone release in the pituitary gland. Besides manufacturing growth hormone and thyroid stimulating hormone, the anterior pituitary also controls ACTH production. ACTH causes production of cortisol by the adrenal glands. Known as "the stress hormone," cortisol levels can be elevated in adults and children diagnosed with depression. In those patients, MSH deficiency is the more likely diagnosis.

Both ACTH and cortisol are tightly controlled. When cortisol is high, ACTH is normally shut down by an order from the hypothalamus. Conversely, when cortisol is low, ACTH production will be revved up by the hypothalamus; the adrenal begins releasing more cortisol. But biotoxic patients with MSH deficiency have problems regulating cortisol and ACTH in more than 40 percent of cases. We call this impaired control, *dysregulation.*

When there's disruption of MSH production, some patients won't have many symptoms. How can this be? MSH is the big controller—no MSH, no control and lots of symptoms. Here's the rub: ACTH overproduction increases cortisol, which can compensate *temporarily* for MSH deficiency. If the biotoxin illness is detected early on, the number of symptoms a patient has might be relatively few if the ACTH response is in "overdrive." We need to treat patients at this stage, before it's too late.

Once the heightened ACTH response "burns out," however, the symptoms we associate with MSH deficiency quickly appear. The hyper-stimulation of adrenal production of cortisol also falls as ACTH levels fall, which results in a rapid crash from high to low cortisol levels in blood. As an omen of impending disaster, however, many doctors will measure cortisol, find it low and mistakenly prescribe extra cortisol without knowing that low ACTH might also be present.

Unfortunately, what this supplemental cortisol does is *further* suppress the already compromised ACTH production. Without ACTH, bad health problems can be made much worse. I say treating biotoxin illnesses with steroids is similar to "throwing gasoline on a fire," and with good reasons. I can't tell you how many biotoxin disasters I've seen, from optic neuritis being turned into blindness to inflammatory joints turning into frozen joints, when a physician prescribed steroids. Don't let anyone give them to you if you're biotoxic!

MSH plays an important role in regulating your sex hormones, too. Gonadotrophins are hormones secreted in pulses by the pituitary, and it's this pulse release that's controlled by MSH. Gonadotrophins are required for normal production of both male and female hormones. The male hormones, called androgens, have a lot to do with biotoxin illnesses. Since androgens control other hormones and their responses to cytokines, when you take away androgens, you multiply the adverse effects of MSH deficiency.

My impression is that the reason that 70 percent of the chronic fatigue population are women is that they're lower in androgens than men. Most men with the syndrome have lower androgen levels than their healthy buddies.

"Umm, this is a little embarrassing," said Jim. "I noticed that when I got sick, I could care less about sex; you know what I mean. And even worse, I started having trouble—I thought I needed Viagra!"

I nodded; it was a symptom I hear frequently.

"You know, Jim, problems with sex hormones, especially androgens, happens a lot," I said. "Women have androgens too, and so they notice a loss of libido, too. And their periods go all haywire. Flooding, clots, cramps, lots of ovarian cysts—gynecologists are looking at lots of MSH deficiency every day without knowing it. If they would do the simple blood tests, they would see it. Even worse, a lot of women will develop chronic vulvar pain and a painful urinary condition called interstitial cystitis. Intimacy just won't happen often when a woman has those two conditions. The symptoms of MSH deficiency break apart more than a few marriages.

"Frankly, the role of MSH in women is being missed too much. Don't forget history: years ago when it was legal to give MSH without the FDA looking over your shoulder, many women with unusual menstrual problems were treated successfully with MSH. Someday, I'm going to do VCS screening in the waiting room of a dysfunctional uterine bleeding clinic or an endometriosis clinic," I said. "I'll bet I could clear out about 30 percent of the patients by treating the undiagnosed biotoxic illness, especially if I had some MSH to use. But now, the FDA is really strict about using medications just because they might work. So I can't give the mold guy the MSH he needs when he comes in for his erectile problems. Fortunately, if you clear out the mold toxins, his male functioning usually returns. And women are no different in this regard. Enjoyment of intimacy requires some MSH!"

And once again, the problem with androgens, gonadotrophins and MSH is failure of normal feedback controlling mechanisms, the recurrent theme of the Biotoxin Pathway.

MSH deficiency nails the posterior pituitary as well. Production of antidiuretic hormone, ADH, will be dysregulated in over

60% of MSH deficient patients. Without the right amount of ADH protecting our body's water content, the kidney can't prevent *losing* free water, so patients begin to urinate more frequently, sometimes to the point of significant dehydration. As affected patients lose water, they begin to be thirsty all the time.

I used to count the number of water bottles my patients would bring with them to our two-hour initial visits. If the patient had two water bottles, they were sure to have major problem with ADH-dysregulation.

As the patient begins to get dehydrated due to ADH deficiency, the level of salt remaining behind in the blood rises. The sweat gland detects the extra salt content and corrects the salt problem by dumping extra salt out of the blood into the sweat. With a thin layer of salt on the skin, even after a morning shower, the patient now becomes an efficient conductor of electricity (electrical ground). The susceptibility to static shocks that follows is amusing at first ("she's the electric eel lady"), but then the symptom becomes a nuisance.

Patients start getting an increased number of static shocks from car doors, light switches, drinking fountains, carpets and loved ones. The first few times biotoxin patients told me about turning on light switches with their elbows or how they stopped watches and Palm Pilots with their extremities, I didn't pay attention. Salty sweat and shocks from biotoxins?

It happens all the time. I hear about static shocks all the time from my patients *because I ask.*

One uninformed medical "expert" testifying for the defense mocked my opinion that susceptibility to static electric shocks was associated with biotoxin illness. I simply asked him what he knew about ADH/osmolality and sweat gland regulation in MSH deficiency. When taking a medical history did he ask patients about increased thirst *without* asking about shocks? He admitted he'd *never* asked those questions, and then he called me "combative."

The fact that he knew nothing about static shocks in biotoxin patients didn't stop him from attacking my point of view. Some expert.

Other extremely important effects of MSH deficiency include changes in gut function and nasal mucus membrane function. Cytokines regulate absorption of many compounds in the gut. MSH-deficient patients will often have problems called "leaky gut," poor absorption of key nutrients or even inflammatory bowel disease.

I mentioned antigliadin antibodies earlier. MSH helps the small intestine *block* any leak of the bulky gliadin molecule *from* the gut into the bloodstream. When gliadin is absorbed into blood, it becomes a foreign invader (antigen). Our protective cytokines signal the invasion and recruit white blood cells to eliminate the antigen. The white blood cells release cytokines to detect and attack gliadin, and those extra cytokines cause even more symptoms. Low MSH lets in more gliadin, which brings more cytokines, more illness and even less MSH. It's a vicious cycle.

We used to measure antigliadin antibodies to diagnose celiac disease, a chronic illness genetically linked to HLA DR. We don't any more because so many people with the antibodies were not true celiac patients. The gliadin antibody people are simply MSH deficient! MSH keeps the intestinal lining healthy. It's certainly possible that the increased number of patients with antigliadin antibodies in biotoxin illnesses is solely due to *initial* MSH deficiency, permitting gliadin to go where it normally shouldn't.

So have your physician measure MSH if you have abdominal pain in an alternating pattern of constipation/diarrhea or secretory diarrhea (the kind of diarrhea that you get when you haven't had anything to eat). Don't accept a diagnosis of "irritable bowel disease" without MSH and gliadin antibody assays! Cytokines also control the "tight junctions" found between the cells that line intestinal membranes and can promote either absorption or release of compounds from the gut itself into the gut cavity.

"Tami was told just after she got sick that she had irritable bowel disease," said Jim. "She had all kinds of tests done, but nothing was ever found. They gave her an anti-depressant. And it didn't work."

Poisoners in Your Nose
Multiply Antibiotic Resistant Coagulase Negative Staphs

MSH protects the mucus membranes in the nose. And the germs that *stay* in your nose only do so by virtue of permission from MSH. So it shouldn't be a surprise that when MSH deficiency has eliminated *protection* from opportunistic invaders in the nose, some new germ will appear. Sure enough, more than *80 percent* of MSH-deficient patients will be victimized by *colonization* (it's *not* an infection) with multiply antibiotic-resistant coagulase negative staphylococci (the acronym for these bacteria is "MARCoNS"). Thirty years ago we were taught in medical school—and I still hear docs today give a knee-jerk response to patients who ask for a deep aerobic culture for MARCoNS—that all the coagulase negative staphylococci (CoNS), usually *Staph epidermidis,* were completely benign. They were regarded as normal residents of skin and nose. That may have been true *years ago*, but not any more! Bacteria are changing all the time; the new MARCoNS have adapted to the new population of MSH deficient biotoxin patients, flourishing in our noses.

Healthy people and biotoxin patients have very different results on deep nasal cultures. When we did those cultures in 100 patients who had normal MSH and no significant number of symptoms as part of a study reported to the American Society for Microbiology recently, we found the benign CoNS in 38%, but only one had the dangerous MARCoNS.

In contrast, when we looked at 471 patients who had low MSH and lots of biotoxin-related symptoms, we found that *80* percent of those patients had MARCoNS. In the MSH-deficient patients, resistance to more than one class of antibiotics was found in more than 98 percent of the isolates, with resistance to 3.4 antibiotic classes, on average. Resistance to *methicillin* was found in 60 percent of the entire group of MARCoNS.

These methicillin-resistant (MRCoNS) organisms, a subset of the MARCoNS, are no longer the benign, normal flora of 30 years ago. They have two separate mechanisms to cause damage to MSH, making us feel worse. The first mechanism comes from the release of hemolysins, proteins with a chain of 22 amino acids that join to form "tetramers," a four-member matrix of hemolysins noted for their ability to destroy red blood cells. These compounds actually make small pores in membranes of red blood cells, releasing some of the interior contents, especially iron, which the bacteria need for life. The hemolysins are absorbed into the bloodstream, where they're recognized as foreign antigens, stimulating a cytokine response, damaging the leptin receptor, and further reducing MSH production.

We think that there's an additional mechanism "used" by MARCoNS to hurt low-MSH patients. Exotoxin A, known to be made by the cousins of the CoNS, namely coagulase *positive* Staph, *Staph aureus*, and now found in MARCoNS isolates, is capable of splitting MSH, totally destroying its function. More research is needed here, but this initial finding could underscore why low-MSH patients never feel better if the nose is still colonized by MARCoNS. Get rid of the toxins first, and then get rid of the MARCoNS.

MARCoNS also present therapeutic problems for physicians because they can rapidly acquire antibiotic resistance. I initially used just a topical preparation, Bactroban, to try to eradicate MARCoNS. Bactroban is commonly used by hospital employees to reduce nasal carriage of *Staph aureus*, especially the feared methicillin-resistant strains. Bactroban resistance in the MARCoNS developed within two weeks! No surprise, then, that when patients began to relapse, when I used oral antibiotics to try to eliminate the organisms, the pattern of rapid relapse with resistant strains was repeated. I now use a triple antibiotic combination, much as we use combinations of multiple classes of antibiotics to help treat tuberculosis and to prevent development of resistant strains.

In a small group of patients with a refractory biotoxin illness, the Post-Lyme Syndrome, in which patients usually have been treated with multiple antibiotics and/or prolonged antibiotics, it's not unusual to find MARCoNS with resistance to up to *eight* different classes of antibiotics.

Our work with researchers like Dr. Phil Domenico of Winthrop-University Hospital in Mineola, N.Y., and Dr. Mark Shirtliff at the University of Maryland School of Dentistry is adding another dimension to the cunning pathogenicity of MARCoNS. These organisms make biofilms. Protected by a thick coat of polysaccharide matrix, these biofilm-formers make antibiotic-resistance factors, differentiate activity, and cooperate in waste removal and a lot more. Our preliminary data points an accusatory finger at prolific biofilm formation as a significant risk factor for chronic fatigue. It may turn out that the "dreaded genotypes" get that way by selectively permitting biofilm formers to grow.

MARCoNS thrives in the presence of low levels of MSH. MARCoNS aggravate the MSH deficiency, perpetuating their own growth by sending out stealth hemolysins to launch cytokine attacks on the leptin receptor and destroying MSH by producing exotoxins. MSH apparently eliminates MARCoNS from its biofilm by stimulating a vigorous, targeted cytokine response against the organisms in their sanctuary. The protection by MSH from colonization with MARCoNS is lost in MSH deficiency.

This is another example of normal regulatory mechanisms gone awry in the Biotoxin Pathway.

When chronically fatigued patients with MARCoNS and MSH deficiency fail to respond to the antibiotic protocol, we've used an experimental Staph vaccine with excellent results. We've asked the FDA for permission to do a double-blinded, placebo-controlled clinical trial to prove efficacy of the vaccine in patients with chronic fatiguing illnesses and low MSH.

MSH is an extraordinarily powerful hormone. More roles for MSH will surface as awareness of its importance grows. When chronic patients ask, "How long is it going to take for me to feel better and how do I know if I *am* getting better?" a simple response is to ask, "What's your MSH?" If the level is going up, that's good. If the level isn't rising, that isn't good. If the MSH level is falling, then there are additional cytokine responses that must be identified and treated immediately.

In biology, control is exerted through negative feedback systems. In the Biotoxin Pathway, we have disrupted normal feedback-control mechanisms. The chronic illness caused by exposure to biotoxins is a result of a failure of normal feedback loops. Instead, we have a disastrous positive feedback loop: the more MSH production is compromised, the more opportunity there is for additional problems to appear and further compromise that production. It's a vicious cycle, one in which a patient's health continuously falls.

In biotoxin illnesses, there is no single magic pill. The goal in treating biotoxin-induced illnesses is to stop the downhill plunge, and to attempt to return health to normal. Often, if all we do is arrest the illness and stop the exposure, patients will have some improvement. That's a worthwhile goal.

But the ultimate goal of treatment must be to take care of all of the aspects of the Biotoxin Pathway. That means removing biotoxins, stopping inappropriate fat cell production of cytokines, stabilizing C3a, correcting VEGF deficiencies, restoring normal levels of MMP9, correcting damage to leptin receptors, restoring normal integrity of mucus membrane defenses, restoring both anterior and posterior pituitary hormone function and protecting peripheral cytokine release by white blood cells. We still don't have the final word in on the effectiveness of shutting off the autoimmune response, but our work is progressing.

"This pathway is incredible," said Jim. "I thought I had it, up until all those feedback loops and MSH kept criss-crossing, going back and forth. There are so many interactions—bacteria that don't invade, but make us sick fighting for their niche in our nose. The gut protecting us from food-induced illness, until MSH goes low. White blood cells, cytokines, hormones, growth factors. And some doctors thought Tami was kooky. Seems to me like they're the kooky ones. They just brushed her off! One physician called her a 'crock' to her face. But look what they didn't even know. We've got a whole lot of teaching to do, Ritchie. Maybe we should just start by making our doctors listen to their patients. I bet there are millions like Tami out there."

• • •

Sir William Osler, often called the Father of American Medicine, suggested during the last century that if the physician understood syphilis, he would know all of medicine. Perhaps today, Osler would revise that statement, because the dynamics of the Biotoxin Pathway—including the genetics, cytokine physiology, differential gene activity, toxicology, endocrinology, immunology and autoimmunity—are as complex as any part of medicine.

In the World of Biotoxins we live in now, knowing the Biotoxin Pathway will be the start to exploring 21st Century Medicine.

CHAPTER 5

Shannon's Story

"This is Alice Johnson. I've had chronic fatigue for 10 years and my doctor thinks I'm sick from my basement. I was told by my doctor to call this number and ask for Shannon."

"That would be me," I said. "How can I help?"

So begins another day for me in Pocomoke, Md. Part of my job is being the friendly, considerate voice on the other end of the telephone. I guess I'm the link to chronically ill patients, who hope that a new approach to diagnosis and treatment will give them back some quality of life. I get to see the results of treatment, both before and after, acting as a support group, a cheerleader and a troubleshooter. For me, all the physiology and biochemistry in the world means nothing without helping people. And in Dr. Ritchie Shoemaker's office, that job starts with me.

My entrance into the world of biotoxin-associated illnesses was accidental. I was job-hunting after leaving my job in Washington, DC and returning home to Pocomoke. I stumbled upon an opening for a receptionist in an office of a local physician.

After being hired for the job, I was given a copy of his book, *Desperation Medicine*, and told to memorize its contents. At first, I thought the task unnecessary—perhaps even an exercise in egocentrism. Why would this man want me to commit to memory his 500-page treatise on chronic neurotoxic illnesses caused by biotoxins? But I read the book, and was surprised that I became fascinated by the tales of illness and recovery within.

I wanted to meet these people and see if their years of pain and suffering really had ended with the treatment methods the book described. Honestly, I was skeptical that these dramatic success stories existed.

My first week remains a blur. The constant ring of the phone, coupled with a steady string of unanswerable questions, combined to create a kind of chaos I had never experienced in my previous employment. I felt as if I had been shipwrecked and no sails could be seen on the horizon.

I felt especially lost when the biotoxin patients called. These patients were so different from patients with clear or familiar problems who knew what was wrong and knew it was fixable, whose colds and sprains were temporary discomforts with treatment methods that would set them back on their feet in a matter of days. In contrast, the biotoxin patients had no idea what was wrong, and so they had no way of knowing what treatment—if any—existed to ease their pain.

After my first week, I began to feel more comfortable on the phone and had compiled the first of a long list of responses to frequently asked questions (FAQs). I was still far from knowledgeable, but felt capable of answering at least some of the more common questions. It was during this time that I realized most of the biotoxin callers not only were searching for treatment, but also for understanding.

As I gained confidence, I took more time to listen to their stories and try to hear what these people were trying so desperately to say. While brain fog, decreased concentration and ability to focus on a topic are most certainly significant parts of neurotoxicity, these symptoms weren't responsible for the anguished sobs and shocked confessions—nearly all of the callers had almost given up hope at some time. Their despair almost seemed like part of the syndrome.

A unifying theme of disinterest and misdiagnosis from the medical profession began to appear. By the time some of these patients found Dr. Shoemaker, they'd already been through many doctors and many thousands of dollars for tests and treatments. Occasional therapies gave temporary relief, but most served only to encourage patients to distrust and dislike doctors, even though some were treated by professionals at centers of academic excellence, like the Mayo Clinic, Johns Hopkins University and UCLA. Every time we saw a patient who had been misdiagnosed by Duke University, Dr. Shoemaker's medical alma mater, he seemed especially distressed. I know it sounds terrible, but after listening to the stories expressed by many biotoxin patients reporting similar behavior from physicians, I began to lose faith in my own assumption that the medical profession really knew what was best. I hear the same kinds of stories almost every day. It really is sad.

At least lepers in this era receive a modicum of consideration far greater than the biotoxin people ever do: leprosy is an interesting illness now.

I don't think I could count how many women have called and said that if Dr. Shoemaker merely patted them on the head, said they were depressed and gave them Prozac, they would scream and run from the building! Who could blame them? Mercifully, that doesn't happen here.

Cancer patients often become depressed during the course of their illness. Does that depression negate their cancer? Of course not. We must ask: Then why are patients with diagnosable diseases such as Lyme disease and Sick Building Syndrome treated as if they're freakish or immature, as if they should be drugged and hidden from public view? Worse than that, however, are the women who are told they're suffering from the mystical "housewife syndrome" and made to feel that they're not only crazy but also somehow inventing their pain and debilitating exhaustion. And then, some are told they just want antibiotics or attention.

Based on the thousands of people I have met or talked to, I can say that they want to feel better. They aren't inventing some phantasmal illness for bizarre gain. They are sick. Doctors should heal the sick.

Men seem to fare better with the medical establishment, and are often told their illness is directly related to overwork and underplay. To think a man with chronic Lyme disease could recover after playing a few rounds of golf is absurd; but many of the men who come into the office have been told just that.

Understanding and sympathizing with people who have been through such ordeals has become a significant part of my job, so empathy is a given. Understanding is difficult without having first-hand knowledge of their experiences, but it grows clearer each day.

The most difficult part is trying to keep the patients on track when they call. Without cutting them off or providing them with the brusque and sometimes patronizing treatment they've received elsewhere, I must somehow coax them into encapsulating their stories into a few minutes on the phone.

So many patients ask me if I have "a few minutes" when they call. Most understand when I explain that I have three other ringing lines and have to frequently put them on hold. I must confess it's a challenge to balance the need to help these people with the need to keep the phone lines clear. It's my hope that I accomplish that goal and that no one hangs up after a call to the office feeling as if they've been dismissed or that their stories are superfluous.

I worry that others in the life of each caller don't understand and have no idea—or won't take the time to understand—what's wrong as well as their loved one's longing to be understood. So many of our patients tell me that just knowing what's wrong, even if it means learning some impenetrable words and terms—helps them. They are no longer dealing with the fears of the unknown; they can see their illness represented on a lab slip, usually for the first time. There it is in black and white, a diagnosis! With that diagnosis comes the verification that they're *not* crazy. Many of our patients cry when they're finally given a definitive diagnosis, because they're so grateful to know they were right all along—that there *really* was and *is something wrong* with them!

Children with biotoxin illnesses remain the most heart-wrenching. Meeting children who thrive in the summer only to become sullen and unable to function while attending mold-riddled schools is infuriating. The parents are inevitably terrified and worried, and have spent months or years wondering which diseases/conditions could be responsible for their child's illness.

Parents of these children come to the office with boxes of medical records and results of tests ranging from spinal taps to PET scans. Some of these children are so accustomed to having blood drawn, they point to the correct place for the phlebotomist to insert the needle. These children are being robbed of their youth by unseen illnesses whose existence is denied by health care systems that are increasingly more concerned with the bottom line.

Maybe it's because I see so much of the damage that mold causes that I can't understand why others don't see it as well. Administrators in schools should suspect mold trouble somewhere within their school system when they see test scores falling, absenteeism increasing and more students with chronic upper respiratory problems.

Instead, these children are ignored, their health is deemed a non-issue and any mention of mold is considered to be a joke. "Mold doesn't hurt anyone," is a phrase I'd like to tear into small pieces. It's wrong, it hurts people and it prevents them from being treated as an individual worthy of respect and dignity.

When I hear that school maintenance budgets are cut, I know it won't be long before I'll be getting more phone calls when the roof leaks aren't fixed, the soggy carpet dries out on its own and the plumbing fixture floods are just mopped up. Then there is the school system that says is trying to fix a mold problem by tearing out carpets, all the while letting the *humidity* levels inside the school rise to over 70 per cent without a second thought. Sometimes I think that the administrators who make those decisions should either work in my chair for a week or work in the fungal habitats they have allowed to be created in public school classrooms. Maybe a TV reality show will pick up on the idea of making higher-up people *responsible* for their actions by having them personally participate in the environmental effects of their decisions, instead of being oblivious to the indescribable misery they cause for others.

While the children suffer, so do the teachers and aides in these schools. One patient who works in a mold-infested school learned about the disregard the hard way. She became sick quite some time ago and ran the gamut of tests and diagnoses, as is usually the case. Specialists from Johns Hopkins told her she had sero-negative rheumatoid arthritis, fibromyalgia and depression, with asthma to boot. The diagnosis she recalls best was that "she was too fat." She doesn't have a fond memory of the compassion or skill of those doctors.

At long last, she came to Dr. Shoemaker's office and was finally given an accurate diagnosis. A strong woman, she continued to work and tried to fix the problem via the proper channels. She informed the administration that she became sick from the mold in her school, and that several students were symptomatic, hoping the school would be remediated and become safe for all. She was barely humored, but mostly, ignored.

This woman became so sick at one point she collapsed on the cold concrete slab floor during a class. The principal was informed, did nothing except to remove the children from the room. She continued to lie on the floor for more than one hour, in and out of consciousness, before her daughter, who was coming for a visit between classes, called an ambulance.

The horrible experience is magnified when you consider that her students witnessed this tragedy from the hallway. What did they think when their teacher lay barely responsive on the carpet and nothing was done? While the hour on the floor must have seemed eternal for the teacher, it must have seemed even longer for the children who could do nothing to help.

Office buildings are also subject to mold infestation and the people who become sick from that exposure are often treated in the same callous manner. Employees at a Social Services building in nearby Accomack County, Va., were upset when they discovered that their common and debilitating illnesses were deemed either psychosomatic or hypochrondiacal. One chronically ill patient whose condition improves when out of the office building can be considered coincidental, or perhaps is even faking the illness to glean sympathy or slack time from an employer; but a dozen patients with the same symptoms, genetic susceptibilities and improvement with treatment are most likely explained by the presence of mold in the building.

When those patients also have photos of black mold growing in the bathrooms, basement and ceiling tiles of that building, there's only one viable conclusion—these employees are being poisoned by their workplace. The response by Social Services was that a new building was scheduled for construction in three years and that the employees who were sick (being poisoned) should simply wait until the new building was complete. Ludicrous!

Even more disheartening was the realization that these patients, whose jobs were to help members of their community, were being allowed to suffer by the leaders of that same community. When one employee became enraged enough about the patronizing attitudes of her supervisors to consider legal action, she was told, "Go ahead and hire a lawyer. Spend lots of money. But you can't sue the government!"

Schools, office buildings and homes are being remediated nationwide because of the presence of mold; millions of dollars are being spent in this effort. Lawsuits result in losses of millions of dollars paid by employers, insurers and builders who knowingly allow mold to thrive in their structures.

Some of the patients I've come to know are involved in litigation to recover financial compensation for their medical bills and lost wages, but most are simply trying to get better, get back to work and on with their lives. These people aren't looking for a lawsuit or piles of money; they're just trying to regain some sense of normalcy in their lives. Hopefully, these lawsuits will prompt employers and builders to use preventive measures in their structures and to remediate at the first sign of toxic mold. It will most certainly be both morally and financially beneficial to do so.

While my introduction to biotoxin-associated illness was accidental, it's been a delightful and rewarding journey. I've met hundreds and hundreds of people who have recovered from their illnesses and who have gone on to help others by telling their stories and relating the stops along the winding road to their recoveries. These people are the true heroes and warriors in the battles against biotoxin illnesses.

As the number of such recoveries grows, I hope the recognition of such illnesses does as well. As I write this in 2004, Lyme disease is the most common vector-borne infectious illness in the U.S., with about 23,000 cases reported in 2002, though the disease is greatly under-reported. Twelve states account for more than 90% of reported cases. The U.S. Department of Health and Human Services says that the number of reported cases of Lyme disease and the number of geographic areas in which they occur is increasing. Mold illness is far more common, possibly causing as many as 25 *million* people to be ill in our country.

My ultimate hope is that none of these patients ever again have to listen to empty words from physicians. Their chronic illness isn't "housewife syndrome," or "all in their heads." I continue to believe that eventually, these patients will be treated with the respect and sympathy they deserve.

CHAPTER 6

Quietly, the Legal Revolution Begins

Susan Donahue, Eastern Correctional Institution

Westover, Maryland
January, 2004

"The state dropped their case. Now I have to negotiate the final settlement." It was my patient, Susan Donahue, with news from Somerset County Court.

"They just dropped their appeal? Incredible," I said. "It's over, and you win? The Worker's Comp decision stands? Surely they know that now the floodgates are wide open for all those *other mold claims* from the workers at the state penitentiary. I don't understand. They fought you for so long, verbally attacked you and me both; delayed your time for justice, threatened you with major financial loss and now they just quit?"

"Well, I don't understand them either, but I don't care," she said. "It's over. Maybe I can get a new life started. I've still got most of my foot left, I'm not completely exhausted and I can even sleep a few hours. Now, I can support my daughter, pay these medical bills."

• • •

Sue's medical record was two-volumes thick. With her attorney's help, she'd won full benefits for life due to her illness, which was caused by the biotoxin forming molds growing in her office in the West Compound at Eastern Correctional Institution (ECI). Her employer, the State of Maryland, had appealed the decision and sent her to *their* doctor for a second opinion. Sue thought she was about to face a lengthy legal challenge, full of more delays that would drain her remaining savings. But instead, the battle was over. No bang, no whimper, just over. She won.

"I'm out of the building, the blood clotting test finally looks real good and the gangrene has stopped," Sue pointed out. "Does this mean I can stop all those blood thinners? The foot surgeon said my surgery site is healing well. How about I stay on the warfarin, but maybe cut down to just one heparin shot a day?"

Just a few weeks ago, Sue was like a wolf caught in a trap that chews off its own leg rather than die in the snow. Her clotting problem from the mold was rotting her toes and her foot, but now she was free. She was pushing hard to start living again—she wanted to enjoy her freedom, even with a limp.

She won.

"Stop heparin? After the mold almost killed you?" I asked. "Look, we've talked about the immune events that your body's response to the mold toxins set off. It's like Pandora's Box. Once your immune response to mold opened that box and turned on production of those anticardiolipin antibodies that cause the clots to form abnormally in small blood vessels, you can't just turn them off by winning some court decision. Cardiolipins don't just stop being produced once their production genes are activated. If the clot was in your toes before, how do we know it won't be in your fingers or brain tomorrow?"

Despite my comments, I sympathized with her desire to cut back on her medications. "We can cut back on the shots," I agreed.

"If the gangrene threatens again, you'll need to restart the heparin shots. I just hope that the next clot isn't inside your head!"

Just two years before, Sue Donahue was working as usual in her new office in the Social Services wing of the ECI, in Princess Anne, Md. She was 39 when she became ill. She'd worked for 15 years for the state; she had a flawless work record, with more than 100 sick days saved toward her anticipated retirement in 15 more years. The new West Compound also housed the educational facilities and two cafeterias, one for the 100 inmates and one for the 400 prison employees.

It was in one of the classrooms near Sue's office that the water leak first appeared in the summer of 2000. Stained ceiling tiles only looked slightly ominous, because the really lush growth of black mold on top of the tiles remained hidden from view. The main air return for the heating and air conditioning system was on the tar-and-gravel roof, next to the areas of greatest water intrusion. Unchecked mold growth—amplified growth as it's called in the trade—provided an invisible biotoxin harvest that the HVAC continuously dispersed throughout the West Compound.

Employees began to have recurrent bouts of coughing, sore throats and muscle aching. Still others suffered severe mood changes and more than a few were given prescriptions for antidepressants and tranquilizers.

The response from ECI officials was predictable: "There is no illness from exposure to areas of water intrusion." The problem, they maintained, was just stress or dry air or fibromyalgia or old computers or blank walls. The warden brought in a team of "experts" who told everyone there were no health problems and the building was fine.

Sue didn't believe him. Normally an energetic woman, Sue was exhausted by mid-morning.

"There's something wrong with me and it's not stress or anything else," she said. "I'm never this tired. I walk around like I'm in some kind of brain fog. This isn't me."

• • •

Sometimes revolutions are quiet. Such a revolution started in Maryland, on January 15, 2004. Sue Donahue was an unlikely heroine on an unlikely battlefield: a Worker's Compensation hearing. She was the first of seven people I treated from 2001 until 2003 who became ill while working in the West Compound. Everyone who worked in the office complex knew about the roof leaks, the dank smells when they turned on the air conditioning and the black mold growing on the ceiling tiles.

When Sue first started getting sick in Autumn 2001, it would've been easy to blame any number of things for her illness—the history of myasthenia gravis, the low blood counts, even the break-up of her marriage—all were possible explanations for her progressive fatigue, sleep disturbances and pain.

At first, everything hurt, but then the foot pain started. I'd never seen such unusual mottling—purplish grey-black blotches all over her left foot. She wasn't getting much blood delivered to the smallest blood vessels (capillaries) in her right foot, either. It was almost like what you might see in an elderly, diabetic smoker with high cholesterol and blood pressure problems, but not in an energetic 39-year old Social Worker. She had been my patient for years; there was nothing new about her ongoing illnesses that could have caused her new symptoms; the only change in her case was the transfer to West Compound earlier that year.

"Ms. Donahue, this capillary circulation problem is so strange," I said. "When unusual illnesses appear without explanation, I often can prove that biotoxins and inflammatory chemicals like cytokines are potential culprits. Let's do a VCS test to see. We'll do all the blood vessel studies too, but I think they'll be normal."

Sure enough, Sue was unable to see the subtle grey/white lines of the specialized eye test, visual contrast sensitivity (VCS). She had all the lab indicators of a "multisystem, multisymptom" illness, including a deficiency of MSH, high levels of leptin and high MMP9. She had the HLA genotypes from Mold Hell: each individually conveyed high susceptibility. Nothing else could explain what was wrong other than her mold exposure. She started on cholestyramine (CSM) and noted quick, remarkable improvement. Her symptoms vanished, VCS deficits disappeared; leptin and MMP9 fell, even if the MSH didn't rise. If MSH doesn't rise, she runs the risk of more illness from mold in the future.

Sue had an interesting way of looking at her need to take medicine constantly to stay well.

"You know, Dr. Shoemaker, if you mix this powdered crap in berry punch, it isn't too bad," she said. "Anyway, I'd swallow pureed dog manure four times a day if it helped me the way CSM does."

"There's only one way to be sure about the mold making you sick," I told her. "When you go on vacation next week, and you're out of the building, it'll be safe to stop the CSM. We'll track your fatigue, muscle aching, cough, brain fog and all the rest and VCS-test you again as well. We'll check your labs too, especially leptin and MMP9—they change almost as fast as VCS does in mold illness with re-exposure and re-acquisition of symptoms. Nothing bad should happen when you're away from the building, off CSM. Then, you'll go back in the building without CSM to see what happens. If you get worse in just a short time, get out of the building and come here immediately. We'll recheck your symptom list, get another VCS score and labs again and get you back on CSM. If you do get worse, you know what that means: the building is harming you, maybe even killing you. We call this a repetitive exposure trial. As it turns out, this trial shows the answer regardless of which toxin-formers are growing indoors."

"But what about mold outside the building?" she asked. "I mean, mold is everywhere, right? So, how can I stay healthy if I'm still exposed to mold?"

"That's the whole point, Sue," I explained. "Just because you're exposed to the 'ubiquitous fungi of the world,' doesn't mean you are exposed to *amplified* mold growth. There's a big difference in toxin production and illness if the mold is growing without much competition for food, water and living space. In a tablespoon of soil from your garden, you'll find hundreds of species of fungi. But they don't make us sick because they're kept in control by all the other living organisms in the soil. Amplified growth doesn't happen in the garden soil.

"In the building, you will only find a few species of molds, but they have almost no competition. They aren't being controlled. Indoors, mold growing in water damaged areas is a biological system without control."

Sue did her repetitive exposure trial on *three* occasions. The pattern was clear. She would be fine if she stayed away from the building, even off CSM, but she would get sick when she went back into the building without the protection of CSM. She'd recover when CSM was begun again. Her series of tests and opinions from vascular specialists didn't help much, but they did rule out a long list of potential other sources of her symptoms. All I had to guide me in managing her medical care after the diagnosis was completely secured was a steady increase in the level of antibodies to cardiolipins with re-exposure.

These antibodies, sometimes seen in lupus or the "antiphospholipid syndrome," were responsible for clotting problems in blood vessels in any part of the body, including the brain and kidneys. The syndrome was responsible for recurrent first trimester spontaneous miscarriages in young women, too. Nobody really knew the cause of this "autoimmune" disorder. Basically, the body began to attack itself, like in lupus or rheumatoid arthritis. In the worst cases, immune suppressant drugs, like those often used in treatment of cancers, were required if blood thinners like warfarin and heparin didn't help, and even then the prognosis was pretty grim.

"Sue, you've got to get out and stay out of that building. Nothing is stopping the anticardiolipins. Every time you go in that building and you're not on CSM, the levels shoot right up," I explained.

"And your leptin is telling you something, too—remember, leptin drives production of MSH, but if there's an immune response to mold toxins, from those cytokines we talked about, the leptin receptor won't work and MSH won't be made. You'll get more tired and be in more pain because of MSH deficiency. And then, when leptin isn't working in the brain, your body will just make even more leptin, trying to compensate. Leptin stores fatty acids like crazy in fat cells, and it won't let fatty acids be burned—and that's why you gain weight and never lose it despite any diet you try."

Sue was trying to understand what was happening inside her traitorous body.

"OK, so the mold toxins are pushing up leptin, making me fat too? And the mold toxins are lowering MSH, making me tired and hurting all the time? Are the mold toxins driving the anti-cardiolipins too?"

It was important that she understand what was happening to her, because once she understood her enemy, she stood a chance of being able to beat it.

"Sue, it's the unopposed cytokine response to the toxins that turn on the genes that make the antibodies. In the end, it's the autoantibodies that are the enemy, but it's the biotoxins that turned on the autoantibodies," I explained for the 10th time. "The language here is new and the concepts are, too. But you don't have the option of not knowing what's going on. The new language of biotoxins is the language of your illness. You need to learn it. One of the results of this poison is a reduction in your learning ability so if it takes you more than one sitting to get it, fine. But you need to understand your enemy."

The breakthrough in Sue's case came from an elevated blood sugar level. I've been using a diabetes drug, Actos®, for years to turn off production of leptin, MMP9, TNF and PAI-1. It's a simple gene therapy. It's no big deal to give the drug, and put patients on a no-amylose diet. Amylose is the complex carbohydrate used by plants to store sugar as starch. Saliva has the enzyme, amylase, which digests amylose into the simple sugar, glucose. All the new fads about low-carb diets miss the importance of the effect of those carbs on blood sugar. Amylose drives blood sugar up fast. If you eat amylose first and your blood sugar skyrockets, all the Actos you take won't knock out all those cytokines. If you stay on the no-amylose diet while taking Actos, you will rake in all the Actos benefits. Then, just watch cytokines—and pounds—fall.

As was described in my book, *Lose the Weight You Hate*, published by Gateway Press, when we use Actos, we turn on genes that help move sugar into cells, so Actos helps our diabetics reach normal blood sugar. But Actos also turns *off* the genes that make all these harmful cytokine products and leptin too. If cytokine excess were driving formation of autoantibodies, would Actos turn off the over-production of Sue's anticardiolipins?

When Sue had a blood sugar of 150, a level high enough to convince someone else she might have diabetes, it was time to give Actos a try. She started avoiding all the starchy, amylose-containing foods and took Actos once a day. The cardiolipins started to fall and her foot pinked up.

• • •

Sue loved her job. Each time I said, "Get out now, before your foot falls off," she would say how well she was doing on the medicine. She couldn't quit. Where else could she find such a high-paying job? And the benefits of state employment were substantial. Sue is a single mom; maybe things would get better.

That was not to be. Despite Actos, warfarin and heparin shots twice each day, the gangrene exploded. She needed amputation of at least one toe. As any diabetic who has suffered loss of an appendage from poor circulation knows, once the surgeons start whittling, the only question is how high do they have to go? If one toe were dead now, when would the entire foot become wooden, blue and cold? And then, what would stop the immune response—amputation just below the knee? Above the knee?

Finally, Sue got out. She had no job, no source of income, no lawsuit, but she got out in time. We knew the mold was starting the biotoxin pathways that clotted her foot, even if there wasn't a case series of 50 patients to convince a judge that the "novel" science was real. The *process* of medical diagnosis in Sue was sound. Traditional concepts of medical science and differential diagnosis needed to establish causation were all in the mix. Surely the courts would agree to listen to something new if the underlying foundation was rock solid.

ECI had "phased out" her position, denied her claim for medical retirement, cut off her benefits, denied her use of a "sick time pool," ended her insurance and then said the building didn't pose a health risk. The supervisors on West Compound didn't treat several of Sue's co-workers real well when word got out they were seeing me, too. Ken Gaudreau, a trial lawyer from Salisbury, Md., agreed to take on her legal claim for damages. Ken wanted specific items in Maryland's Worker's Compensation law identified in Sue's case. Even if the claim were for something new, the mechanism of law applied to Sue's case.

The first venue was a Worker's Compensation Commission hearing, and things didn't look good. The state-appointed physician took less than 10 minutes to examine Sue. He stated that she was depressed, overweight and had myasthenia; mold was not a factor in her illness.

What an outrage. His superficial exam showed his "superior" powers of deduction, apparently far more incisive than mine. I had hundreds of hours of clinical work in Sue's case. Her claim was denied by the state, largely based on the State expert's opinion.

What about all those multiple re-exposures and proof that the building exposures were the only possible cause? And what about those prospective exposure data documenting that just three days in the building, without CSM, was enough to cause a measurable increase in leptin, MMP9 and symptoms, as well as fall in VCS?

No comment.

Gaudreau remained firm in his soft-spoken way. "This case is solid, let's get it in front of a judge," he said. "Shoemaker's report is solid, the evidence is solid. The dissenting medical opinion is an empty piece of fluff. Hang in there, Sue."

What he didn't tell Sue was that the Worker's Comp board in Maryland had never ruled in favor of the plaintiff on a Sick Building Syndrome case. Gaudreau did a lot of Worker's Comp work: he had begun to take more than just a passing interest in Sue's case. There's always a first time, he thought.

For Sue, the Worker's Comp hearing was her last hope to maintain health benefits, not to mention an income, however small. She had other worries, too: How much would a hospitalization cost? How many dollars per toe amputation? Who would hire a 40-year-old with a multisystem, multisymptom illness who was on record blaming her employer for her illness? What insurance company would underwrite a patient with progressive gangrene of unknown cause?

The case was scheduled for a hearing two days later, but it was postponed. Months went by, and it was postponed again. ECI offered Sue a deal, one that would give her 50 percent of her benefits for a couple of years, but they offered no health insurance. Gaudreau was firm: the case was solid; stay the course. She turned down the offer, even though she was broke.

Finally, the hearing was held. Sue talked with the state's attorney outside of court for almost an hour before the case was heard. He knew me from a few years before when Maryland had contested a Worker's Comp claim for a worker in the Department of the Environment who had acquired her own chronic fatiguing illness while harvesting fish, as directed, in a massive fish kill from *Pfiesteria*, in the Pocomoke River. The attorney had lost that case, too. Maybe there was hope for Sue.

The judge, a commissioner actually, had a reputation for being tough but fair. Worker's Comp litigation is known for the high frequency of claims that aren't based in medical reality. But in the docket full of patients who might be presenting with fraudulent claims or exaggerated injury and feigned pain, there were patients like Sue, honest, hard-working and until now, never complaining.

Sue was Gaudreau's only witness. The mold reports from the blackened ceiling tile in the education room confirmed exposure. The health data was solid, even if lab tests like HLA genotypes, leptin, MSH, cytokines and anticardiolipins weren't part of everyday judicial vernacular. The opposing medical opinion didn't address any of the pertinent health issues, simply saying that mold doesn't cause illness and introducing the statement from the CDC that said there were no studies that proved mold makes people sick. The judge had to see the truth through the smokescreen.

Sue testified for more than an hour. Over and over again, she was asked to backtrack. Each detail of her story was scrutinized. Talk about a fine-toothed comb!

Shannon Brown took the call from Sue two days later.

"Shannon, we won, we really won!" she screamed.

Gaudreau had done it—legal precedent was established. The repetitive exposure trial carried the day. Surely, it will hold for the series of patients to follow.

Sue now had insurance and income adequate to support herself and her child without going on medical assistance. She would be allowed to convert her temporary disability to permanent disability. All in all, the court got the benefits right.

ECI didn't comment, though their risk manager had called me previously to talk about the hypothetical "possibility of mold trouble." They did nothing to fix the roof leaks, nothing to try to reach known affected workers and nothing to screen all exposed workers and inmates with VCS, HLA and MSH. They did nothing to protect employees at all. Despite the fact that several other workers have become disabled by the mold exposure, and have brought their own lawsuits, the State still hasn't done anything to reach out to affected workers.

Sue Donahue is doing better now, and her imminent risk of amputation seems a little further away. Who knows what complications and side effects her current medical program and those in the future will bring?

There are many lessons to be learned from this case. Human illness resulting from exposure to toxin-forming mold can be positively identified. The illness is treatable; reliable screening tests can be done inexpensively. Detecting the illness before it becomes life-and-limb-threatening, as in Sue Donahue's case, prevents morbidity and prevents costly lawsuits.

The time has passed for "benign neglect" to be a valid management policy regarding mold and health effects. Now that a legal precedent has been set, even in just a narrow venue like Worker's Compensation, it is likely that more cases will follow. No longer will solid, hard working, good-hearted people like Sue have to give up their health to work in a dangerous work environment, without an opportunity to recover lost compensation.

This case marked the beginning of the post-revolutionary mold era. It probably won't be a quiet revolution.

CHAPTER 7

Martha Knight

Some Coincidences Are Just Too Coincidental

Shannon was politely, but assertively, knocking at my office door just before Christmas 2002.

"Dr. Shoemaker, I think you need to take this call."

"In a minute, Shannon, can you ask them to hold? I'm almost done talking with Dr. Hudnell," I said.

"It's Martha Knight. It sounds important. I'll tell her to call back, in what, five minutes?"

• • •

"Ritch, what we really need is an entire building cohort of patients from one building each exposed to mold, with complete data on everyone," said Ken Hudnell. "Check HLA, MSH, all the hormones you're looking at, symptoms, VCS, MMP9, everything. Show who's sick and who isn't. Then show that with treatment, the sick ones become biochemically the same as the well patients. There would be nothing experimental about seeing everyone at one time, as each would be treated as an individual and we know that treatment of an individual works. The concept is to have a *complete* group of exposed patients."

We'd already done prevalence studies on all employees in a few small buildings, but not with *all* the labs and biomarkers—it was like Thanksgiving with turkey but no stuffing, gravy or cranberry sauce. Sure, the roast bird was truly delicious, and the data were excellent as well, but where were all the trimmings? It was one thing to know that the Biotoxin Pathway applied to individuals and small groups, but comparing individuals to a big group, with each evaluated simultaneously had never been done.

"That cohort would really demonstrate the validity of your diagnosis and your model of illness," said Ken. "Your treatment data are very convincing, but we need a complete population study as well."

Shannon interrupted my talk with Ken *again*, "This call sounds important."

• • •

It was Martha Knight from the Baltimore Washington Conference of the United Methodist Church.

"Bishop Felton May asked me to call you about a mold problem in our office. We've just evacuated our offices at 9720 Patuxent Woods Road in Columbia, Md., because of toxic molds," she said. "A lot of us have been sick since we found the leaks, but it wasn't until a couple of weeks ago that we had the building inspected. They found *Stachybotrys* and *Aspergillus*. Bishop May's physician, Dr. Warren Ross, said we should contact you about possibly looking at all of us. The Bishop himself would like to talk to you about our case soon."

"There are 55 people who worked in the office," she continued. "We handle administration, pension, health plans and more for all the Methodist churches in this area. To tell you the truth, we need your help right away."

"What can you tell me about the building?" I asked. "Mold needs moisture and food; cellulose and paper products are its favorite."

"Well, we've got plenty of both," she said. "There have been problems with the building from the very first day I started working there seven years ago. There were leaks in the windows and roof every time it rained. The water would come in over the window tops and run down the walls, particularly in the lunchroom.

"The Bishop knew from the time he first entered the building —and he was there about 6–8 months before I got there—that something was wrong. He has a background in biology; he was going to be a doctor, but he decided to become a minister instead.

"I first knew there was a problem in the building some time in the first few years. There were three or four of us, including the Bishop, who complained of problems—from sinus headaches to excessive fatigue. But there was nothing obvious, other than moisture that the landlord continually tried to contain and correct.

"We had one property management group during the first three years, and although they were very friendly on the phone, they never really fixed anything. And the other problem we had was that there were several tenants here prior to our occupying the whole building. The heating and air conditioning system was chopped up into different generations of equipment. It was all pretty obsolete, to the point where if something broke, they couldn't find parts to fix it except in the junkyard. There was quite a challenge, keeping up with the HVAC system during those first few years; nobody was really happy with it. We never even knew to look for mold."

"Of course, I'd like to help," I replied, "We've done a few clinics for a number of building owners. Basically, we'll go to a site away from the building, interview everyone, do our testing and come up with some conclusions about whether or not there was a health issue. The problem is that the testing is expensive and well, there's an expense associated with getting our staff up there, too."

Martha replied immediately, "The Bishop's first concern is not the cost. He wants everyone who's sick identified, treated and returned to work as if this nightmare didn't take place at all. Somehow, I've got to get us moved into a safe environment, files and all, cleaning everything first, take care of the year-end business details, and now also get you here before New Year's. And Christmas is next Wednesday."

Talk about everything coming down on one person.

"Martha, let's see how my office can help. I hear you about the time pressures, but if you have sick people, they can be treated if the exposures haven't gone on too long, but we need to get started," I said. "Maybe we could set up a clinic next Saturday. My staff could come up, we could screen everyone, draw the necessary blood tests, do Visual Contrast Sensitivity tests, physical exams, nasal cultures and pulmonary functions. If we find proof of illness—and with the combo of molds you have, that's likely—at least 25% of your staff will be sick, given the statistics on genetic susceptibility I have. We could have a report ready by mid-January. But one thing you need to know is that I have to be able to evaluate *everyone*."

"You could do that on short notice?" Martha seemed to be jumping up and down on the other end of the line. "The Bishop insisted that everyone get tested unless they had good reasons not to. I'll get started on the arrangements, get everyone scheduled and make sure any paperwork is completed ahead of time."

And so, on December 27, 2002, we worked up 42 people from a moldy building. The others came independently shortly thereafter to Pocomoke for their work-ups. The results of the BWC workers exactly mirrored information we'd already collected from small groups of people and individuals we'd worked up. Sure enough, those without HLA susceptibility to mold were rarely sick. Those with illness had VCS deficits, MMP9 elevation, high leptin and low MSH, just like the model of smaller groups suggested.

Here was Ken's building cohort, with the initial contact made within five minutes of his asking for it. We also had a clear, defined source of water intrusion along with unequivocal proof of amplified mold growth. Even better, this was a population without a lot of confounding health problems—only two people were smokers and no one had alcohol or drug problems.

More importantly, these were caring people whose jobs were based on helping others. Now they were working together to help themselves. Everyone had a story. And most importantly, the responsible party, the Bishop, put everyone's health at the top of the list. Martha said that when he decided on his career that he felt he could do more to help people by ministering to their spirit than as a physician attending to their health. Looking back on this case, he showed an example of caring and responsible leadership that is an example for all others to follow. The Bishop wasn't cheap and defensive like so many building administrators or owners or employers are. He didn't argue about whether doing a blood test was necessary or denied that the illness existed, even though there was the possibility that the BWC itself could bear some responsibility for causing the employees' illnesses.

As it turned out, the BWC story contains all the themes you will meet in *Mold Warriors*. There were many sick people, some with short-term, mild illnesses; others ended up with what appears to be *permanent* impairment. We now know the spectrum of illness could be predicted at initial diagnosis, but that information only began to solidify by clinical work with this group.

Martha herself had quite a story to tell:

"I went to work there July 1997, and in September, I started with a series of symptoms that were unusual for me. The first thing that happened was an abdominal pain on the left side, profound fatigue and pain down my legs. That started a pattern of the fatigue and pain that would come about twice a year, lasting about five months. This went on for a few years, but no one could diagnose it.

"A lot of other diagnoses were researched; I had many medical evaluations. Physicians would say, 'yes, you have some spinal disc degeneration, but it's not pinching a nerve and it's not on the left side, so we don't really know what's wrong with you.' But that started my journey from doctor to doctor, trying to figure out the cause of my pain and fatigue.

"I actually had several operations, trying to solve things that the doctors thought were wrong. I would be better at home recovering from my surgery then when I went back to work, within a couple of months, the pain would return. They fixed things that they thought were wrong during these surgeries, but it didn't work.

"I was having considerable back pain, and finally an incidental cyst that had been growing in my spine grew to the point that they thought *that* might be the problem. So in September 2001, they took a cyst about the size of my thumb out of the nerve canal in my spine. Most of the doctors were telling me that those cysts didn't normally cause pain, but I finally found a person at Johns Hopkins who'd seen a couple of cases like that where pain *had* occurred. And they were willing to take it out to see if that would help.

"At that time, I was having trouble sitting up for more than a few hours at a time. And there was some compression on the nerve because of that cyst. So they went in and removed it, which is a fairly serious operation. I was healing well and the pain was subsiding. Then in February of 2002, something else happened which was odd. This time I had a post-nasal drip—not a cold and not the flu. But the fatigue was so great that I was in bed for three days without even getting up.

"And when I finally did get up, I had pain from the top of my head to the bottom of my toes, and everywhere in between. I also had tingling in my feet and hands, and both felt asleep and cold.

My feet would fall asleep while I was driving in the car. My body began to mistake cold and hot for pain. If I put my hand under the cold-water faucet, it would hurt. And on the left side of my body, the skin was extremely tender, and you really couldn't even touch the top of my leg.

"At that point, I got depressed. I mean, after that many years of pain-cycles, and then going through this elaborate back operation trying to solve it, and then to have the pain come back all over my body—that was very discouraging. And again, nobody seemed to have an answer. My primary care physician examined me and she said, 'There is no neurological syndrome that follows the path of pain you're talking about.' But I said, 'I don't care if there's no syndrome. I'm here so that you can tell me what's wrong with me—or send me to someone who can tell me.' But she had no answers.

"So I finally went back to the pain doctor that I'd seen when the pain was getting worse in my back, and she said, 'OK, I'm going to pretend like I've never seen you before, because your symptoms are different this time.' And she gave me a full neurological workup, and she said she still had no idea of a cause. She sent me for every test in the world; I had an MRI for every part of my body. I had brain scans, thyroid scans, endless blood tests and hormonal workups. But everything came back normal, and finally she said, 'Have you been exposed to toxic mold?'

"And I said, 'Not that I know of.' Nothing more was said or done. Another year went by before we found the mold."

"She said, 'Well, I don't really know what's going on with you, but I'm going to give you some medicine. It sounds like the pain is just being amplified in your brain more than the actual stimulus that you're experiencing. So I'm going to give you some medicine to turn off the pain.' And she gave me an anti-seizure drug to try and control the pain for me.

"The other symptom that I had during that time period was a 'sensory echo.' I would be walking, for example, and I'd put one foot down—and my body would experience a second step, even though I hadn't put my foot down again. And if a car passed me on the highway, it was like I was seeing it go by twice. It was like an old film that got off the sprockets on the wheels, and it would jerk. Everything was like that, and she also tried to turn that off for me.

"Then last summer we had a drought here, and I don't know whether it was because of the drought or the medicine she gave me got these symptoms under control, but the summer was pretty good. Towards the end of the summer, however, the symptoms began to come back. Mostly, it was the pain at first. Then in early October, I got a headache over my left eye every day for a month. And it triggered a couple of migraines for me during that time.

"Coincidentally, that's when it started to rain!"

"For two weeks at the end of October, it seemed like it rained every single day. A coworker reported a musty smell in his office. And he began to miss work with a sinus infection. When he asked me to smell his office, it smelled terrible.

"Eventually, we found a leaking window in his office in 2002. We called in the property manager. They started tearing plaster out of the wall, and they found mold on the drywall underneath. The year before the building had been sold to a new owner, and the property management company caulked all the windows right before the sale. But the offending window in the coworker's office hadn't been caulked, and when they finally looked beneath it, behind the wall, it was November 2002. I stuck my nose in there and it was the most God-awful smell I'd ever smelled.

"The following day they came in and tore down that part of the wall—and it turned out that there was a rodent's nest in there, with a tremendous amount of excrement, and mold growing on top of the excrement. And there was actually water in the wall, between the wallboard and the cinder block and water stains showing the cinder blocks had been completely saturated. It seemed to extend along the wall into the next office. So they went into that office and did the same thing—and they found another shrew's nest there. In that office, the water had actually rusted out the metal studs in the wall and had pooled under a cylindrical table. I guess that's why we didn't see the water. When the table was turned over, it was covered with mold.

"The inspectors determined that the builders had never sealed the wall to the floor, and also that they'd never filled in the control-joints all the way around the building. This created holes that allowed the shrews to come in and build nests. Water had been seeping through the cinder blocks for years, and the blocks were only a couple of feet away from the workers' desks.

"An inspector poked a hole in the ground at ground level, and a tremendous amount of water gushed out from under the building. We took plenty of photographs that showed the water damage just in case we needed them.

"I'm pretty certain that when they opened up the wall cavities, the building contamination took off. They didn't isolate the room or use plastic containment or anything. I'm sure that's when the mold spores got into the air conditioning system and everything. The spores went all over the building, and everyone in the building began to have respiratory problems. Within a few days, many employees were complaining of sinus pain. I started being really short of breath, something I never have. I don't have asthma, though I have wheezed intermittently from allergies in the past. I think that exposure created the 'point-source' you and I talked about on the phone.

"Looking back on it, many of us had been feeling bad for a while. It was the sheer horror of seeing the mold that made us start to talk to each other about our health problems. Because I was the administrator in charge, I began to consult with various environmental experts about the problems. I finally contacted a physician in Annapolis who knew about environmental medicine. From my symptoms, he diagnosed me with a neurotoxic injury. And he first refused to let me go back in the building ever, though he relented and said I had to stay out for 30 days."

"Finally, the landlord hired a remediation team with full protection garb and respirators to clean up the entire building. They found that this entire wall, about three-quarters of the way up, had been soaked over the years with water from the outside. And they sealed that wall off completely and put drains on the outside of the building. And they revamped the entire drainage system. We closed the building for a couple of days while that was going on, and then we tried to open the building again the following week.

"But the air quality was so bad that people couldn't work in there, so the building was closed throughout Thanksgiving week.

"At this point, the Bishop went to his physician in Columbia, Md., and he failed the visual contrast test—and so he decided to move everybody out. But unfortunately, some of the mold was on the papers we carried out of the building. We just didn't know much about mold back then. We wiped down cabinets and other furniture but no one realized the paperwork was contaminated. And so the headaches and sinus aches and asthma continued in the new space.

"At this point, 40 of the 55 people who worked in the building were sick, most with sinus problems and allergic reactions. And some of them had been complaining of symptoms for more than a year. I started having new symptoms—muscle cramps, body aches and sinus congestion.

"Then on Dec. 27, Dr. Shoemaker came here and brought many of us to a local church, where he set up a temporary clinic. And it turned out that only eight employees tested were still healthy. The landlord wasn't interested in testing for toxic mold. But the church hired an environmental remediation company that found both *Stachybotrys* and *Aspergillus*. The landlord didn't agree that the building had anything wrong with it, but after we moved out, he gutted the inside of the building.

"I'm better now, but I'm worried about long-term problems. We just don't know enough about the future effects of this illness.

I'm getting new neurological symptoms and I still struggle with a level of fatigue that doesn't allow me to function normally. Another new group of symptoms comes after I'm exposed to chemicals. I have to be very careful not to get exposed to any kind of chemical, because it will just send me to bed for a couple of days.

"When I started on CSM, oh my goodness, I felt bad at first. Every pain that I've had for the past six years came back. But I stayed with the treatment because no one else knew what was wrong. I'd been to many doctors, but no one had been able to help me. I'd gotten to the point where I just thought I was crazy. But the fact was, the other doctors didn't know enough science to help me.

"Fortunately, Dr. Shoemaker explained to me how the neurological pathways are damaged in a biotoxin illness, and that was the first thing that made sense to me, because I knew that I had experienced those kinds of reactions. And nobody else had found a solution, other than to medicate me."

• • •

Mold illness symptoms can vary around a common theme. All people with the illness will have multiple symptoms from multiple body systems. The most common symptoms are fatigue and chronic pain. To me, these two symptoms simply underscore just how important MSH deficiency is. There are so many other symptoms, though; the respiratory people blame problems on allergy, cigarettes and sleep apnea. The gastrointestinal specialists blame problems on reflux, diverticulosis, gall bladder disease and the big one—irritable bowel disease. When the weight gain from high leptin kicks in, well, here comes the "too fat, out-of-shape" group of diagnoses. My patients tell me that this diagnosis is made really quickly when they see a new physician. Of course, it isn't too long before the stress and depression labels are glued onto the foreheads of mold people with letters that never are removed.

Imagine telling a physician who's supposed to help you but who's unaware of what mold does to people, that you have chronic fatigue, pain, and 18 more symptoms, with all standard blood tests coming back normal. It doesn't take too long before the hurried physician puts on his best solemn face and asks, "Are there some stresses in your life?" And if the individual listening to you is a loved one, well, let's just say it promotes marital break-up. "The Spouse's Tale," is just too common in biotoxin patients.

I used to count the number of diagnoses patient had when they came for their first visit—all the diagnoses were simply unproven assumptions. The number exceeded an *average* of four. And that was before lupus, possible multiple sclerosis, sero-negative Lyme disease, sero-negative rheumatoid arthritis and hypersensitivity pneumonitis diagnoses, among others, were trundled out by physicians for public consumption.

The worst, and most insulting, diagnoses were the "all-in-your-head" suggestions either made by the physician or loved ones. So many times, patients have told me that just having their abnormal lab tests recorded in black and white took away an enormous emotional load.

"I just knew something was wrong," said Martha to me. "But until I had your tests done, nothing showed up. And here it is. Now I can face my illness and start getting better."

• • •

The Methodist group had those stories and many more too. Some people had neurologic problems, like unusual numbness, tingling or tremors. Others primarily had cognitive problems—"brain fog" is a pretty good description. Respiratory symptoms were the most dominant in another group. A couple patients did fairly well on VCS, as no test for mold illness is 100% accurate. What we showed once again was that there were markers for mold illness that we could use to define accurately who was a "case" and who wasn't.

In other words, how do we define what mold illness is? Thanks to BWC and so many other patients, we had that answer. The definition of a "case" has been shown repeatedly to be correct ever since Christmas 2002.

Once we had a validated case definition, we could target specific tests to follow in treatment. If a person had too little anti-diuretic hormone (ADH) for the amount of salt (osmolality) in blood (60% of mold patients will), for example, then treatment should continue until the ADH/osmolality mismatch was corrected. The same idea applied to ACTH/cortisol, androgens and MMP9. The illness wasn't fixed until symptoms were gone, VCS deficits gone and abnormal labs corrected.

For some, MSH didn't improve and symptoms stayed. This finding helped us find out why the multiply antibiotic-resistant, coagulase-negative Staph germs (MARCoNS) living in patients' noses were so important. We rarely found MARCoNS in well people who had few symptoms and normal MSH, but found it nearly all the time in the sick ones. We already knew that the MARCoNS showed up *after* the mold illness began and wasn't the *cause* of the illness. The large cohort of BWC patients would let us follow those who were treated. If MARCoNS was the pathogen I believed it was, we'd be able to follow treated patients' relapse without re-exposure to mold and show the MARCoNS had now arrived.

It sure did, but only in MSH-deficient patients.

Many more insights came from treating the BWC people. There were a small group of HLA-susceptible patients with low MSH, VCS deficits, but no increase in symptoms. What was up with them? It looked like the Biotoxin Pathway didn't apply to them. How can people who *should* be sick, *not* be sick?

The answer was ACTH. All of these patients had sky-high cortisol. They were driving the second most important anti-inflammatory hormone after MSH like crazy. Their high ACTH was making their cortisol shoot through the roof. At first I wondered if they had some kind of rare illness, like a tumor making ACTH or another unusual condition, Cushing's disease. High cortisol should shut down ACTH production, but not in these patients.

The answer came from the normal passage of time. There was no demand for treatment from these people; they felt fine. So what if VCS and MSH and HLA said they should be sick! They weren't.

Their health did not last.

One by one, the high ACTH/low MSH patients turned into low ACTH/low MSH patients. It was almost like the pituitary was "burning out" its main defense against the stress of biotoxin attack. As soon as the ACTH fell, cortisol levels fell and symptoms took an elevator ride through the roof.

So, the model of Biotoxin Illness became confirmed. The role of ACTH in the presence of symptoms in mold illness was proof of why treatment with steroids, including prednisone, was like "throwing gasoline on the fire." I can't tell you how many disasters I've seen following misdiagnosis of biotoxin illness that led the attending physician to conclude that steroid treatment was necessary. From blindness in *Pfiesteria* and Lyme patients to dramatic worsening of everything in mold victims, steroids are the greatest poison.

Why? The answer was so obvious in retrospect. I was reading about differential effects of steroids on cytokines, changes in immune function and inhibition of white cell migration, hoping to find out that while steroids were manna from the heavens for my asthmatic or lupus patients, they were a growling Cerberus from Hades for biotoxin patients.

The steroid use shut down the ACTH response. With MSH deficiency, there was a really thin line to keep ACTH going. The jolt of 40 mg. of prednisone a day smashed the ACTH like an elephant walking on eggshells.

Please doctor; never give prednisone to a patient without considering a biotoxin illness first. Please patient; if your doctor wants to give you prednisone, ask him to rule out a biotoxin illness first. No, come to think of it, don't *ask* him; *make* him think of biotoxins first.

BWC gave us a lot more lessons about mold illnesses than just the physiology. Martha Knight sent me a copy of the directive from Montgomery Insurance Co. It said that no expense associated with a diagnosis made by Dr. Shoemaker would be covered. The letter itself would intimidate many administrators.

Not Bishop May. He was adamant that if the insurance didn't cover "my bills," then BWC itself would. Costs didn't come first: his staff did. In all my previous dealings with those with the potential for financial liability, Bishop May stands alone as the most committed to seeing his employees return to health.

Martha didn't have much time to find a new location. Fortunately, she found adequate space just next door; the move was made in a hurry. Even with all the computers used by BWC, they still had to keep hard copies of thousands of documents. How can you move a file cabinet full of papers from one office to another without possibly cross-contaminating the second building with mold spores (bearing toxins) from the first building? Did it really matter that the air samples didn't show all that much mold floating in the offices? If the airborne spores fell into the papers in the filing cabinets, then the files were the moldy Trojan horse.

BWC couldn't take time to try to answer the questions about possible cross-contamination. Eight patients came to wish they had.

Martha and seven previously treated co-workers went to work on papers from contaminated files just before tax time in 2003. The files had been sealed since the move, but now it was time for the annual IRS reporting.

Open the files, people. Let's get to work.

Martha was one of the first to get sick again, followed quickly by her co-workers. Each reacquired a typical mold illness. They'd all stopped their CSM, why should they take it? They were out of the sick building, so why put up with the side effects? Lesson from BWC: Re-exposure in susceptible patients not taking CSM prophylactically will result in illness again.

The second time around, each of the eight knew what was happening, so they all were re-treated quickly. Subsequently, Peter Steinmetz, an environmental biologist, was able to prove that mycotoxins were in the file cabinets, as well as showing that there was a different fungal toxin-former present in office #2.

When would this nightmare end? Martha organized a *second* move; this time only after *every* document in the office had been copied in a sealed room under negative pressure. Imagine three copy machines going full tilt for a month just to make a new toxin-free set of documents. But what about the documents that staffers had taken home for work? When would the potential for cross-contamination end?

In the middle of all this upheaval, there are 55 stories that could be told. The lady with a dual dreaded genotype, profound illness and MMP9 too high to believe who shouldn't be improved, but was. The lady with joint pain and muscle spasm so bad that her fingers and toes clawed uncontrollably. The lady with lesions in her brain that looked like MS but wasn't. The lady who developed catastrophic lung, joint and gastrointestinal problems that CSM couldn't fix. Included in the group were a couple of people who weren't sick but acted like they were. They jumped on the lawsuit bandwagon quickly.

And Martha Knight. She greeted us on clinic day, looking like a normal person. All around her were obviously affected people, people with the "hollow eyes," I call them. Near the end of the day, while the lab crew struggled to finish processing 15 tubes of blood from 42 patients going to three labs, with three different temperature requirements thrown in, Martha told me more of her story. The illness truly had changed her. She wasn't the same at work or at home.

Martha's story isn't unusual. She started to break down, tears clouding at first, but then they flowed in sobs of pain. The one part of her life that was most precious, her marriage, was in trouble. Later, her husband readily agreed that they were having trouble. But once the illness was identified, the "problems" disappeared.

It's the rare spouse who doesn't have second thoughts about continuing a relationship when their loved one becomes neurotoxic. People just aren't the same after suffering from exhaustion, mood swings, cognitive impairment and many more symptoms. Having symptoms every day adds too much strain to "for better or worse." It doesn't matter if the spouse has an illness from toxin-forming mold, Lyme disease, Chronic Fatigue Syndrome or has a rock bottom-low MSH from an unidentified cause. The number of symptoms and the unremitting effects of so many hormone and cytokine systems not working ruins relationships. In fact, more than 70 percent of marriages in which one spouse is ill end in divorce. God only knows what the figure would be if they added non-married couples. I think the number would be even higher.

The spouses that stand by their loved ones are the unspoken heroes of the mold illnesses. Not all do. Adding the pain of spousal wandering, infidelity or inattentiveness to chronic symptoms from biotoxin illnesses is emotionally crushing to many patients. It isn't a matter of *whether* these people consider suicide, but *when*.

When the patient is a child, and Mom spends all her time on the child's illness to the exclusion of the husband, is it any wonder that the home life becomes strained?

Not all the spouses of BWC patients were as understanding as Martha's husband. The lack of support from loved ones is a barrier to healing.

The legal battles at BWC began soon after Sue Donahue won the first Worker's Compensation case for mold illness. The stakes were high for the insurance company. If they agreed with all I said, then they would be paying out huge amounts of money. The stone wall directed against me was particularly maddening, but in a way, completely understandable. That didn't make it any easier to swallow, but the insurance company was being attacked legally, so they began to use all their legal weapons to fight back.

They brought in their experts, including Dr. Cheung in some cases, to say that nothing was wrong with the building and that mold illness didn't exist. The experts cited allergy, depression, arthritis, psychosomatic illness and more, as the real reason the BWC people were ill. They ignored the data and attacked me. They said we should forget the facts that proved the illness real. The list of experts who testified for the insurance company included the same group of med school graduates without hands-on experience that we've already seen in earlier chapters.

In court, the insurance companies lost. The insurance company experts tried to put forth ever-increasingly more detailed opinion papers, but they still lost. Naturally, the insurance companies appealed their losses to higher courts, delaying their obligation to pay for the illness.

I guess there will be appeals and postponements of the eventual Pay-out Day. The longer the insurance companies wait, however, the more likely it will be that more academic papers on mold will be published. Perhaps it will be more likely that the final verdicts will have punitive damages attached. The lessons we've learned in medical malpractice are that juries don't like it when big companies deliberately ignore obvious facts to deny the rights of the little guy. And juries also don't like it when big companies or government bureaucrats act with incredible arrogance and abuse of power to cover up obvious facts.

People like Martha Knight and Bishop Felton May fought in their one way to prove that mold illness is real and that it must be diagnosed and treated. They fought to protect their own health and that of their fellow co-workers, without ever stopping to consider the cost in work or money to reach their goal. They are true Mold Warriors.

CHAPTER 8

Mold at the Ritz

"What Doesn't Kill You Will Make You Stronger!"

Summary: At the tender age of 56, former high-powered corporate attorney and successful venture capitalist Carol Anderson is struggling to re-build her life, after a devastating confrontation with toxic mold. Her mold castle is one of America's most exclusive and expensive locales: a super-luxury apartment complex in Washington, D.C.

By Carol Anderson

Several years after the disaster began, I can't stop asking myself the endless "What if?" questions.

- What if my husband and I had never heard about the plans for a brand-new luxury apartment complex in the heart of Washington's "West End" section near George Washington University? Would we still be married?
- What if we hadn't been so impressed with the "elegant architectural design" and the breathless brochure-descriptions of the "five-star hotel amenities" and the "100,000-square-foot sports facility and spa?"
- What if we hadn't settled on the purchase of the apartment back in November of 2000, and then moved in. The Ritz was just for the two of us. The five kids were grown and would be

coming to visit us at our home in Florida.

- And here's the toughest question of all: If *those* things had never happened, would I still be raising venture capital, serving on boards and financial committees—instead of being chronically ill and virtually disabled because of the multiple toxic molds that were allowed to flourish inside the five-star health threat known as the "Ritz Carlton Residences?"

But let me back up for a moment, and tell you how it all started.

• • •

After more than three decades as a corporate lawyer and entrepreneur working with "start-up" businesses, I was thrilled by the prospect of living in an elegant and tasteful *pied-a-terre* located only eight minutes by car from Washington's Ronald Reagan Airport. Working with the developer's architects, I worked up a plan they implemented to combine two units into a single, perfect residence.

Although the apartment complex project was beset with extraordinary delays from the outset, I was pleased with the eventual results. The elegant, brightly decorated apartment was loaded with custom features and upgrades. Beautiful, carved "built-ins" were installed throughout. The kitchen displayed the latest in luxury appointments, along with such high-profile, high-tech equipment as SubZero and Poggenpohl appliances, and the bathrooms were lined with beige Italian marble. Silk curtains, imported wool carpeting, antique wallpaper and custom beveled mirrors throughout completed the effect. To cap everything off, the front hallway featured a hand-painted fresco of a Florida tropical garden—a reminder of our permanent home in Palm Beach.

Unfortunately, these glittering accoutrements would be of little value. Behind the ornate trappings and invisible to the eye, moisture seeped along interior walls, pipes leaked steadily, water poured in from the roof through the HVAC system, and mold spores began to reproduce at an alarming rate. I knew none of this, of course, as I worked diligently to make this apartment our home away from home. When in DC, I was there constantly, talking on the phones and working on my computer.

Once I'd worked to fix up the inside of the apartment, I thought things would settle down, but curiously, the building seemed to have a will of its own, and there was a constant sense of crisis. Workmen scurried around with their walkie-talkies, shouting to each other and dropping whatever they were doing in our unit to handle a bigger problem somewhere else. It was all very mysterious. We knew there were problems but we didn't know what they were or how serious. Things seemed to go from bad to worse.

Six months went by and I thought it was going to improve. Again and again, we were startled to discover that we were without electricity or heat. The water was shut down frequently and when the water did flow, it was a strange, cloudy color. When the water would go off, that would shut off the heat pumps, cutting off the heat. We could smell sewage odors emerging from the toilets and sinks. Food smells from other apartments were common and conversations were surprisingly clear in the bathrooms as owners discussed their problems at the start of the day.

Construction on nearby units often began at 3 a.m. and continued until well after dark. The official language on the site was Spanish (fortunately, I was fluent!) The workers seemed eager to please, but they had very little supervision. They would often barge into our unit without knocking, in search of spare tools or equipment, or simply by mistake. At times we spotted rats running loose in the hallways. When we complained, the managers would shrug and tell us they'd lay more traps.

As the chaos mounted, my husband and I asked each: "Who's in charge of this project, anyway?" The developer—an outfit called Millennium Partners—were headquartered in New York, and we rarely heard from them. Meanwhile, the rate of turnover seemed unusually high. As soon as we got to know a "project manager," he'd be transferred, never to be seen again. And, of course, each new manager felt entitled to ignore all the promises made by his predecessor.

And so it went, week after week and month after month. Was this the "legendary luxury" we'd been promised? Of course, we weren't the only couple in the building to feel the stress. After six months, dozens of occupants were complaining. And the worst part of our helplessness was our gradual realization that complaining was pointless. You'd inform the concierge about a problem—no action. You'd tell the engineers, and nothing would be done. You'd beg for help from the project manager: Forget it.

At one point I even wrote to the investment partners who'd funded the project in New York, to no avail.

All complaints fell on unhearing ears. Things were bad, but we had no inkling that the construction problems we could see would end up to be trivial compared to the health effects that gradually emerged. The ongoing, inexorable release of toxin-laden spores made by fungi growing behind walls and deep in the heating and air conditioning system was still unknown, at least to the residents. I had no knowledge of anyone else being sick until well into August 2002 when I became violently ill myself.

Through the Doctor's Eyes ...

When Carol Anderson first visited my office in November of 2002, she was weak, distracted and exhausted. She had the look I so often recognize in biotoxin patients; I call it "the hollow eyes." The patient is alive, but she looks like life is completely gone. Early in the two-hour office visit, she had the nervous energy that comes from seeing a new doctor. By the end of the visit, I might have called for a wheelchair to help her to her car. She was worn out.

She explained that she'd been sick for months, and that she wasn't getting any better even after she left her dream condominium at the Ritz. Carol had been referred to me by another resident of the Ritz Carlton, a woman who lived on the floor directly above her. This second patient, along with her son, had become ill two months before Carol got sick. Until November of 2002, however, the two women had never met.

Increasingly frustrated by her illness, Carol's neighbor had already consulted physicians in California, Arizona, New York and New Mexico, without success. The daughter of a physician, she'd already done extensive research on toxic molds, and her exhaustive studies had led her to my clinic. She soon referred Carol to me.

"I'm so grateful to my neighbor," Carol told me early on. "She showed me the power of support groups to help in these ordeals. Without her, I'd still be floundering all alone."

She went on to describe her bewilderment that no two residents had the same set of symptoms. Her friend's son had chemical sensitivity and unusual seizure-like activity, she pointed out, but others hadn't. Her neighbor could drive an automobile, but Carol was unable to manage that simple task because of depth-perception difficulties and blurred vision.

While a few of the neighbors weren't sick at all, most of the residents seemed to be in denial, or attributed their fatigue and confusion to old age. Others, especially the TV celebrities and the sports stars, don't want to admit they're sick because it would affect their jobs, their insurance rates or some other factor.

As I requested, Carol had sent a copy of her records and the environmental reports ahead of her visit. I was especially interested in the comments from an Occupational Medicine specialist from George Washington University. He would be on an invited panel with me in just a few days to define national criteria for mold illness.

Looking at his report, I saw that he wouldn't be able to diagnose a mold case if a ton of *Stachybotrys* drywall fell on his head.

The day may soon come when he's held responsible for failure to diagnose mold illness. I wish he could avoid the harsh impact on his professional life that will bring by learning how to hone his skills and to recognize what mold did to people like Carol.

The environmental report she gave me from the developer's experts was worse. Leaving aside the issue of "sham" or "selective" sampling for a moment, the report seemed reassuring, but their testing made a mockery of the word, "comprehensive." For all Carol knew, they might as well have tested the bottom of the toilet tank for mold when they showed results from the "bathroom." Carol only found out later that the experts had supplied her information on a selective basis. They didn't disclose until much later that they'd found *Stachybotrys* and *Aspergillus* in her apartment. The chain of deception had begun.

"It didn't seem important," they later admitted. "We didn't want to alarm anyone."

Of course, when Carol had her own tests done, the results were startlingly different, though the same two species of toxin-forming molds were again found. Who was she to believe? But this was all after the fact. Sadly the extent of the deception would take months to unravel.

Some toxic molds make their spores and toxins fly everywhere. We call them "bioaerosols." Others, like the nasty *Stachybotrys* species, produce gummy, moist spores that rarely get up into the air. And yet, because a worthlessly trivial five-minute air sample didn't show anything particularly terrible or anything toxic, Carol's condo was called remediated.

Wrong.

Just five minutes was all it took for re-exposure to Carol's suite before she became deathly ill, after the "remediation" was complete.

Carol Says That Symptoms Speak Louder Than Words

Dr. Shoemaker asked me a series of questions about my health and possible environmental exposures. He especially wanted to know about other potential toxins. After all, the entire state of Florida is home to all kinds of toxin-producers. I couldn't blame the Ritz for my illness if I had acquired my toxin illness from a source in Florida. Just look at the opportunities for toxin illness in the Sunshine State: I might have lived in a moldy home in Palm Beach, swam in red tides off Sanibel, visited nursery sites with a history of crop damage from fungicides all over the state or fished in the Harris Chain of Lakes in Central Florida just north of Orlando.

No, those were not activities that I'd enjoyed. I never got sick from eating the fish from tropical reefs and I never had Lyme disease, either. On every one of the possible exposures on his biotoxin list, I just didn't have it, except the Ritz condo home. My case was mold from the Ritz, period.

I think even he was surprised at my symptoms. I was sicker than my appearance. Most people say I look pretty good. I probably could fool the age-guessers at the county fairs; most of them would say I'm in my mid-forties, but I am somewhat older. So when he asked me about biotoxin symptoms, my list of symptoms was pretty long. Here are the categories:

General: fatigue and weakness. Had both.

Muscle and joint: muscle aches, cramps, delayed recovery from normal activity, unusual sharp stabbing pains, lightning bolt pains, morning stiffness (lasted one hour) and joint pain in large and small joints. Had all every day except the lightning bolt pains; I get those so rarely but they are over quickly so I don't complain.

Headache: Oh, my, I had real blasters. I held my temples with my palms to describe them. "Ah, yes, the *Pfiesteria* salute," he said.

Eye: sensitivity to bright light (the fluorescents in the exam room were killing me), red eyes, especially with chemical exposure, tearing, blurred vision. Check on all of those.

Respiratory: Sinus congestion, cough and shortness of breath (more on these later). Right to all.

Gastrointestinal: Abdominal pain, diarrhea, often secretory. Oh no, I thought you'd never ask.

Cognitive: Here I started to cry. Recent memory, gone. Difficulty assimilating new knowledge, you bet. I have to read two legal pages about six times each before I can begin to remember them. Word-finding problems, yes. Abstract handling of numbers. He asked me to divide 7 into 91. I couldn't. He just sat there waiting, "Go ahead, and just see the numbers in your mind." But I couldn't. Eight something? 12 and a remainder? The worst part is that I'd read about his abstract math question before my visit. He asked the same question and I still couldn't do it. Confusion? Yes. Decreased ability to focus and concentrate. Yes. Disorientation? No, you can't get me on that one. I never get lost, except the other day and then again last week. And then, well maybe I did have some trouble there, too.

Hypothalamic: mood swings, right. Appetite swings? Some days I was ravenous and other days I'd look at food with no thought of eating. Sweats? Night time? No, that wasn't menopause. Temperature regulation? OK, so I was always cold.

ADH: thirst and frequent urination, I had, but never any static shocks in Florida. In DC, sure, but never in Florida.

Neuro: Numbness and tingling in my hands and feet. At first one day it would go away, but now it stays, only to appear in one or the other side. No, I rarely got vertigo—see, I'm not so bad. Metallic taste? Sometimes everything tasted like copper pennies —about 80 percent of the time.

Adding to the Diagnostic Process

Carol certainly had multiple symptoms in multiple systems. She only needed symptoms from four of the different system categories to raise concern about a mold illness. She had them all.

So we had the potential for exposure, lack of confounders and presence of the typical grouping of symptoms not found in other illnesses that are included in the differential diagnosis of a mold case.

Next we needed a VCS test. While her eyesight was 20/25 (great acuity), she flunked the VCS test badly. Her physical exam was normal. And she was right; to most observers she was the picture of health. By the way, I hear the comment made to mold patients repeatedly, "You just don't look sick, how can you have so many symptoms?"

Her labs were next. I would start her on the CSM protocol, even before the HLA gene test, MSH, leptin, MMP9 and all the rest of the lab results came in. She had biotoxin mold, no question about it. My only question was how many of her lab results would be correctable. Carol had been hit really hard. I would do my best to describe her illness and what we could do about it.

"Carol, this is the Biotoxin Pathway," I said, as I drew out on a large piece of white table paper the scientific explanation for the various effects that biotoxin exposure can have on the body. At a first visit, I'm not sure how much information a biotoxin patient can retain, so I keep my explanations simple. She caught on quickly, however, and asked me many questions. She especially wanted to know how I'd been able to make the connection between exposure to mold and other biotoxin-related illnesses.

I gave her the "snapshot explanation," but I could see she was intrigued. I talked about the importance of the unusual Staph that might be living deep in her nasal cavities, releasing its various poisons.

No Illness Can Do This Much

I heard what Dr. Shoemaker was saying, but I felt so numb. Genetics? What about my daughters? MSH? I'd never heard that a hormone could be involved in so many things. What hormone class was MSH in? Hypothalamic? I tried to spell neurohormonal-immunomodulatory. Twenty-nine letters—that's the longest word in the dictionary!

What if my MSH were as low as he said it probably was? I didn't know that I had 34 health symptoms. And that fellow at George Washington acting as if he were an informed physician had said I was depressed. What an outrage. Depression? I wonder how many times he's said that in court when he testifies for insurance companies?

Well, what else is there? How much more can I find out about getting better? What accounted for the unusual neurologic findings? Was there an autoimmune problem here? Dr. Shoemaker didn't know, but he said he would look at anything I could find out. I believed him when he said that there was so much he had learned about mold illness, but there was so much more to find out.

Looking back on that day, as I am writing this in May of 2004, it seems like far more than just 18 months ago. Yes, Dr. Shoemaker did look at the link I found from mold exposure to antibodies to myelin basic protein. As soon as he found the link in me, he promptly sent off blood samples he had saved from more than 150 patients to prove the antibody situation was real.

Myelin is the "insulation" of nerve fibers. Antibodies to the myelin could disrupt the function of nerves and damage them without killing them. In a way, I wish I hadn't asked for the testing. My myelin antibodies are off the charts. Later on, when we talked about other autoimmune problems he found out I had, without any other known cause, Dr. Shoemaker told me it was my insight that started him looking further. He talked about the cytokine response to mold toxins as "opening Pandora's Box of autoimmunity." How true that was. I wanted that box closed forever, but that was wishful thinking, I guess.

Face to Face with Black and White

The photocopies of my lab results sat on my dining table for a couple of days. I'd pick them up, read the grim news and put them down. Later, when I picked up the labs again, I ached to see that the news was still grim. My genes, the HLA analyzed by the fancy PCR testing, stared at me. I had brought my daughters into this world to be healthy and vigorous. And now, I'd probably given them a sentence of life imprisonment from mold toxins because when they'd come to help me, they brought their genes—my genes—into the mold. God, if only I'd met Dr. Shoemaker sooner and we'd known that there could be a risk to them because of their genetics. If only I'd known, I'd never have let them set foot at the Ritz.

But I was the one who begged my youngest daughter to leave California to look after me—I was desperate, thinking of getting a court-appointed guardian at the time because I knew I met the legal definition of incompetent. To think that I'd compelled my daughter to come into this Tar Pit at the Ritz to become a Tar Baby, caught in the sticky mess of mold illness that further traps you whenever you struggle against it. I can bear a lot, but this guilt will never leave me.

My MSH is dead low. My ADH is dysregulated and my ACTH/cortisol is shot, too. But the good news is that the MMP9 isn't too bad, or so Dr. Shoemaker tells me, probably to try to lift my spirits. Poor man, he doesn't like to give his patients bad news, but maybe there is some hope for me after all.

Why Carol Felt So Bad

Carol's labs confirmed the extent of the hormonal hit she'd taken. There was no way to get around the fact that she was in big trouble. Even worse than all the hormone disasters was the presence of the biofilm-forming, multiply antibiotic-resistant coagulase-negative Staph in her nose.

I already knew that she wouldn't get better until that once-benign organism was eradicated. It wouldn't be until May 24, 2004, however, at the annual American Society of Microbiology meetings, that I'd prove to an audience of scientists why the unusual Staphs were so harmful. As further research into the Staphs and their biofilms and their toxins progresses, Carol's case stands out as one that could have been treated more aggressively if only I had known then what I know now.

We could fix her ADH problem by using a nasal spray that we give to little boys who wet the bed. We could fix the ACTH problem by fixing the toxin problem, but the FDA still wouldn't budge to let me give Carol replacement MSH.

What a tragedy. It was just like standing by, watching some kid with juvenile onset (Type I) diabetes waste away, doing nothing, when we could safely give him a simple shot of insulin. Carol was like one of hundreds of patients I had waiting for the FDA approval I'd asked for in January 2002.

Nope, the FDA wouldn't do it.

A Toxic Odyssey

What makes some patients accept their fate and others fight with every breath of their being? What makes some bitter and others energized to seek answers? This is a question I often ask myself in the practice of medicine. Sometimes it's religious faith that makes people strong. In others it's their upbringing. Carol seems to have a strong sense of right and wrong, but does not appear to be a serious follower of organized religion.

"I'm a believer, but not a zealot," she tells me. She's the first to admit that she's lived a highly privileged life.

"We lived all over the world. My father worked for a multinational pharmaceutical company and I spent a large part of my early life in Brazil. I speak Portuguese, Spanish and French. At the age of 11, I went to boarding school in England for six years. Then I studied in Switzerland. The first time I lived in the U.S. was when I came here to college. My parents had moved to New York."

How does this early background bear on her recent adversity? I asked.

"I do think going to a British boarding school for somebody who grew up in Latin America was a pretty daunting experience. But it certainly gave me the tools to deal with tough situations later on. I learned to take care of myself at a young age. My mother is English and I learned the expression, 'Stiff upper lip' early on. Long odds and impossible tasks don't usually daunt the British, and I guess some of that rubbed off on me," Carol admits.

"Fortunately, I was well prepared for the battleground that is biotoxin-associated illness—prepared by adversity, that is, during a challenging career that tested me to the limit. I graduated from Vassar College in three years, Phi Beta Kappa and Magna Cum Laude. My father-in law responded in characteristic fashion; he suggested that I immediately 'learn to type'—a vital skill, he suggested, for any woman who wanted a career. Instead of going to typing school, however, I enrolled as a graduate student in Economics at NYU. Then I worked at the Federal Reserve before my husband, a doctor, got drafted during the Vietnam War and we moved to Fort Benning, Georgia.

"I had three children under the age of three, and soon after my first husband and I moved back to Washington, the marriage floundered. I worked on Capitol Hill for the Joint Economic Committee while attending Georgetown Law School and raised the kids on my own. Once I graduated, I went to work for the high-powered New York law firm of Milbank, Tweed. It was a struggle, but I managed to make Law Review at Georgetown, and the children and I enjoyed a relatively stable period in our lives. Soon after my divorce, I met and married my second husband.

And then things got really hectic. We traveled a great deal. At one point, we had four teenagers living at home. For a while, I specialized in energy project development for Merrill International and had clients all over the world. Later still, I did a lot of work with start-up companies and young entrepreneurs.

"And then I got sick. Although many wrong-headed federal researchers and profit-hungry insurance executives insist that mold illnesses are simply 'allergies,' the fact is, I never had allergies before moving into the Ritz. I certainly didn't have a compromised autoimmune system. Nonetheless, I got walloped with an attack on my central nervous system—I lost sensation in my feet; I had weakness in my left hand. I couldn't do the simplest, most routine tasks, and I couldn't remember new facts no matter how hard I tried. My vision was constantly blurred; my speech was stilted. I couldn't find the right words and I'd frequently stop in mid-sentence with no clue of what I wanted to say." For someone who'd relied on her brain in order to do her work, this was a frightening experience. Only after I learned that these problems were typical of a biotoxin illness did I begin to learn to try to deal with them.

"The only 'good news' in this poisoned nightmare was that I knew how to locate the right specialists who could diagnose me and then treat me effectively. But I never stop wondering: Where do the people who don't know what's wrong with them, and can't afford treatment, go for help?"

The Doctor's Analysis: "It's the Mold, Stupid!"

When I ask Carol to describe her first exposure, it's a harrowing tale.

"Gradually we began to learn about leaks and floods going on in the building," she recalls, "but it was all after the fact, and it was a slow, drawn-out process. By the time we understood the extent of the structural problems, it was too late for my health. The harm had already been done."

Then she described some horrific symptoms from her earlier exposure: "I had bloodshot eyes and was constantly congested whenever I went to the construction site in 2000 and 2001. I began to snore for the first time in my life, and I was getting more and more tired. I was also becoming increasingly irritable—picking fights with people and getting upset over the small things. Growing more and more absent-minded. Losing credit cards and keys. Losing my car in parking garages."

I told her what she had already guessed. She was a classic example of the mold-exposed patient.

"But the worst was still to come," she said with a sigh. "Amazingly, I started putting on weight!"

I nodded. Sudden weight-gain was a familiar part of the mold saga.

"How much did you put on?" I asked.

"About thirty pounds, I guess. You know, I've read your books and what you say about insulin resistance makes a lot of sense. Frankly, you're the first physician who's ever asked me where I live or how much weight I've gained. I think those two questions are so important, but most physicians don't seem to care. It's pretty frustrating."

"What about other people in the building? Are there others who are sick?"

"Oh, sure. The developer made me feel like an isolated freak and yet knew other people were sick."

I spent the next half hour explaining how some people are genetically susceptible to molds and other toxins; others seem invulnerable. Carol nodded.

"That makes good sense. I've never understood why one person gets sick and another can tolerate the exact same environment. My husband looked at me as if I had a screw loose. As far as he was concerned, if he didn't have a reaction, then it couldn't be real for someone else.

"And it doesn't help that most of the people who are sick are women—that just breeds testosterone backlash. One of the owners calls us 'malcontents,' and others refer to us as 'hysterics.' Even though a number of the workers at the site have experienced bleeding—and there were several cases of pneumonia, even a few hospitalizations. But the workers are afraid to lose their jobs, so they try to hide their symptoms. Some of the security guards were sick.

"Even worse, both my daughters complained about getting headaches, rashes and colds when they came to visit me in the apartment. My youngest told me, 'The air smells strange in here.' What an understatement that was," she sighed.

"I never thought I'd have to start again from scratch at my age. For the first time in my life, I was ready to devote myself exclusively to my career—instead of worrying all day about car pools or getting some child to sports practice."

Too bad. My new patient is the first to admit that her life has been irreparably and radically altered by her recent illness. Example: She's had to move 11 different times since she first left the Ritz Carlton Residences on July 30, 2002.

She remembers a particularly grueling evening in July of that year: "Within two hours of moving in, I got a terrible headache and took to my bed. The tightness returned to my chest—only later would I learn it's routinely used name: 'asthma'—but it isn't asthma at all. I struggled with violent stomach cramps, diarrhea and nausea. Burning with thirst, I had to pee every half hour. I started bleeding from my gums, and then I found blood on my pillow from one ear. By morning, I lacked the strength to walk to the bathroom and crawled there on my hands and knees.

"Somehow, I staggered outside and managed to drive to my general practitioner's office in Virginia. He told me not to return to the apartment. He then put me on prednisone (a synthetic cortisone) for three weeks. Within a few days, he sent me to several specialists (an allergist, a pulmonologist, an environmental doctor), each of whom prescribed different medications. By now I'd forgotten how to put paper in the fax machine and how to turn on the coffeemaker. A disaster.

"Thank God my daughter soon arrived to take care of me—even though she had to call the doctor immediately, because she found me sitting beside a pile of medications with not a clue about when I should take them.

"For the next five days, I stayed home alone, getting progressively sicker. Finally, at 5 a.m. on the fifth day, I experienced an epiphany: This was not the flu. This wasn't old age, or fatigue, or depression. It's the mold, stupid. That's what was making me sick—or maybe it was from chemicals they'd used to clean it up (or both?).

"Somewhere around dawn, I crawled from bed and wrote a letter to the Ritz developer. They never responded, of course. But I was determined to prove my point. I don't know how I did it, but on the fifth day, I got dressed and went out to meet some friends for dinner. I was blotched and swollen (I'd gained 17 pounds of fluid and must have been a sight from the looks of horror that I elicited around the table. They wanted to take me to the emergency room right away, but I have a horror of hospitals and promised to go to the doctor the next day. As I tried to make it through the meal, I realized that I felt better for the first time in days.

"The next day, I picked up the phone and called the building manager and asked that they re-test my apartment unit. Finally, I came to my senses and called my own Certified Industrial Hygienist expert to do some independent testing. It took five weeks to get all the results back, and there was obviously no relationship between what he found and what the building's experts reported. I knew the deck had been stacked.

"My doctor at the time ordered me to move out immediately, which I did, into a hotel (Four Seasons), but the physical damage had already been done. From that point on, I was allergic to everything—perfume, soap, feathers, pets, toothpaste, detergents, chlorine in the water, the plastic in the cups. I couldn't brush my teeth without getting a metallic taste in my mouth. If I rinsed out my mouth with tap water, my tongue would swell up. If I went outside and breathed car fumes, I'd come close to passing out. I couldn't understand what was wrong with me. I slept all the time and was constantly confused, but during lucid moments, I started reading anything I could get my hands on. I started reading everything about neurotoxins, and I soon came across a book on Multiple Chemical Sensitivity (MCS). That book was a godsend, because it described what I had to a T. More importantly, it gave me concrete suggestions on how to cope with my new condition. And it taught me about the various fungi and the mycotoxins they emit.

"After that I read books on chemical brain injury, on autoimmune disease, on excitotoxins, on the fungal link, on inflammation. Bit by bit, I found descriptions of every symptom I had! For two months, I never left the hotel room except to go to the doctor's or the sauna (I learned early about the importance of detoxifying your body). I went on a very limited, low-carbohydrate diet, much like the one Dr. Shoemaker has been prescribing for patients for years.

"My daughter and sister began to care for me in shifts. They installed HEPA filters in every corner. Food was left at the door; the hotel cleaning ladies were required to be perfume-free and to use nothing but vinegar for cleaning. Sheets and towels were washed with special detergents.

"After seven long weeks, it was pretty clear that staying in a hotel wasn't going to be an acceptable long-term solution. But where could I turn? The developers were refusing to pay for alternative accommodations and told me I should move back into my apartment. When I showed them the results of the environmental study I'd commissioned, they angrily denied that the remediation had been inadequate. For four months, the apartment stayed vacant. Mold continued to fester—but after two more leaks from the apartment above, the developer finally agreed to come back in and 'redo' the work.

"Meanwhile, the chemical assays I'd authorized just before their second round of remediation work were beginning to tell the tale, and with a very loud voice. Those tests showed massive levels of *Stachybotrys* and other toxic molds including *Chaetomium*, a demyelinating mold. They also showed that I'd been subjected to astonishing toxin-levels. When they checked my furniture, some items had levels of *Aspergillus* and *Penicillium* at 43 million colonies/cubic meter. And yet the developers insisted that there was no need to clean the carpets, the furniture or my clothes!

"Caught in the middle of an expensive and painful divorce, I suddenly found myself totally strapped for cash. To survive, I began borrowing money from friends and family. I was impoverished now—and that experience was unprecedented for me. It taught me a great deal."

The Failed Remediation–A Cure Worse Than the Disease

"After a month of wrangling between my insurance company and the Ritz developer, the latter finally agreed to begin repairing at least some of the water damage to my apartment caused by the leaks.

"A few weeks later, I got a surprise. The phone rang, and the voice on the other end—it belonged to the latest harried-sounding 'project manager'—quickly announced: 'Uhh ... we've found a few spots of mold in your unit, behind the baseboards. But don't worry about this minor problem—we're cleaning it up as we speak.'

"Foolishly, I believed him. But a month or so later, when I returned to the Ritz to inspect the repairs, I got a nasty shock. It happened when I moved a sofa away from a wall and spotted a hairy-looking patch of creeping black mold. A moment later, I was pulling open a closet door and gasping at the furry patches of gray and pink mold advancing along the walls.

"Once again, the fumbling project manager assured me that the 'cleanup and eradication' would be swift. It wasn't. As the months passed, I returned to the apartment again and again and each time I arrived, the developers told me they had discovered new areas that needed remediation. I asked them if there was any risk from the mold and they told me it was harmless as long as I didn't touch it! That comment only seemed odd to me after the fact. At the time, I took them at their word.

"Although they never articulated their reasoning, toward the end of the remediation, they no longer wanted me in the unit until they'd finished the clean-up and I was suddenly banned.

"Finally, in June, they give the apartment the 'good housekeeping seal of approval.' The mold is gone, I'm told. The place is clean. I spend two days in June there, get a few headaches, but don't think anything of it. In July, I come back to the apartment. Everything is covered in pink dust and the air conditioning is off. The air has a heavy, stale smell to it. The developer is now frantically remediating all 161 units of the Ritz condos, as well as remediating the Sports Club and Hotel. They shut off the water to do the repairs and this, in turn, shuts off the heat pumps in the units, so that the air conditioning doesn't work. The engineers are supposed to go around and reset the heat pumps after the shutdown, but somehow my two heat pumps have not been reset and I didn't know how long the air-conditioning had been off in my apartment—I now know the mold had been having a field day in my absence."

Dr. Shoemaker's Assessment: If at First You Don't Succeed, Try, Try Again

"Where are you living now?" I ask Carol on her next visit. She'd brought along stacks of environmental studies that detailed the species of molds identified in her apartment. I know she's been moving around, living with friends, staying in hotels, trying to find a safe place to live. While she has a home in Florida, she's been too sick to fly and cannot make the 20-hour car ride.

"My daughter and I are living in an apartment the developer has provided us," she mumbled after a bit.

"What?" I ask in surprise.

"Well, after leaving my apartment vacant for 4 months, the developer has now agreed to do a thorough cleaning. They sent me a letter the other day saying they'd sent movers into my unit, had packed up the contents and moved them to another location in the building. For the first time, they're offering me an alternative place to live—an apartment in the South building. They assure me it's clean and has never had mold. We've been there for three weeks now, but we think that one may also have mold because I'm getting symptoms again and my daughter has been sick."

After listening to Carol's epic story of biotoxin-linked illness, I gave her the most useful advice: "Get out of that complex and don't ever go back!"

Let's face it: Even though the NY developer "remediated" her apartment twice, Carol's hypersensitivity to molds and chemicals is such that I would never recommend she live there again. Carol tells me that even though they agreed to remediate her apartment again, they've refused to clean her furniture and clothes and reinstalled them into her unit once they completed the second clean-up, thereby cross-contaminating the unit and necessitating further clean up. It's bad enough that Carol has to return to the Ritz periodically to check on her unit, in order to prevent the developer from accusing her of being 'negligent.'

But these 15-minute stays have undermined her progress and caused her VCS test scores to plummet.

My reaction: "We've discussed this, Carol. These toxins are deadly. At least we have concrete proof of cause and effect—but you simply cannot risk further exposures. Given your genetic susceptibility, you must avoid the apartment completely."

Carol has been coming to see me on a regular basis for more than a year. She takes her medications properly, yet her test results are repeatedly disappointing. While there has been slow, steady improvement (with the exception of two relapses that we can tie directly to re-exposures), Carol's MSH levels are stubbornly low. It's clear that she will not improve until she receives MSH supplements; unfortunately, they're not FDA-approved yet. It'll be years before the trials in humans are complete and by then the damage in Carol's case may be irreversible.

Her most painful question, during this period of agony: How can one building make a person so sick?

Because her sharp sensitivity to chemicals, detergents, pesticides, fertilizers, asphalt, gasoline and household paint—even in low quantities—will trigger nausea, dizziness, swelling, chills, stomach cramps and fatigue. She's become a prisoner in her own smell-protected world. Typically, after an exposure, it takes her three to four days in bed before she recovers enough to function again.

On to the Next Generation of Mold Warriors!

The next time Carol comes to my office, she brings along her beautiful, slim and energetic younger daughter.

"My mother has been a handful to look after," the young woman tells me, her voice vibrating with sadness. "It was scary to see how quickly she deteriorated. I hadn't seen her for a while and when I arrived in DC, I was shocked at her condition. She was getting lost in the car. She couldn't balance her checkbook or work the computer. She couldn't tell north from south or east from west. She'd go in the kitchen, turn on the oven, walk away and forget what she'd done."

"She's doing better now, don't you think?" I ask. How am I going to broach the subject of her own health, I wonder?

"Oh, yes. There's been some real improvement, but she's still incredibly forgetful and seems to have trouble making up her mind. The divorce probably hasn't helped. Of course, she does have chemical sensitivity to a heap of products and that seems worse. Come to think of it, I haven't felt too well lately myself."

There it is—my chance to leap onto the subject of her own health. "How about you? Do you think her apartment also made you sick?"

"Well, it's true that I've had a lot of colds lately, and sore throats. Bronchitis, too, but its winter, and I just figured it was the flu season. I haven't been able to run like I used to. I get short of breath."

I nod diplomatically. I'm so used to my mold patients finding a hundred and one excuses for their multiple symptoms. Again and again, as I walk them through the questionnaire during their first office visit, they're shocked to see how many symptoms they actually have.

"How about joint pain and muscle aches?" I ask.

She looks at me in surprise.

"Well, now that you mention it, I have been achy lately. And I'm having problems with my menstrual cycle."

Bingo! She's laying out a fairly typical case of biotoxin exposure, yet the case isn't completely clear yet. She does look extremely fit.

"Look, you seem quite healthy to me," I point out, "but your Mom is worried about you. How would you feel if we did the tests she went through? Carol is worried that maybe you were harmed by the mold, too. Of all the things she doesn't need now, it's worry about you. The best defense from worry and fear is fact. We'll just do the tests to make sure you're OK, and then we can give her some peace of mind."

Reluctantly, the daughter agrees to have her blood drawn, but we're rushed now, so I don't do a VCS.

Three weeks later, the daughter's test results are back and I'm shocked. Her genetic biomarkers, inherited from her father and from her mother, make her highly susceptible to molds. Her lab results are poor: low MSH, low ACTH, and presence of the bad acting Staph, elevated liver enzymes. Not at all what I expected, but what I had feared.

Her mother is overwhelmed with guilt when she learns that her daughter is ill.

"Oh, my God. What have I done? Here she is in the best of health and she comes to look after me? How could I have known that apartment would make both my daughters sick!"

• • •

Months later, when I asked myself a key question, "What personal qualities combine to make a 'Mold Warrior?'" Carol Anderson was the first patient who came to mind. Carol had been courageous enough to admit a basic fact: Her life had been irretrievably altered by her illness. She hadn't worked for two years. Her financial situation was dramatically changed. She endured a bitter divorce after 18 years of marriage. Her husband's two children by a former marriage, whom she'd raised as her own, no longer spoke to her. Her own natural daughters now have symptoms of exposure—especially the daughter who came back East to care for her mother when she became ill.

She fights for her rights, despite an unbelievable opposition. She's a Mold Warrior.

But Carol Anderson doesn't sound like a victim. She hasn't been intimidated by her experience, and she's without self-pity. Somehow, this Warrior has even managed to keep her sense of humor. Carol had documented neurological difficulties and central nervous system damage by the time she came to my office. She was also about to undergo her second of five depositions in the divorce. When I suggested that an anti-anxiety medication might help her cope with the enormous stress she was facing daily, she laughed out loud.

"No thanks, Dr. Shoemaker. I need to keep my head straight,

as much as I can, anyway!"

Carol's battle cry ought to energize every biotoxin-disease researcher in the land:

"I intend to fight with what little energy I have left to increase awareness of the serious harm to health that molds can cause. When I think about how many schools have mold problems and children being diagnosed with ADD, I shudder. People worry about lead or radon. Mold is as pernicious as asbestos—or maybe even more so, because it's so common. Every flat-roofed school is going to have leaks sometime. Mold can grow invisibly and change its chemical composition without warning. We certainly have a massive educational undertaking ahead of us!

"Now that I know the harm toxic mold can do, I won't accept mealy-mouthed excuses from people who should know better—for example, remediation company officials who don't protect workers or the public and who aggravate the problem by spreading mold spores throughout a building rather than cleaning them up; school administration officials who don't properly maintain schools, putting teachers and students at risk; hospital officials who fail to protect patients and staff; government officials with mold-infested public buildings who cover-up the problem rather than remediate it; and insurance companies officials who try to 'cap recovery' or refuse to pay altogether.

"These organizations have the medical records right in front of them. They review the claims, and ask their physicians to come up with reasons to not pay for legitimate expenses. The doctors see how sick patients can become after mold-exposure, yet they come up with a smokescreen of confabulated diagnoses—like the George Washington guy who said I was depressed—just to avoid paying what they're required to do by contract. The insurance companies should be asking themselves why the number of mold-related claims has risen so dramatically over the last five years! Instead, they usually deny the problem by resorting to such pejoratives as 'junk science' and 'mass hysteria.'"

Anderson, the Lawyer on the Mold Case

"A few months after my consultations with Dr. Shoemaker, there are 9 lawsuits that have been filed. Mine was the third. Seven of the nine are now using my lawyers. I discovered that there were two unexplained cancer deaths at the Ritz this year; one involved a 43-year-old woman who died of cancer of the sternum. Another was my neighbor—a 63-year-old gentleman, the picture of health, who developed pancreatic cancer and was dead nine days after diagnosis. People in the building still don't see a connection, but time will tell.

"Of course, our various lawsuits will be an uphill battle. The developer has at least 20 lawyers on its team and seems to have a bottomless supply of funds. To them, it's a game. It's all legal strategies and theories. They forget that these are real live people whose health is at stake.

"Indeed, the developer and its lawyers show an extraordinary contempt for anyone who claims to have become ill in the building.

"I personally think the stalling tactics being used by defense counsel are not in the developer's favor. We may be at the beginning of the mold saga—where we were with asbestos and lead 20 years ago—but the science is developing rapidly. Just think what you've been able to accomplish scientifically in the last year. It's only slightly more than a year ago that the labs started being able to identify the mycotoxins that are being emitted from the molds. The ability to classify the molds has become increasingly refined and they're able to measure smaller and smaller particles. All this will help with proof in the courts.

"But these cases aren't easy to bring. Some lawyers don't like to lose in court and are predisposed to settle. Settlement may be appropriate at some point, but you have to have someone on your side who's prepared to go all the way or you'll lose.

"Typically, lawyers don't like to push the envelope with new legal theories, especially where cutting-edge medical science is concerned. They prefer the tried-and-true claims and those are not always the ones that apply to you. For instance, in my case, a number of lawyers told me to forget MCS and brain injury. 'It's just not provable. Not like allergies and asthma.' That wasn't acceptable for me. I wasn't going to wrap up my ailments in some ill-fitting suit for the ease or convenience of some lawyer. I wanted someone who was prepared to do the work and take the tough road, if called for.

"Successful mold litigation requires a combined expertise in building defects and personal injury. There are only a handful of firms today in the United States that have this combined expertise and they are hopelessly overworked.

"I hate to pry, but aren't these cases done on a contingency basis?" I ask, "If you're paying expenses, won't the other side just bleed you dry?"

"Yes," Carol admits, "but these are very costly to bring and typically firms will require the plaintiff to pay a portion or all the up-front costs. This becomes a battle of experts and can cost hundreds of thousands of dollars. You put on a battery of experts and the other side puts on an equally convincing group of luminaries with polar opposite views. In the meantime, the patient is sick and financially disadvantaged, faced with overwhelming decisions on jobs, insurance, you name it, trying to cope with the basics of where they're going to live and how they're going to support themselves. Often they have children who are sick and having trouble in school ... well, you know the rest. I don't mean to be a pessimist, Dr. Shoemaker, but you have to think long and hard once you go down this litigation path and then you have to have unbelievable resolve to put up with what's going to be dealt you.

"Although some readers may find this difficult to believe, I'm not a litigious person at heart. Throughout my legal career, I've always been considered a skillful negotiator—someone who brought parties together, found the common interest and forged the creative solution. Two years ago, if the developer had simply acknowledged that I was sick and offered to buy my apartment back at cost, I would have taken the offer in a flash! But today, after two failed remediations and the exposure of my daughters, my position is much more intransigent. *What doesn't kill you makes you stronger*—and I've just begun to fight back! I've lost so much that I have little *left* to lose. In short, I've become the 'Plaintiff from Hell.'

"I know we'll suffer some defeats along the way. But those of us who were injured at the Ritz are certain we have the facts on our side. With the help of physicians who are scientifically adept and courageous enough to challenge the conventional wisdom about toxic molds—and dedicated enough to do the painstaking work of *documenting* their findings—we're going to turn the tide in this country. Of course, you're going to have to find some Mold Warriors in the legal profession and move them along to adopt bold new strategies in court as well.

"It won't be easy. In fact, it might even get harder for a bit as the insurance companies, developers and other building owners join ranks to fight the tide. But we can fight back. And eventually, they'll fold in the face of the growing ranks of Mold Warriors!"

CHAPTER 9

Romo and the Roadblocks In Hampton Bays

The School, Truth and the Teacher

Pat "Romo" Romanosky and her husband Joe had driven all night to get to Pocomoke for her first office visit. She didn't mince words.

"I'm sick of being sick and I'm sick of being told I'm *not* sick," Romo said.

"Most of all, I'm sick of being told stories by those who should be fixing our school. Our building is notoriously filthy," she said. "I'm sure some of the dirt from when I started 29 years ago is still there. We have carpeting in halls and classrooms that gets shampooed once a year in August. They won't take up the carpet because it's covering the asbestos in the floor tiles underneath. The roof in my work area has leaked for years. It's so hot, it's worse than a sauna. At least in Finland, you can jump into a snow bank or something to cool off; not here. I have to open windows and put on fans or the heat suffocates us. Many of the teachers have demanded fans because it's impossible to teach in that heat. I guess the heat is a mixed blessing, though—it probably dries out all the soaked ceiling tiles.

"And I have a lot more to tell you about the moisture problems in the school.

"I'm done just talking about this issue. There are kids at risk and the school is doing squat to protect them. I don't know if you need to know all the details, but I brought a time line for you that summarizes name-by-name what we're up against. I've documented everything so that no one can come back to me deny the truth. And I brought a copy of my speech to the PTA from February. We're just not going to be put off by the administration any longer."

Romo's summary and the text of her address to the PTA clearly and heartbreakingly tell the story.

• • •

"Romo, I'm seeing that your concerns about water intrusion, mold growth, failure to maintain a clean environment and much more have not been addressed by the people who have the duty to take care of these issues," I said. "When did the health issue begin?"

"I can't be sure exactly. There are so many symptoms and to tell you bluntly, my memory isn't good. That's why I brought the time line. Those facts are secure. I don't believe the school really cares. Just look at the environmental report done on the Hampton Bays Union Free Elementary School," she said. "What they're passing off as a finished product is a farce. They didn't sample any walls with visible mold, the carpets, the stained ceiling tiles or any of the classrooms where the smell is so bad. I don't know what to call the smell—propane? Something dead?"

• • •

Romo's office visit wasn't much different from so many others: an informed worker sees water intrusion and evidence of amplified mold growth, or smells musty smells. The worker sees sick co-workers, including students in this case, and then asks the building owner to make changes to clean up the mold and correct the water intrusion. Claims of illness usually are ignored as 'allergy,' but when the complaints don't stop, the illness is blamed on 'asthma,' 'cigarettes,' or 'stress and depression.' The worker usually becomes so frustrated that she leaves or gives up the fight.

But quitting isn't part of Pat Romanosky's genes.

"Look, I know I'm sick and I know the building is doing it to me," she said matter of factly. "I just don't know how to prove it. I've got to get better. I have a new grandson and one more on the way. I'm so tired, I can barely take care of myself, much less Joe."

Romo had started feeling bad several years ago. At first it was just her voice—her throat got so bad, she was told to whisper, and then to use a bullhorn in class. Next, her arm hurt, for no reason. An orthopedist said she had an "overuse syndrome" of her left elbow, but she rarely used her left arm for anything strenuous. Not too long after that, she developed weird neurologic problems—numbness and tingling that appeared out of the blue and left without a trace. The neurologist ordered nerve tests, neck X-Rays and a brain MRI for her symptoms, but no one found anything wrong. She saw a long series of doctors, but got no diagnosis and no relief.

In Hampton Bays, on the southeastern shore of Long Island, NY, the Lyme disease capital of America, if someone has multiple health symptoms and no reason is found, it won't be too long before someone suggests Lyme disease as the diagnosis.

Romo knew Dr. Joe Burrascano from his treatment of Lyme disease in one of Romo's family members. For years, Dr. Burrascano, practicing just down the pike from Romo, has been treating Lyme patients from all over the world. He listened to her story, suggested a trial of antibiotics, but he wasn't convinced that Lyme was the problem. And just as he suspected would happen, Romo didn't respond. When Romo started to tell him about the mold in the school, he wrote my name down for her and told her to go to Pocomoke.

And so she did, in April 2003.

• • •

Romo had the typical symptoms that all mold patients know so well. I'm going to tell you all of them, since indoor toxic mold is not just a little nuisance that causes a little cough. She had fatigue that put her on the couch, weakness that didn't fit her phys ed teacher's physical fitness-buff body, constant aching and charley horses from Hell.

A little bit of physical activity put her in bed for a couple of days (remember the jargon from Chapter 4: "delayed recovery from normal activity?") How about those sharp, stabbing pains and then the "cattle prods," the ones that felt like a lightning bolt thrown by Zeus right through her? Her eyes were the worst; bright light bothered her; they were bloodshot and tearing constantly, and blurred vision was an unwelcome garnish.

How did she suddenly become so short of breath, with abdominal pain that resembled so many gastrointestinal illnesses, when all her tests were negative? Diarrhea that woke her from a restless sleep became the norm—that is, diarrhea got her out of bed at night, since she usually couldn't sleep. And even if she did grab a few hours of sack time, she never was refreshed in the morning when she struggled to get out of bed. It was her joint pains and morning stiffness that made her think she had Lyme disease at first.

When I asked her to divide 7 into 91, without using paper, she laughed at first, then broke down. She couldn't see the numbers in her brain! Joe now did the checkbook, a job Romo had performed for years.

Memory for her was just a song from the Broadway musical "Cats." Other cognitive problems—the ability to concentrate, assimilate new knowledge, find words in conversation, confusion and disorientation—turned her into some new person who looked like Romo, but didn't think like Romo.

The "Spouse's Tale," is the story of the healthy partner in the marriage. After listening to the same story repeatedly, I know that mold illness shows just how solid or weak the marital relationship is. When mold steals a loved one's health, leaving behind a shell of hollow eyes, relationships either wither or survive. There's no more of life as it was before the mold illness. The surprise to me in my practice is to find a spouse like Joe, who stands behind his woman, for better or for worse. Sometimes that's hard for spouses, when one is so tired that intimacy can mean nothing more than saying goodnight tenderly.

Joe was right there for Romo; she hadn't been the same for several years.

"She can't think like she used to," confirmed Joe. "When we get her out of the building, she's a little better, but not much. If the mold can hurt her this way, what is it doing to someone who isn't as strong as she is?"

Her arms were numb from time to time, but the tingling never stopped. Mood swings, appetite swings, sweats and temperature instability were present, showing how besieged her hypothalamus was. Her greatest joy, holding her grandchild, was now a problem, as she shocked him repeatedly. Imagine having to turn on light switches with your elbow because you got a static zap every time you got close to a properly grounded electrical box.

Pat didn't have Lyme or any other confounding illnesses, though she knew the list of other biotoxin illnesses. Joe had picked up ciguatera in Hawaii years ago.

Romo's physical exam was pretty normal. Her VCS test was positive. She had the potential for exposure, symptoms and no other explanation. There was enough reason to give her a label: she was a mold patient. It was just that simple.

• • •

"How can I prove that the building is causing my illness?" she asked.

"Let's get your labs done," I suggested. "The abnormalities will profile your illness and provide a platform for an attack on the real cause of your symptoms. Once we know the physiology, we can deal with it. Anyone with 30 health symptoms of more than a few years' duration, like you, needs to know what to do to feel better. I use a sequential approach to treatment which takes five separate steps. First, we eliminate the toxins with cholestyramine (CSM). Then we continue with CSM and begin to knock out the unusual colonizing Staphs. Then we correct the hormone abnormalities, growth factor abnormalities, high MMP9 and the rest. After you're better, we'll prove that you have symptom resolution, correction of the VCS deficits and reversal of all the biochemical problems your baseline labs will show.

"Then, we keep you away from the school and stop all medications. You'll use your own exposure to tell us what the source of the mold exposure is. If you get sick at home, well, we can't blame the school for being the source of the illness. But if you don't get sick from your home, car, church, supermarket or anything else, including that wetland near your town, then we put you back in the school for ***three days***. If the school is the source of the problem, you'll get sick. Then you get treated again. And then you don't go back into the school."

Romo paused. "You're telling me I have to get sick to prove my point? And what about the dose of mold I'll be exposed to? Isn't the amount of toxin I'd see in three days so much smaller compared to 29 years of exposure that I won't get sick?"

That's my favorite slow-pitch softball question. Wham, hit that one right out of the park.

"That's the whole point of the genetics of susceptibility, Romo," I agreed. "In a way, the idea is similar to a pregnancy test. The woman who has missed her period and who has a positive pregnancy test isn't a *little bit* pregnant, she's pregnant! You're exposed to mold and you get sick from mold. No 'little bit' about it.

"The toxicologists go nuts about these amounts. They don't understand that health effects from mold toxins aren't dependent on the number of teaspoons of toxins over so much time. Once you are 'primed,' that means your immune response genes are revved up, so new exposure is like flicking a switch. If you have a vulnerable HLA DR genotype you could be hurt if you're then exposed to indoor mold toxins. Now don't get me wrong—the greater the dose of toxin or the longer the exposure, and the longer you stay exposed and untreated, the longer your immune response will start to recruit *additional* problems, including those autoimmunity illnesses we talked about."

"So, if I've got the mold genes, then my children might have them too," said Romo. "And they could give them to my grandkids. Oh, my God, I can't let the children get sick!

"OK, Dr. Shoe, we're going to Florida for a couple of weeks for the long Spring break. We could stop by here on the way back to Long Island, get the labs done again, and prove I'm improved. Then I could wait a few days, not taking CSM before I went back to school, prove I'm still improved. Then I make myself sick again."

She paused again for a moment. "I'm sorry to ask this, but do you know what you're doing? I mean, it looks like there are about five steps I have to take and you're going to record all of them. But I don't want to feel better and then turn into a 30-symptom queen again, either. I don't want to give away my life for a school."

"Romo, you don't have to do anything," I explained. "Any of the treatments I use are ones you have to give consent to. If you want to know if the building is the culprit, this is what you have to do. If all you want is to get better, then we don't have to do either of the two prospective exposure trials. Your option, either way."

"But if I don't do this, then we can't prove that the building is the real problem," she said. "And we can't show anything about all the others if I don't take the first step. I can't believe that there are so many kids at risk."

Romo stopped for a minute, as if to breathe. "And not just my school—how many others are worse? You know, I see the kids with their coughs and their asthma inhalers and kids that just act like they aren't there. Could mold be hurting them like I think it is?"

"Romo, we have data to present in a paper on 156 patients from 150 buildings all ready for the big Mold Conference up in Saratoga Springs this fall. I haven't even *applied* to submit my paper yet, but the 5-step, repetitive exposure approach is one I've been using for several years now. The lawyers and epidemiologists love it, because *we can pinpoint exactly where the illness comes from*, provided the patient is willing to give informed consent. We use *11* different biomarkers to show how the illness changes with each step."

"Oh, this is incredible—*proof.* Finally, solid proof for the School Board battles? You know the School Board will fight me," said Romo. "They're caught with an old building with water problems and no money to put up a new school. If I get sick from the school and then we have others show they get sick too from the school in the 5-step protocol, well, the Board would need a new school. Hampton Bays isn't Southampton, where all the rich people live. I mean, we're blue collar."

• • •

Pat Romanosky followed the 5-step protocol exactly as prescribed. She quickly responded to CSM, with resolution of nearly all of her symptoms. She stayed well after she returned home, and after stopping the CSM. This meant that no exposure away from the school, including shopping, the inside of her car, her church, even the nearby wetland, was making her ill. I could feel her fear over the phone lines after she went back into the school, unprotected by CSM.

"It's back. The illness is back!" she told me. "The red eyes, muscle aches, joint aches, confusion, memory—I can't believe it. It's only been 36 hours since I went back in. What am I going to do?"

She had tracked her exposures carefully during her diagnostic re-exposure. In animal labs, they call the repetitive exposure studies "ABAB." Do something once and measure an effect. Do the same thing to the same animal a second time and document the same result. The ABAB studies are the most powerful kinds of study in medicine, pharmacology and toxicology.

When we add in the *exact time* Romo became ill, and the objective changes in her health, complete with all the changes in markers, especially the changes in blood levels, the cause was certain. The special labs, ones that change rapidly with exposure, like MMP9, leptin and VEGF all exploded almost immediately after contact with the school. As they say in nearby Rabbit Gnaw, Maryland, we had our possum in a burlap sack.

• • •

Romo Revs Up

Romo was ready. She had the proof she needed. As she later said, protecting the children was her driving force. The school was a threat to their health, all right. And the threat wasn't small: exposure to the school was hazardous for those genetically primed for mold illness. How many students and staff, present and past, were affected? Her appointment to the School Safety Committee gave her the duty to act to safeguard the health of students and staff of Hampton Bays Union Free Elementary School. It wasn't just an option for her to speak out about health effects from mold exposure; it was her responsibility!

From my previous work, we already knew that 25 percent of adults exposed to amplified growth of indoor fungi or their bacterial and other microbial companions could develop a mold illness like Romo's. Children aren't little adults, but symptom screening in schools has shown that after mold exposure, there are two groups of kids, almost like "yes" or "no." One group will have no more than two health symptoms and the children in the other group will average about 12 symptoms. Each group's parents think the other parents are crazy.

We also knew from doing lots of screening clinics, using physician-recorded symptoms and VCS, that we could demonstrate abnormalities in children that left little room for alternative diagnoses. Eliminating alternative diagnoses cleared the decks for confirmation of the illness: mold exposure from the school. The rest of the labs showed some children were in very bad shape from the mold. With these labs, we established a baseline. Don't forget that the labs themselves don't make the diagnosis; they *support* the diagnosis. And the labs are ones that can be ordered from just about any laboratory.

Romo stayed on aggressive CSM treatment as prescribed and continued to work. After all, she only had another 18 months and she could retire. As she started to ask which teachers and staff members were ill, some weren't willing to talk to her, but most did. Many of the staff agreed to fill out a standard symptom survey.

"Dr. Shoe, this is scary. Nearly 30 percent of the staff has more than 15 symptoms out of the 37 on the list. Can you imagine walking around with 15 symptoms and not doing anything about it? Wait, I had 30 symptoms! I forgot, again. It amazes me to see what all the mold in the building does to your ability to think," Romo laughed.

"Like we should give away our brains as part of our loyalty to our employer."

"Romo, those numbers shouldn't surprise you," I said, "One

quarter of the country has mold-susceptible genes. So we would expect that in any moldy school or business place you will see illness in 25 percent of staff.

"You know, I have to wonder about the kids. Do we have any way of seeing what the statistics on learning disability are in your school? Maybe the percentage of kids on Ritalin or Adderal or an antidepressant? And how about a five-year follow-up report? Wouldn't the PTA want to know if attending school risked the future health or ability to learn of about 25 percent of the graduates?"

"What we would like to do, Dr. Shoemaker, is bring you here to do a formal screening clinic," she said. "The teacher's union would cover your expenses. You could bring your wife, stay with us, and maybe catch a show in the City. And don't forget, I'm well known for my secret recipe for cooking lobster from the Long Island Sound!"

Doing a clinic in a state where I am not licensed can be sticky. You can't just walk in and start seeing patients and ordering lab tests and treatments. Every state has different statutes covering visiting consultants. New York is particularly sticky, since its Medical Board has a reputation for being really aggressive about going after the medical licenses of some of its own physicians in the state. From what I'd heard, New York's Board had questioned a number of physicians who treat Lyme disease, including Dr. Burrascano, and some of those docs had been stripped of their medical licenses. Tom Monaghan, executive director of the NY Medical Board, was kind enough to clarify the law for me. As long as a physician licensed in NY attended me, and that physician was the one who ordered the labs and treatments, I would be allowed to be a consultant.

"Romo, I would be happy to do a clinic," I told her. "But we'll need to do the clinic thoroughly and precisely. As you know, the power-players sure won't want the school to be proven to be a 'Sick Building.' The clinic will be a threat to the School Board; any physicians who were negligent in *not* making the diagnosis;

the insurance companies; and politicians dealing with budgets. They will try anything to attack you and me. If that old school is so hopelessly contaminated, and it sounds that way, we're talking a new school and some lawsuits over health effects. Maybe even a class action suit, since hundreds of children could be still suffering needlessly. When the parents of Johnny, class of 92, has been sick for years and isn't as big a professional success as he says he should be, who is going to compensate him? This issue is politically and economically explosive.

"Since most of the voters in a school district won't have kids who are sick, they're not likely to be happy about funding a new school construction bond. Who will vote to raise taxes? And what will the politicians say? They will say the CDC doesn't agree that mold makes kids sick, so why should they support a new school project?

"You better believe that the school will bring in some doctor who will say the school is just fine and the children aren't sick. When the school has that guy's testimony, just see how quickly they refuse to pay for tuition vouchers or cover costs to transport sick kids to a different school.

"Wouldn't it be wonderful if the school took the problem to heart, though, and did a health survey of all the kids, staff and graduates, just to find out? They could fix up the school, set an example for the whole country," I said.

"Dr. Shoe, the school is really bad," said Romo. "The cheapest thing would be to close it. There's a new Superintendent coming in the fall. I don't see any way that the current temporary Superintendent will make any meaningful decisions.

"What about a legal option?" she asked. "If I'm sick from my job—and I am, and so are a bunch of others—with a legal threat, wouldn't that make the school correct the problem?"

"Romo, I don't know anything about the legal situation," I said. "I've talked with a number of attorneys who would be willing to talk to you. Who knows, if enough parents banded together in some sort of class action, maybe that would be the answer. I mean, the majority who aren't ill can't just run roughshod over the minority who are ill.

"While someone is doing the legal homework—and your union should be doing this, not you—I'll check to see if Dr. Jo Feingold will be my supervising physician. She has a practice outside New York City, so maybe she would be willing to make the trip to Hampton Bays."

• • •

On May 30, 2003, a health clinic was held in St. Mary's Church in Hampton Bays, NY. Forty-eight people attended, including staff in the school and teachers from other schools in the district; children of all ages; Dr. Jo Feingold, an internist from New York City; Dr. Jeff Kopelson, a family physician, also from New York; and George Burns, a member of the law firm, Lewis and Roberts; and myself. Romo had done a bang-up job of organization. We worked all day long, finding the expected distribution of people. Some who were exposed were ill; others with exposure weren't. All of the people who didn't go into the school were well. Eight of the children were too young to take the VCS test; the six who took the test included four with symptoms and each of those four were VCS positive.

The results of the clinic were clear. As Shannon Brown logged in the symptoms, VCS scores and lab results, the conclusions were obvious: The school was sick and the severity of the illness was incredibly high.

One of the worst was five-year-old Fisher Haines, a precocious child whose "allergies" never left him alone. His mother had tracked his symptoms over the past year, showing a direct relationship between exposure to the school and illness. As the year progressed, however, Fisher no longer became symptom-free with simple removal from exposure.

"Dr. Shoe, what should we do now?" Romo asked. "The staff is sick. We all want to get better and we all want the school to be cleaned up. Do we all have to do the 5-step protocol? There are bunches willing to take CSM, either from Dr. Feingold or their own doctor, and a bunch who would be willing to do the same thing I did."

"Every person needs to know that there are risks involved with taking the medication, and then feeling better, staying well off medication at home and then re-challenging with exposure back in the school," I explained. "Face it: in the paper we're preparing, there was one person who became irreversibly ill with re-exposure. The fact that some of your colleagues may become irreversibly ill as well by continuing to work in the school isn't the point. As soon as you see people with this illness get better, then you never want to see them relapse."

"Dr. Shoe, wait a minute. These people are making their own choices," said Romo. "If we don't have overwhelming proof of health effects in a group of people, then what will happen? Nothing. The school will continue to be an environmental nightmare and no one will ever do anything to fix the problem."

• • •

In the end, 12 patients, including Fisher Haines, followed the cholestyramine protocol as the first step of the 5-step repetitive exposure protocol. Each had dramatic reduction in symptoms and improvement in VCS scores. Of that group, nine continued on with the prospective exposures, first off medication at home and then without medication but at the school. Each relapsed promptly with exposure to the school, but not away from it. Finally, they were treated again; completing the fifth stage of the repetitive exposure and all remained well.

In response, the School Board denied there was an illness. They commissioned an environmental study of the school which showed that all was well. The report might as well have said: blah, blah, blah. I see owners hiding behind obviously flawed building reports all the time. The roof, desperately in need of repair, was replaced at a cost of $1 million. The school was declared clean and ready for the fall term.

In the fall, six of the nine patients had their symptoms and VCS recorded off medication before school began. All were well. Within four days of re-entry into the cleaned school, the illness reappeared, symptoms increased and VCS scores plummeted. All the cleaning, the money for the new roof, visiting defense medical personnel and environmental studies meant nothing: the school still made people sick.

The symptoms and VCS scores stood as obvious, screaming testimony to the power of fungi in buildings to defeat modest attempts at clean up. What the school did to remediate the building was of no help.

At Romo's request, on Oct. 27, 2003, I wrote a letter to the Hampton Bays School Board, which read:

> "I have completed my review of the health data accumulated on the cohort of patients with occupational exposure to the indoor air environment in the Hampton Bays Union Free Elementary School. Based on information provided to me the school was found to have contamination with at least four separate genera of toxigenic fungi, including *Stachybotrys*, *Acremonium*, *Aspergillus* and *Penicillium*. I have been informed of public statements made by Hampton Bays School District officials alleging that all of the toxigenic fungi have been removed and that the school is totally safe for all staff and students.
>
> "Our data are in direct and complete opposition to that statement. Not only was the school the source of illness for members of the cohort of patients commissioned to see me in association with a NY State-licensed physician in June 2003, but the school continues to be a source of illness for

the six cohort members who completed an additional leg of our repetitive exposure trials. This trial firmly establishes that the remediation attempts performed over the summer have had no impact on adverse health effects in susceptible patients.

"I was given several samples from the school to be sent to P and K Microbiology. A chain of custody was established. The rug from Lucy Dumont's classroom is contaminated with two separate toxigenic species of fungi, *Chaetomium* and *Stachybotrys*. From notes sent to me, your Board is apparently aware that lack of funding to remove the contaminated carpets does not change the fact that the carpets are contaminated. The potential to create bioaerosols from walking on the carpet remains a factor that must be considered in whether or not the indoor environment in the school is safe.

"I have included a table, Time Series VCS, dated Oct. 17, 2003, and a series of four graphs that summarize the health factors that support my medical opinion that the school presents an ongoing health hazard to those people with genetic susceptibility to fungal toxins and who have the potential for exposure to the indoor air in the school. The historical control group is the one I used in my peer-reviewed paper on human illness acquired following exposure to water-damaged buildings, presented 9/10/03, at the 5th International Conference on Bioaerosols, Fungi, Bacteria, Mycotoxins and Human Health, at Saratoga Springs, NY. This paper studied 156 patients from 150 water-damaged buildings, with eleven different biomarkers, a careful case definition and a robust study design that enables us to prove causation of illness by exposure. I have included five of the figures presented in the paper. I would suggest that you review the content of the figures from the paper, as well as the additional figures presenting the Hampton Bays cohort in their repetitive exposure trial, as they effectively render irrelevant any building environmental report purporting to show that absence of fungal contamination is sufficient to ensure absence of illness. As has been stated frequently in environmental studies, absence of proof is not proof of absence.

"The power of the study design stems from the documented ability of cholestyramine (CSM), a non-absorbable anion-binding resin, FDA approved for use in treatment of hypercholesterolemia, to treat the illness caused by exposure to indoor air environments with resident toxigenic fungi. Previous studies on human health and so-called 'Sick Building Syndrome (SBS),' suffered from various methodological flaws; each study fails to convince the reader of the existence of SBS because there is no documentation of prospective acquisition of illness, even if the methods were satisfactory. What we do is to take patients with the potential for exposure, assess symptoms according to published lists, rule out confounding exposures, accumulate a database of biomarkers, including visual contrast sensitivity (VCS) and initiate treatment with CSM according to a published protocol.

"At this stage of the study design **(Base)**, we don't know what caused the illness. At the end of the treatment phase, usually about two weeks, we re-assess the patients using VCS and symptoms **(AC-1).** As you see, in your cohort, 14 patients completed the **AC-1** arm of the study. Next we counter the argument that fungi are 'ubiquitous,' and that fungi anywhere could cause the illness. In this arm, **HOC**, home off cholestyramine, we stop medication and keep the patients away from the known affected environment, in this case the school. We show that despite exposure to the home, car, grocery store and all other potential exposures, but with no exposure to the indoor environment at the school, that there is no relapse in symptoms or recrudescence of VCS deficits. Because the initial study was performed at the end of the school year, 2003, only three patients are documented in HOC. The others simply gave verbal reporting that they stayed fine away from school, without taking medication, CSM. Next, while still not taking CSM, the patients are returned to the school and are re-evaluated in three days **(BOC)**. Prompt deterioration, with relapse of symptoms and development of the same deficits in VCS are noted. It is this prospective exposure of known susceptible patients who are asymptomatic, evaluated with all potential confounders accounted for and with time to develop illness as an important factor, which confirms causation. The patients are then re-treated with no relapse without re-exposure to a fungal-contaminated environment.

"In this case, we had five patients who relapsed at the end of the school year, 2003 (BOC). We also have six patients **(BOC-1)** who relapsed at the end of the school year, 2003, who were treated successfully a second time **(AC-2)** and then taken off CSM over the summer. They elected to not take CSM with re-entry into the school at the beginning of the school in the fall 2003. They functionally test the hypothesis that the remediation performed by the school over the summer was satisfactory and complete. If those patients relapse, we can conclude that the exposure to air in the school causes their illness. Six patients completed this arm **(BOC-2)** of the study, with identical results to both the BOC cohort and using themselves as controls, BOC-1.

"In the table, SEM is standard error of the mean. The numbers in columns A, B, C, D, and E are the final contrast sensitivity values. Please note that the numbers for both control groups, AC-1, HOC and AC-2 are not statistically different. The Base scores are also not statistically different from BOC, BOC-1 and BOC-2. These data are completely consistent with those presented at the Saratoga Springs conference and the 103 patients from 43 buildings presented by US EPA neurotoxicologist, H. Kenneth Hudnell, PhD, at the 8th International Conference of Neurotoxicology, June 2002, at Brescia, Italy.

"The symptoms listed are also completely consistent with the expected findings of controls equaling AC-1, AC-2 and HOC; and Base for cases equaling BOC, BOC-1, and BOC-2.

"I am accustomed to hearing that 'SBS isn't real,' and that patients who say they are sickened from exposure are whiners, complainers, troublemakers and worse. As in this case, the patients' complaints are validated with a rigorous scientific process. It is the scientific process that proves that the allegations of illness from exposure to the indoor air of the school are well founded. The patients in this case provided informed consent to basically put them in harm's way, some repeatedly, to make a point that the school poses definite health risks to staff and students alike. I regard them as heroes, and defenders of the students' health, rather than troublemakers.

"Of note in this case is the presence of one student, age 6, in the BOC-2 cohort. We know that there are predictable adverse metabolic/autoimmune effects caused by persistent exposure to fungal elements and that these effects aren't always evident at initial screening. It is my firm opinion that those with documented illness following exposure be removed from the affected environment. It is also my opinion that because of the severity of the illness in these cohorts, that an intensive school-wide epidemiologic survey be performed according to my protocols (do not substitute anything less rigorous), as case finding can be reasonably expected to include at least 24 percent of the population as a whole.

"Given that there already has been an alarming amount of misinformation provided to the public by the school authorities about the health risks posed by attendance at the school, it may be useful for you to have a public meeting to discuss the reality of SBS and what should be done about it. Videoconferencing may be a good way to accomplish this meeting, given the distance between my town and yours.

"I am aware that there are multiple ramifications that follow the realization that your school poses a predictable risk to health and learning. Many of those ramifications potentially could involve a significant expenditure of time and money. I would suggest that your school follow the precedent set in many other counties across the U.S. by commissioning an aggressive sampling program, a comprehensive remediation plan (i.e., one that doesn't leave contaminated carpets in place, for example) and a thorough health assessment, with treatment provided for identified cases. If you need assistance with the diagnostic and treatment protocol, I can help you and your designated physicians.

"Based upon my review of the sampling information provided to me, my physical exam and laboratory findings in the various cohorts of exposed patients, it is my opinion to a reasonable degree of medical certainty that the patients were suffering from a readily identifiable, chronic illness, often called Sick Building Syndrome, due to their extensive exposure to multiple species of toxigenic fungi and, potentially, other contributing factors associated with water damage in the school. It is also my opinion that the continued exposure of individuals to the toxigenic fungi present in the school poses a continued health risk and should be prevented.

"Please contact me with comments. I will answer as quickly as time permits."

• • •

As we'll see, the battles for safety in Hampton Bays were just beginning, and the next round would become vicious.

CHAPTER 10

Saratoga Springs, NY. Birthplace of the American Mold Revolution, 9/10/03

Certain days you just remember. July 4, 1776, *American Independence Day*. June 6, 1944, *D-Day*. December 7, 1941, *Pearl Harbor*. September 11, 2001, *World Trade Center*. Those who were there never forget; the lessons of history are there for the rest of us.

Other dates don't live in infamy and aren't recorded as yellowing pages of newsprint cut from *The New York Times*. But if our day in Saratoga Springs had been recorded, maybe this is what you might have read:

Special to The Times

> ***Saratoga Springs, NY***—After studying more than 150 "Sick Building Syndrome" (SBS) patients for several years, two U.S. researchers warned yesterday that the nation faces "an unprecedented epidemic" of toxic mold-related disease in the years immediately ahead.
>
> The controversial findings by "biotoxin-mediated illness" experts Ritchie Shoemaker, M.D., and Kenneth Hudnell, Ph.D., also suggested that the U.S. medical establishment has so far failed to identify a public health threat already affecting more than 10 million Americans.

Many of these victims of mold-triggered SBS will be permanently disabled and unable to resume normal lives and careers, according to epidemiological data released by Dr. Shoemaker at the Fifth International Conference on Bioaerosols, Fungi, Bacteria, Mycotoxins and Human Health. ...

• • •

Maybe September 10, 2003, won't be remembered with the same passion associated with the better known dates, but for Mold Warriors the two academic talks given in Saratoga Springs, NY, at the 5th International Conference on Bioaerosols, Fungi, Bacteria, Mycotoxins and Human Health were shots heard around the world. That conference marked the *end* of the era that was characterized by, "We declare that there are no illnesses caused by exposure to toxigenic fungi because *no academic papers prove* there's illness from mold," and the *beginning* of the era characterized with, "This is how you prove mold causes human illness."

Take just a minute to ask the question: How many people were affected by the American Revolution, World War II and September 11th? And now ponder the similar question, "How many people are affected by mold illness?"

• • •

When it comes to mold and human illness, one name looms above the rest: Johanning.

Dr. Eckardt Johanning has organized a series of international conferences on mold; the September 10th date was the 5th such gathering. From my point of view, he was the first researcher to really *try* to pin down the elements of human illness caused by mold. Don't get me wrong, Dr. Johanning was hardly the first observer to make a connection between toxic fungal spores and illness in humans. Just read your Bible: "A house desecrated by mildew, mold, or fungus would be a defiled place to live in, so drastic measures had to be taken." *Leviticus 13:47–50*. In later chapters, Biblical authorities sternly recommend that homeowners should; "Watch the plague and if the plague spreads, the unclean item or property must be removed and destroyed."

That's still pretty good advice for building owners today.

In my view, modern researchers have taken a back seat to Dr. Johanning as being the first to bring public attention to potential health effects acquired following exposure to water-damaged buildings. Some other researchers, Dr. Dorr Dearborn, Dr. Steve Vesper and Dr. Harriett Amman, among others, deserve substantial credit for helping to make a connection between human illness and exposure to water-damaged buildings; they have been doing excellent work on mold and human illness for many years. When I talk to reporters or lawyers, however, Dr. Johanning has the name recognition most associated with mold.

So when the announcement came out that Dr. Johanning's scientific organization, the Fungal Research Group, was going to put on another international mold conference, this was the place for Ken Hudnell and I to present our first two "mold paradigm" papers.

Why was this one such an important conference? We had data to prove that our use of VCS as a diagnostic tool. First identified by Ken as a reliable diagnostic test in people made ill following exposure to a *different* biotoxin-forming organism, *Pfiesteria*, it was equally valuable as a screening test and monitoring device in mold-exposed patients. We had data to prove that the biological markers we knew were important in other biotoxin illnesses, like HLA DR, MSH, MMP9 and all the rest, were abnormal in mold patients as well. All we needed was a high-level conference, preferably one with peer review, to get the word out. And here was just that kind of conference.

To understand why presenting these two papers was so important to "changing the paradigm," of mold illness, let's look at the rigorous and microscopically detailed process that takes place whenever scientists attempt to prove a new concept. In addition to being near the Birthplace of Winning the American Revolution, Saratoga Springs isn't too far from the ancient hunting grounds of the Iroquois, who some historians say maintained a tradition of running the gauntlet to test the mettle of men. Just like in the Medieval times, when throwing the gauntlet (a glove, really) was a challenge to a duel, the scientific gauntlet that tests men and ideas involves a challenge from all sides.

The gauntlet begins with crystallizing new ideas into a working hypothesis—in this case, "Mold illness is a chronic biotoxin-associated illness marked by specific symptoms and markers that can be evaluated and measured in order to confirm its presence."

The next step calls for the investigator to test the idea. If the idea stands up to initial evaluation (including statistical analysis), then the study supporting the hypothesis is prepared for presentation to a group of scientists involved in the same or a similar field. This scientific peer group will then review the work, making suggestions or corrections. If the work can withstand the peer criticism, then the scientific work is ready for publication.

But the initial gauntlet of peers is only the beginning. After publication, the ideas in the paper can be challenged in letters to the publication, followed by rejoinders in which the authors defend their work against the criticisms. Finally, the results published in the paper are usually tested by an entirely different group of investigators who will attempt to duplicate the published results from the first paper.

If all goes well with the repeat studies, then the ideas are incorporated into generally accepted knowledge, although that process usually takes a while. And given the economic implications of mold illness, no one will expect the world to agree overnight that mold does all the things to people we've already observed.

A good example of this scientific approach to securing new knowledge can be seen in the evolution that led to today's widespread use of DNA as evidence in criminal cases. I rarely see a cop show anymore that doesn't have some reference to DNA "found at the scene of the crime." It wasn't always like that, however; the first proponents of use of DNA testing faced years of skeptical and rigorous scrutiny, during which they were required to prove again and again that the new evidentiary tool could stand up to every possible challenge from their peers in forensic medicine.

Who is a peer? Are *all* physicians my peers regarding mold illnesses simply because they have a license to practice medicine? No. Are researchers like Dr. Amman *not* my peers because they aren't physicians and they don't treat patients? Of course not. What about physicians who don't treat patients but who testify for defense firms on mold cases because they hold prestigious positions in the scientific community? What if their "knowledge" arises only from some remote past relationship to mold illness? Ridiculous. We often find this kind of doctor on consensus panels, rendering opinions on others' work, despite their lack of knowledge of the subject matter. Surprisingly, lack of experience in patient care often isn't a disqualifier in health panels. Common sense tell us that what should count is generalized knowledge of the scientific process, specific knowledge of current technology and strong experience in all aspects of the field. Sufficient experience of hands-on patient care to validate that knowledge and experience would be a plus.

We have all seen "authoritative" consensus panels dominated by experts with tainted hands: if you have been paid by a drug company that sells cholesterol pills, you should be excluded from a panel that recommends taking those pills. Or, if you feel you can legitimately recommend taking more drugs for control of hypertension when you are being paid by ten different drug companies, your opinion should have a big asterisk on it. And if you have silver in your pocket from corporations that don't want mold to be a health issue, you should never be allowed to sit on any panel that presents authoritative opinion on mold and health.

If we had full disclosure of conflicts of interest required from defense experts in mold cases like we are now starting to require from authors of scientific papers, I suspect the juries across the country would see that expert opinion is bought and sold every day. As George Bernard Shaw might have said, "We know what the defense experts are; we are just haggling over the price."

• • •

Dr. Johanning's conference would be attended by an international collection of experts, all with extensive working knowledge of the mold issue—even if they weren't physicians and even if they didn't take care of patients. This was a classic example of a peer group!

Keep in mind, the reader also needs to know that not all scientists take kindly to new ideas. We call the scientific process a gauntlet for a very good reason! As you might imagine, the motivations that influence peer-reviewer scientists also vary greatly. In many ways, scientists are no different than other people, and they are quite capable of being swayed by feelings of anger, jealousy, distrust and all of the other emotions that make us human. Some people might think that all scientists are part of a happy family looking for truthful answers to life's mysteries, but the reality is a bit darker. Don't forget that the ongoing search for truth is frequently complicated by the pressure on scientists to "produce." Whether the product involves a patentable discovery or a published paper in an influential journal, scientists never stray too far from the dictum, "**publish or perish**."

And as you can imagine, given the grubby realities described above, a fair amount of "career advancement" that takes place in science is actually accomplished "on the backs" of others and not just from collective good works of cooperating colleagues. The unsuspecting reader might not realize that all sorts of dubious chicanery and conflicts of interest that would never withstand public scrutiny are involved in peer-reviewed science.

Do I need to say something sad about hired science again:

scientists serve at the whim of who pays them for their work. If you want to publish something that vindicates the tobacco industry, for example, I don't think you would obtain big grants from foundations aligned against tobacco companies. When the *Pfiesteria* Wars in Maryland were entering what I referred to as "the Fourth Stage of the Appearance of Good Science," (research =burial), there was no money made available by the government to study *non-nutrient* sources of the estuarine plague. Just imagine the limitations on freedom experienced by the scientists who work for an arm of the U.S. government research consortium: do you really think they operate independent of "management?"

So when we hear our legislators calling for definitive research on mold from the CDC and EPA, who are they trying to kid? Those agencies have budgets that are "managed"; don't think for a minute that the CDC will research mold and human health unless they are given *money* to look at mold!

Let's face it: the national scientific agencies are only going to do research when funding is provided by management. Both management and funding are subject to *political* influences. Political influences might desire to defend or expand some unstated purpose of those in power. What better way to protect the interests of those in power from "new science" than to *attack* that newcomer with the established opinion from the defenders of and feeders from the politically-based scientific trough? And if the bastion's opinions are solicited by mega-corporations and not by politicians, is the result any different?

So let's not have any fantasy-based illusions about the objectivity of the end result of the scientific process.

Those of us with new information have felt the knives of fellow scientists more than once.

• • •

To tell you the truth, I didn't expect a warm welcome from Eckardt Johanning at his conference. I knew that in many ways, I might represent competition for him—I'm the "new kid on the block." He and I approach the mold-human health problem differently so in this international conference of academics from many countries, I thought I'd be greeted politely, but with reserve and sent on my way.

Instead, Dr. Johanning went *out of his way* to welcome me personally, and discussed his reading of my book, *Desperation Medicine*, along with his thoughts for the upcoming conference. Even better, he assigned Ken and me to give our talks during the most desirable time slots.

• • •

In the middle of the run-up to the conference, I got a nice surprise; Pat Romanosky called from Hampton Bays. As Safety Officer for her school, she was being sent to Saratoga Springs to learn from all the experts. And she was bursting with news from the additional exposure trial her co-workers and Fisher Haines had successfully completed. Unremarkably, the data they'd gathered once again proved that the school was contaminated. No surprises there, as the carpets hadn't even been removed in the areas of greatest contamination. Nor had the "summer clean-up" and expensive new roof corrected the problems left behind by years of mold contamination and water intrusion.

Located near Albany, NY, Saratoga Springs is a resort town that features lovely Victorian homes and streets rich with racetrack history and stirring reminders of the Revolutionary War. The battle of Saratoga Springs was one of the early turning points that showed that the rebels had a chance to win against the superior firepower of the British and their mercenary Hessian forces. Driving into the conference center, I could almost feel the presence of times gone by. Back then, long-vanished rebels had dared take on the "Establishment" of their day and had eventually prevailed in their struggle for what became the American Way.

Curious coincidence, wasn't it?

In order to be allowed to give a talk at the conference, Ken and I had each submitted a summary of our talks and then sent in further material that was scrutinized by the conference academic staff of 25 reviewers. When the email came informing me that my talk had been accepted, I whooped and hollered ... bringing Shannon Brown and office manager Debbie Waidner back to my office on the run.

"What's all the ruckus about?" they asked.

"Look at this, ladies, we're going to Saratoga!"

• • •

My paper was a long one. It detailed the exhaustive research in which I'd reviewed the records of 156 patients from a total of 150 buildings; the patients had consented to be in a study reporting treatment results of the 5-step, repetitive exposure protocol.

The group of patients in my study could be divided into sub-groups ("cohorts" is the correct jargon). Some had experienced occupational exposure, others residential exposure and there were those with "exposure unknown." Some had exposure to buildings in which the presence of particular species of fungi had been precisely identified. Others, however, had only noted visible mold, but without precise documentation of the type of fungus involved. The third cohort had been exposed to buildings with water damage, but no identification of fungi or visible mold growth.

We had assembled a large control group to use for comparison, an essential factor missing from most of the reports about mold and human illness.

What was surprising was that there was little difference between those patients with exposure to *known* species of mold and those with exposure to areas with *visible* mold growth. Both of those groups of patients actually were a little *better off* than those patients exposed to water-damaged buildings with *musty smells*. It looked like those who knew they had a problem with mold did better than those who didn't know it! The problem with this observation is that the logical next step is to say, "Don't bother testing the building for mold; just test the people in the building." It is true; a marker for a sick building *isn't* an environmental report so much as it is the sickened people. The indoor hygienists might agree that finding sick people was the gold standard for diagnosis of a sick building, but their business depended on doing the cultures and analysis of contaminants indoors. They weren't likely to agree that their testing was of the limited clinical value my data suggested!

I *always* want to know what species of mold is present (if possible), but the absence of a formal report proving the presence of toxin-forming molds doesn't mean the report proves the absence of those molds in water-damaged buildings. But if a building inspection ignores bacteria or actinomycetes species, for example, or only samples the air for mold for *a few minutes*, what is it worth? NOTHING.

And such sloppy work, often accepted by so-called authoritative organizations, can destroy lives by purporting to show that all is safe indoors in a contaminated environment. Take the short-term sampling tests and throw the results away. What a waste of money!

Every one of the patients had recorded symptoms (the history was taken by the same person—me), a physical exam, pulmonary function tests, VCS tests and blood tests. Some patients had come from a great distance for treatment but wouldn't be returning for the repeated testing which others closer to Pocomoke would receive.

Once the patient was diagnosed, treatment with cholestyramine (CSM) was begun. After all the complicating factors often found in mold patients were corrected, the patient was re-evaluated. In a way, they were now like a slate blackboard that's been wiped clean of chalk. To no great surprise, the patients now resembled the control group of patients, with nothing to suggest an illness. We then asked them to stop the CSM, but stay away from the known source of contamination. Once again, they looked like the control group of patients. Then, after lots of discussion about risk and the like, the patients re-entered the contaminated building, not taking CSM.

We looked at them again after they had been exposed for *just three days*. The illness was back again, with the patients looking like they did when they were first evaluated. They were then re-treated, with confirmation that afterwards they looked like control patients again. Now, if they were going to stay exposed, they took CSM or a similar medication, Welchol, to prevent relapse. This group looked like the controls and stayed looking like controls for up to six months.

The report showed the consistent association of illness with HLA DR genotype, analyzed by PCR, with relative risk significantly increased for the same genotypes I was seeing as being susceptible in multiple other cohorts. We could watch leptin and MMP9 change with treatment, but there was an ominous group whose MSH just wouldn't respond positively to treatment because of their genotype.

Once again, we were able to see that the low-MSH patients—and they were more than 95 percent of the group—were saddled with a series of metabolic lodestones. MSH controls pituitary function, and both anterior and posterior pituitary hormone production was compromised in a significant number of patients, but not the control group. VCS and MMP9 were abnormal in so many MSH-deficient patients as well.

Before and after levels of MMP9 and leptin, combined with improvement in VCS led to a reasonable question: How quickly does leptin, MMP9 and VCS change with re-exposure and re-acquisition of illness? It was one thing to be able to show what the mold illness had as markers, but truly understanding the physiology of the illness meant seeing changes in both hyper-acute and chronic illness, as well as hyperacute changes with "new" illness. The changes from normal suggested by the Bio-toxin Pathway were present in spades in mold illness.

With all this information on what patients with the illness had, compared to what patients who didn't have the illness didn't have, we should be able to develop a case definition. No one had ever developed a *case definition* of mold illness because no one had ever been able to treat it, with abnormalities returning to normal, serving as identifiers. The case definition would be one that would have to detect all cases (epidemiologists worry about sensitivity, no false negatives) and to rule out all non-cases (called specificity, no false positives). The reality in medicine is that there are no tests that are 100 percent accurate, with 100 percent sensitivity and 100 percent specificity. What I wanted was to set 100 percent as a target and see what biomarkers identified that target. That meant that the case definition was developed *after* the cases had been identified.

And as *that* turned out, once we identified the markers of a case from the large number of 5-step repetitive exposure protocol studies, we could *apply* what we knew to identify a case just as accurately with a simple before and after treatment protocol. That meant we could define causation in an individual patient even without use of the prospective exposure arms of our studies.

As it turned out, a modification of the original case definition of biotoxin illness from *Pfiesteria* days was perfect. I had defined the illness from *Pfiesteria*, the fish killer, in my paper in Maryland Medical Journal in 1997 as "*Pfiesteria* Human Illness Syndrome (PHIS)." Believe me, if there were a good word beginning with the letter "H" to add at the end of the title, I would have used it so the acronym for the illness would spell PHISH. The CDC's idea was that my illness title wasn't a laughing matter. PHIS would be called, "Possible Estuarine Associated Syndrome," or PEAS. I didn't care whether it was PEAS, PHIS or PHISH, just so long as the illness was recognized as real.

The CDC experts, as one should expect by now, didn't include any primary care physicians. The list of symptoms they included as important in PEAS showed that the consensus panel had never taken a history from a real person who was a real case. The constellation of verbosity from science and government that accompanied the birth of PEAS into the real world of medicine was excellent preparation for the Mold Wars!

What the CDC did that *was* very helpful, was to suggest a case definition that included the potential for exposure, the presence of symptoms and the absence of confounders. They had a good idea, even if it were based on information filtered from primary care physicians through layers of bureaucrats. Mold illness fit the CDC criteria well; I adopted the CDC guidelines as the first tier of the case definition.

The second tier of the case definition had to include the metabolic foul-ups, along with HLA and VCS testing. There was no reason to exclude what we knew to be true. Using a long process of elimination, it was really exciting to find that the presence of 3 of 6 findings, after the first tier was met, included all of the cases and none of the controls! Here was my 100 percent case definition!

The reality was that most patients with mold illness had 5 or all 6 of the secondary criteria, but finding 3 criteria did perfectly. Making the definition more stringent eliminated true cases and making it less stringent included patients with other illnesses.

Tier 2 included:

- HLA susceptibility
- VCS deficits
- MSH deficiency
- MMP9 > 300

- ACTH/cortisol dysregulation
- ADH/osmolality dysregulation

With these 6 criteria, we had more than enough to show the CDC and the experts of the world what a true case of mold illness was.

And we did.

• • •

Ken Hudnell's talk was slated to go just before mine. As it turned out, Ken's talk on the "fungal paradigm," was the first from our group to be published in high-level peer-reviewed literature. He had rewritten my manuscript on groups of patients from 5 buildings near my office in Pocomoke to become the "landmark" paper on mold, accepted for publication in the prestigious journal, *Neurotoxicology and Teratology*, in July 2004. My fungal paradigm paper, showing that mold illness was a chronic biotoxin-associated illness, was published in a book edited by Dr. Johanning, a compendium of papers from that conference that passed additional layers of peer review.

Hudnell's lecture was the first to show use of VCS and symptoms, combined with leptin and MSH, in mold illness. This paper needed no controls, as the patients served as their own controls as they progressed through the 5-step protocol. Every time I look back on that paper, it was truly a landmark, in that it established MSH deficiency as a consequence of biotoxin exposure for the first time. Incredible as it was back then, when I had data on more than 3000 biotoxin patients, many of whom needed MSH replacement to return to functional life, there was no way I could convince any funding group of the validity of my work if I couldn't answer their first requirement: "Please send us a list of the papers you have published on the subject." Conference talks, like those at Saratoga Springs didn't "not-count," but they weren't enough to sway anyone thinking about investing in research on MSH replacement to actually fund the project.

If it isn't published, it doesn't exist. We're published; we now exist.

• • •

Romo wasn't happy. She and husband Joe had once again driven all through the night to arrive in Saratoga in time for the conference. She was talking like she was still on the New York Thruway.

"I can only stay for one day," she said. "I'm glad they let me stay for your and Ken's talk, but we have to get back tonight. You wouldn't believe what they're doing at the school now. Remember they were going to correct all the problems? Well, they didn't. They went through the motions, washing mold off here, replacing some drywall there, but nothing has changed. All six of us who went back into the school off cholestyramine at the beginning of the school year got sick right away. Poor Fisher is the worst. I just don't know how long I can continue. This illness is killing me.

"And the worst thing is that it's like no one else cares about the mold any more. I'm being called all kinds of names, maybe because of the lawsuit, but no one will do anything. I've written letters to the editor, talked with hundreds of people, talked with my Congressman, nothing works. It's like I'm the wacko out there and I just can't take it any longer."

I thought she was going to break down. All that work, all the heartache and now, the problem was being buried.

"Everything you said was going to happen has come true," she said unhappily. "I can't believe what's going on. I guess I shouldn't expect any crusading from others. The new superintendent is nice enough, but now she's bringing in some medical experts to give her a second opinion. Maybe you know Dr. Cheung?"

"Dr. Cheung, the one who used to work for the State of Maryland?" I asked. "So he's working for the School Board now, is he?"

She nodded.

"Carolyn Haines is really upset," said Romo. "What's she going to do with Fisher? He's sick as can be in the school and well after treatment with CSM out of the school. They're saying all kinds of bad things about her, me and you too, Dr. Shoe. One administration person told me they wanted to make you into a target of a Federal investigation. And they're hauling Dr. Feingold before the Office of Professional Medical Conduct (OPMC) on some kind of trumped-up charges. She has to hire an expensive lawyer to defend her license and her right to practice medicine just because the School Board is so upset. I don't know if the School is behind the charges against her, maybe it was the group of pediatricians, the insurance company or someone else, but Dr. Feingold is in trouble."

"Wait a minute, Pat," I said. "You mean to tell me that the OPMC can convene a kangaroo court to attack Dr. Feingold, but she isn't allowed to confront her accusers? Isn't that against the law? Didn't the Constitution or Bill of Rights prevent that? Can't her accusers be cross-examined? Aren't any civil rights groups aware of Jo's secret attackers? Isn't there some Congressman who will say this is immoral and unethical and illegal and stop it?"

I'd remembered hearing about how Dr. Burrascano had to spend $250,000 to clear his name in an OPMC investigation, and he still didn't even know who was attacking him. Are there any OPMC investigations that aren't clouded by secondary motivations? I wondered. From what I heard, Dr. Feingold would probably have a better chance of success winning a McCarthy hearing in 1953 than exposing the OPMC.

"You and I have done nothing wrong, yet here you are warning me of deep plots and conspiracies to hurt us and Dr. Feingold, too. Where's the crusading Attorney General of New York? Come on, he's taking down the crooks on Wall Street. Isn't he willing to get involved?" I asked Romo.

"Look, Mrs. Romanosky, all you have is a sick building, and you're not breaking any laws by saying so! On the other hand ... remember that poor guy who blew the whistle on Big Tobacco? Didn't they try to bury him under a mountain of lawsuits? Didn't he run into plenty of legal trouble and need personal bodyguards? Is he even still alive?"

Totally angered, I took a deep breath. How can so much human viciousness come from just little drops of water and little mold spores and little mold toxins?

"Dr. Shoe, let me tell you. I've had enough. Now that some improvement is going on in the school and a number of teachers have stayed working and healthy while they are taking CSM, they aren't fighting any more. And some of the parents who used to be so gung-ho aren't walking tall with me any more, either. What more do I have to do? When I think of all those kids at risk, these are my kids too, you know. I taught them and their parents and look at what is happening!"

"Romo, listen, it sounds to me like you're getting a big case of battle fatigue. Just forget trying to fix the problem by yourself. Dr. Hudnell and I have the academic basis of our work to support what you're doing. You don't need to do anything more except to secure your right to a *safe* workplace."

She seemed a bit better, as if my support had eased her pain.

"Dr. Shoe, I've looked into getting out," Romo admitted. "They moved me to the high school after you wrote that letter, but I think the high school is even more contaminated than the elementary school, if that's possible. Honestly, I'd like to take early retirement; I only have a few more months to go to get my pension. Joe works as the mayor of Southampton, but that job doesn't pay anything more than a pittance. For all the time he puts in, it's less than minimum wage.

"And without my pension, which I won't get if I quit now, I wouldn't be able to make ends meet without working somewhere else. I love to work, always have, but face it, who would hire me now? I'm over 50, I have no other credentials other than being a PE teacher and now, I'm being called a trouble-maker.

"If I make more trouble at the school, they can make things miserable for me. Teachers don't have many rights, you know. The administration has all the power. Not us, not our union. The list of what they can do to you is—well, it's really long."

Romo's worries seemed to be stacked up high and her face showed the stress she felt, and the sadness.

"Romo, I know lots of teachers. I've seen what its like for them when they have support and when they don't. Somehow, find the strength to carry on. Someday, Hampton Bays will recognize you for having told the truth. You are a true Mold Warrior!"

• • •

What a day. Within the space of several hours, I'd plunged from the euphoria of academic success into the sewer of realizing that my opponents would stop at nothing in order to prevent the truth about mold illness in *their schools* from going public. Back to back, here was the most successful day of my academic life, and here was the most unnerving day of my life. Federal investigation? Anyone want to fight the superpower FBI-type organizations we see on TV? You're kidding! All at once, I felt a growing awareness of the truly formidable powers that can be arrayed against anyone who challenges the status quo in this country.

All because I could show that mold makes some people sick.

"Discredit the witness"—wasn't that the strategy of defense attorneys trying to win a case? They wouldn't hesitate to do anything to make the witness look bad. And what would it be like, on some not too distant future day, to find myself listening to the question, "Tell me, Dr. Shoemaker: isn't it true that you faced a Federal probe of your role in the Hampton Bays mold case? And will you please tell the Court why we should believe your opinion now, when the government felt that there was enough evidence to investigate you back then?"

It wouldn't be the first time that a witness was intimidated by threats from the defense.

Mold Warriors don't back down from intimidation, not when they're right.

And we *are* right.

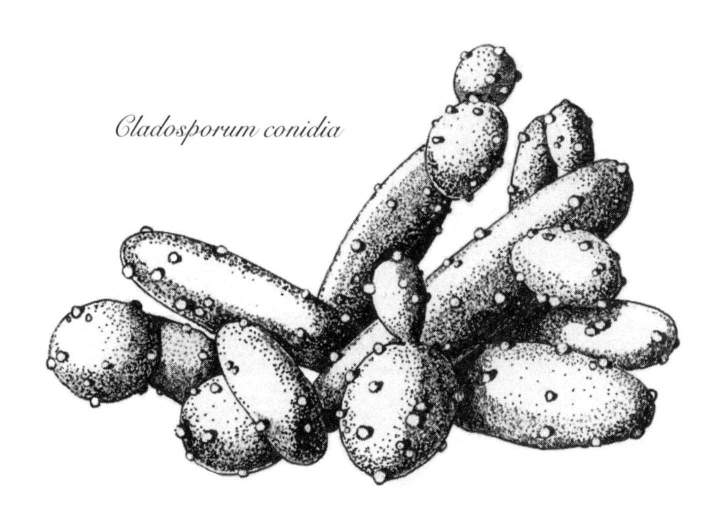
Cladosporum conidia

CHAPTER 11

For Want of a Nail

At first, all Marty Bernstein wanted was a little justice. The time she lost from her mortgage brokerage wasn't a lot, but her feelings were hurt by her boss' attitude and that of his insurance company. They were wrong and should have said so. Payback for her time, maybe something as simple as a verbal apology was all she really needed. The building made her sick and she could prove it. At that point, twelve hundred dollars was all it would have cost her employer to fix things.

Really, she wasn't mad at the beginning—she was working in a different building now and probably could make more in commissions in the resorts near Ocean City, Md., than she could in downtown Salisbury, Md. And she was feeling so much better. But when her boss started telling her she was just making up her symptoms to miss work, and when her big company denied her Worker's Compensation claim, Marty started getting upset and then angry enough to call a lawyer.

She was just one person. Who cared what one person did? And just a few bucks were at stake. No big deal. Stonewall her, deny her claim. She'd probably get mad and leave. There was no legal threat. So, if she got a little mad, who cared?

The big company learned the hard way not to make Marty mad.

Looking back on it, failure to provide just a little money to take care of a small claim ended up costing the company's insurance carrier a whole bundle, and made Marty a Mold Warrior.

Marty's case was just like the old adage, "For want of a nail, the horse was lost. For want of a horse, the rider was lost. For want of a rider, the battle was lost."

For want of a few dollars and a touch of understanding, another legal precedent was set in Maryland supporting use of the 5-step repetitive exposure protocol. It was clear the insurance company decided they wouldn't back down in this case, no matter how little money the case involved. These battles over mold illness were attracting the attention of insurance companies, no doubt about it. Nip this one in the bud now, seemed to be the attitude.

Bad decision.

• • •

One of the advantages I have as a primary care physician in a small town is that a lot of people around here know I work with unusual illnesses; mold illness is one. If someone's worried about mold, there's usually someone who they know, who knows me. Several years before Marty came to see me in August 2002, another employee from the building where Marty worked had come for evaluation of her chronic symptoms that never made sense to her attending physicians. As part of that work-up, I knew the building often had water leaks and that *Aspergillus, Penicillium* and *Acremonium* grew there. Each of those genera are toxin-formers.

Marty had been sick for eight months. She kept working and didn't think the building was the problem, even though six of her co-workers were sick and water-soaked ceiling tiles were part of the office decor. At first, her symptoms were fatigue, cough, shortness of breath and memory problems. Maybe she was working too hard or maybe it was time to quit smoking cigarettes. And she was carrying a heavy stress load, too.

Then the neurological problems began. Numbness and tingling in her left arm and left leg came and went. Her back hurt constantly, but not in the same places from day to day. When she went blind temporarily, she arranged an appointment at Johns Hopkins University Hospital.

Her symptoms said multiple sclerosis (MS), but without proof, she was given the diagnosis "myelopathy of uncertain etiology." MRI scans of brain, neck and thoracic spine didn't show the plaques typical of MS. The Hopkins neurologist was stumped.

Meanwhile, Marty didn't get better; the illness wasn't truly MS. More symptoms appeared. Cognitive trouble, muscle aches, headaches and more fatigue followed like grains of sand piling up in an hourglass. Abdominal pain, cramping, joint pain and morning stiffness all began slowly, adding their weight to a crushing illness. Nothing helped—B12 shots, pain pills, antidepressants, you name it. It wasn't Lyme disease, although she took antibiotics for the possibility. What do you do when the best hospital on the East Coast (Duke University will argue with the "best" label, as will Harvard) can't come up with an answer?

Then they told her she had fibromyalgia. She said, "Bologna, I'm going to Shoemaker."

• • •

I really despise the diagnosis of fibromyalgia. And I can't stand the lack of intellectual rigor used by doctors who make the diagnosis, either. *Gag.*

I used to think that the fibro term was a medical code for, "We don't know what the Hell is wrong." Now I believe the diagnosis of fibromyalgia is the ultimate example of how physicians patronize patients. Fibro usually means a biotoxin illness; certainly more than 98 percent of patients I've seen who've been diagnosed with fibromyalgia actually have MSH deficiency. Fibromyalgia diagnosis: *Gag.*

I've heard more than once that the real reason fibromyalgia was coined as a diagnosis was to provide "intellectual cover" for all the rheumatologists who made their reputation publishing papers on Lyme disease. We now know that the Lyme-susceptible genotypes comprise more than 19 per cent of the total population. That means that 19 percent of all patients with Lyme are likely to develop a biotoxin illness following the tick bite from which they get the Lyme, the infectious disease. So when those patients are treated with antibiotics, there's no way they're ever going to improve until the biotoxic basis for their chronic illness, complete with all the Biotoxin Pathway, is treated.

If you have been diagnosed with Lyme disease and someone is suggesting treatment with lots of antibiotics, but you have never even had a VCS test and never have gone through the diagnostic labs in Chapter 4, and never have been treated with Actos/CSM, you haven't been treated properly yet.

So, what do you do to treat fibromyalgia? First, you eliminate the absurd lack of definition of physiology. Second, you get a real dose of skepticism about how fibro came to life as the bastard child of failed treatments of big name medical people. Third, you say, "We aren't going to take this any more."

I have repeatedly complained that consensus panels are a ruse, a smokescreen of pseudoscience that simply puts the validation of an entrenched scientific bureaucracy on whatever idea is acceptable to the *payors* who organized the "impartial" consensus panel.

Anyway, the rheumatologists, all of whom had documented that 20–30 per cent of patients with Lyme Disease had persistent symptoms after the Lyme had been treated with antibiotics, and all of whom had a ton of patients with multiple symptoms from multiple body systems, got together to create a consensus panel to cover their failed model for Lyme. In that big net that focused at first on Post-Lyme were huge numbers of mold patients. The birth of fibro was a curse for those with chronic biotoxin illness.

Can't you just hear the rheumatology consensus boys; "Aha! Yes, many Lyme patients develop 'fibromyalgia' after their illness. Nothing is wrong with our antibiotic ideas. Our reputations and funding are safe. Yes, that blasted fibro was what was wrong. Give the poor sufferer some antidepressants and tell them they'll get better unless they really want to be ill."

Talk to 100 fibro patients, and see what they've been told about their fibromyalgia.

Fibro saved a few reputations of a few prominent scientists but the popularity of the diagnosis, made so often in biotoxin patients, condemned those sufferers to unnecessary illness.

• • •

Marty didn't have fibromyalgia. She had a mold illness.

Her VCS test was positive. She had 28 symptoms; to this day I still don't understand why the neurologist didn't document her symptoms. It only takes a few minutes to ask about the grouping of symptoms biotoxin patients have. If the neurologist had taken a history instead of passing Marty off as having some fantasy diagnosis, he might have seen how ill she was.

She had a dual mold-susceptible genotype, MSH deficiency, dysregulation of ACTH/cortisol, dysregulation of ADH/osmolality and MMP9 over 1400. All any person has to do to get a boatload of sympathy from me is to walk in with a lab report showing MMP9 over 600 (*see, Chapter 16 Turquoise Water*).

Marty started on cholestyramine (CSM), with only slight improvement. Actos to the rescue! Sky-high MMP9 must come down for improvement to occur. Sure enough, Marty followed the no-amylose diet, took her Actos and the MMP9 barriers to health came crashing down.

Actually, at first we didn't know that Marty's illness was from her workplace. She was better, but now we had to use her as a human arrow to point to the source of the illness. Her home had a basement, maybe it had mold. She had multiple different sites she visited or spent time in since the illness began. What ruled them out as being the cause of her illness?

The answer was simple. Stop CSM and let her go where she wanted, just not into the building with the known fungal contamination. No surprise, she didn't get sick at home, in the supermarket, at her place of worship or anywhere else. Now let her go back into the mold-filled building, and she becomes sick again in three days. MMP9 shoots through the roof, VCS falls, symptoms return, especially her numbness, tingling and pain. Fortunately, the blindness doesn't recur.

Actos and CSM are re-started and Marty does fine. And if you ever wonder if the super-low risks of taking Actos to lower MMP9 were worth taking, just ask Marty. Without Actos, CSM couldn't have done the job of lowering MMP9 quickly enough.

• • •

Along the way, Marty asked for treatment for her rosacea.

"My face always gets red when I go to work. It's worse on my cheeks. My first doctor said I had lupus, because of where the rash was on my face. He said it was a malar rash. The dermatologist said I had rosacea. I don't get the little pustules they talk about on the rosacea Internet sites, but I just have this redness that won't stop. If I go into a restaurant or a grocery store and start to get more flushing, I'll look for mold. And I know now that I'll find it."

"Marty, rosacea has no known cause. I call your rash, and it is totally due to mold exposure, 'mold facies.' It might be due to release in your skin of complement or VEGF, or maybe it's both, I'm not sure yet, but it's from mold exposure, for sure. It isn't as commonly seen in patients with your brown hair and darker complexion. The fair-skinned patients, especially the blondes, really get hammered. And it's far more common in women than men. We're having excellent results with a special topical preparation of CSM that a compounding pharmacist, Dennis Katz makes, but nothing is a better cure than prevention of exposure. And no, it's not lupus."

• • •

Marty finished her 5-step trial, repeating the exposure arm to the building just to be sure that there would be no mistake where her sickening exposure came from. She still couldn't get any satisfaction from the building owner covering her two weeks of sick time. He didn't believe his building was the source of her illness. The sick time was hers to take and he wasn't going to replace it, not to mention paying for her two weeks of time missed from work. And forget about him covering any of the big co-pays she had to pay to Johns Hopkins for all those MRI exams. The rest of Marty's medical bills were hers alone, too.

• • •

Off to Court she goes. She won her initial Worker's Compensation claim; now the insurance company wanted to contest the decision. They would use their formidable resources to overturn the judgment. They would attack Marty and me with the best experts they could buy.

Her medical evidence was so solid; I couldn't imagine how the insurance company was going to defend the case. They could only win by throwing stones at Marty and at me, of course. If anyone looked at the facts, there was no reason for argument. But court cases aren't decided on facts alone.

Dr. Ronald Gots entered the fray in April of 2003, hired by the insurance company. I had heard of Dr. Gots. More than one person wrote me about Dr. Gots and how they disliked him. He was often asked by insurance companies to provide assistance for the defense in litigation. After reading one of his prior reports about one of my mold patients, a report that lambasted me without basis, I could understand why he was so despised by so many. By the time I finished reading his unsubstantiated report about my opinions, I was ready to believe I was some kind of wacko doctor (he called me "unconventional"), who said only bizarre things about the purported illness.

The problem was that Dr. Gots hadn't commented on the basis for therapy. He made no comment on my scientific method (if he did, it would have been immediately fatal to his employer's pocket book) and he clearly didn't have any experience with seeing patients in the last *20 or more* years.

Who is this guy, anyway? He slams everyone that files a suit against his clients, the insurance companies. He has no recent experience in clinical medicine (he hasn't treated patients for years, it seems). I don't understand why these companies use him so much. He's no closer to clinical insights than the man in the street who happens to read about clinical cases.

One might ask, how is this guy continuing to live off the dole of insurance companies?

I logged Dr. Gots' name into a Google search (you can too) and found out he had been performing independent medical exams for years for many insurance companies. He was included in a Dateline NBC (broadcast 6/23/00 The Paper Chase) show, when he was apparently hired by State Farm to perform medical exams on auto accident victims. On the Google search, he was listed as working for a number of corporations to provide expert medical testimony for the defense.

In a later legal case, this time in Colorado, the plaintiff's attorney Joseph Coleman wrote into the public record about the standard for testimony in that state ("Shreck") and what Dr. Gots and his group of comrades hired by State Farm insurance had to say in lock-step, to prevent me from testifying in a clear case of health damage to the plaintiff:

> "State Farm's untimely claim that Dr. Shoemaker suddenly does not meet *Shreck* qualifications is 'supported' by a barrage of attorney argument and self-serving, contrary opinions.* Anyway, State Farm's expert opinion adverse to Dr. Shoemaker's opinion existed before December 2003 and the *Shreck* standard has not changed. With no changed conditions, either different facts or 'a significant change in controlling law made by a higher court or by statute,' *Buckley Powder Co. v. State*, 70 P.3d 547, 557 (Colo. App. 2002), courts generally adhere to the law of the case doctrine 'which protects parties from re-litigating settled issues,' Id. The doctrine applies to trial court rulings and 'gives preclusive effect to decisions involving the same parties in the same case,' Id at 557. See also *S.O.V. v. People in Interest of M.C.*, 914 P. 2d 355 (Colo. 1996). The propriety of Dr. Shoemaker being accepted as an expert and entitled to offer expert opinions was resolved by prior order and should be adhered to.
>
> "State Farm's purported complaint against Dr. Shoemaker is that State Farm, and its various experts, claim Dr. Shoemaker's approach is 'anecdotal,' not 'scientifically proven' or widely accepted. Admittedly, Dr. Shoemaker is one of the leading experts in the field and has treated hundreds of patients demonstrating medical conditions analogous to the Plaintiff. State Farm's 'expert' falls into two different categories: e.g., Dr. Gots, a professional witness who regularly recites the position advanced by insurance and industry which supports Dr. Gots' organization, but Dr. Gots fails to have time to treat even a single patient, and State Farm's other doctors do not claim to treat patients from a biotoxin exposure perspective. Dr. Shoemaker evaluates many patients who come to him for treatment and the doctor has provided medical services.

"In light of State Farm's Motion: Is the issue boiling down to the validity of a medical doctor who actually sees patients and tries to resolve persisting problems (Plaintiff's Dr. Shoemaker) versus the opinions of a doctor who has not treated patients during the past couple decades, while he spends his time behind a desk offering litigation support for insurers and corporate America (State Farm's Dr. Gots)? Obviously, this issue is a credibility argument involving the 'weight' to be given to opinion testimony, not and issue over admissibility.

"Much like his opinions, none of Dr. Gots' information on Plaintiffs' conditions are original or based on real experience. Instead, Gots relies on a summary of Plaintiffs' medical records (wherein no answer for plaintiff's persistent problem has been identified) and regurgitation of industry publications which are surprisingly similar to the industry defense to cigarette illness claims, i.e., 'cigarette science' and junk journalism which deceived courts and the public for decades about the harm of cigarettes. Fortunately, with the rejection of blind adherence to the 'pigeon-hole' analysis of *Frye,* the disproportionately large (but erroneous, corporate financed publications) 'scientific' literature was ultimately successfully challenged by new, correct science.

"The first of the 'new information' is from the Institute of Medicine of the National Academies 'IOM' (a great name, too bad State Farm fails to identify its corporate financiers) (Motion, p.3). State Farm attached as Appendix 6 to its motion the IOM's 'pre-publication version' of a report titled 'Damp Indoor Spaces and Health.' This 'new information' is not yet complete nor has it been published for public consumption. In fact, apparently only State Farm, through its expert Dr. Gots (who has association with IOM) has been able to gain an 'advance' copy. 'New,' yes; reliable 'information,' no. Maybe Dr. Gots should read State Farm's own internal memoranda wherein State Farm issued 02/07/01

Operation Guide, Section V which stated: *'If mold is discovered, a claim representative sensitive to mold should remove themselves from the exposure. Appropriate personal protective equipment should then be used to continue the inspection of the loss.'* See Appendix 2. Before State Farm's attorney tries to convince the Court that exposure to mold and their mycotoxins have not been 'proven' to be harmful, maybe the attorney should read State Farm's own 2001 memo which recognizes that the health risk is so severe as to require either human removal from the mold area or use of personal protection equipment."

* Whether the experts relied upon by State Farm are qualified to provide the opinions they express in their various written reports or "affidavits" should be noted, before the court mistakenly believes that State Farm somehow is the "source of all truth." Dr. Gots is addressed infra. As for Dr. Miller, he always claims to be an expert, regardless of the actual medical issue and Dr. Scott appears to express opinions regarding Dr. Shoemaker's reports and opinions, which are outside the realm of his own expertise. Anyway, determining the admissibility of an expert is not based on a popularity contest involving opposing experts.

The judge threw out the unsubstantiated and false arguments of Dr. Gots and State Farm. Let the legal battles begin.

• • •

But what was the basis for Dr. Gots' attacks on me—and why were those attacks delivered with such intensity—when he didn't have any personal experience with treating patients?

As time has passed, I have been opposed by Dr. Gots in many mold cases. Fortunately, his work doesn't sway many judicial opinions, but his novel attacks on me, each more shrill and each without basis, now increasingly fed through defense attorneys taking a deposition and then in court, show he is a resourceful, intelligent man. While I disagree with him on almost everything legal, and I feel he's ethically and morally wrong to stand in the way of justice for patients made ill following exposure to water-damaged buildings and mold, I must tip my cap to him as a man who does well what he has chosen to do.

As a matter of fact, Dr. Gots was hired to attend one of my depositions (it lasted 7 hours!). I think the aggressive, in-my-face defense attorney wanted the mere presence of Dr. Gots to intimidate me (fat chance), but during the occasional break from the Inquisition, I did have the opportunity to talk to him about health problems in our respective families. He showed a side of compassion and concern for personal issues that wasn't paralleled by his opinion letters. Maybe if we had met some years ago, our battles wouldn't have the same bitterness they do now. By the way, the offensive defense attorney in that case settled the case; the claim was paid.

While I respect Dr. Gots' ability to win cases for his employers, I still would question him about the basic elements of being a physician. He isn't a doctor in my operational definition. He is a paid commentator who has a medical degree.

My feelings about Dr. Gots are not the issue. He does well what he is paid extremely well (as documented in the public record of the Colorado case) to do. Similarly, as much as I don't like other college basketball teams performing well against my beloved Duke Blue Devils, you have to respect the opponents (even UNC) for what they do.

But when I find out that Dr. Gots is a behind-the-scenes participant in the verbal attacks I have to hear from defense attorneys, well, it's time for the Duke team to emerge victorious! And we do.

• • •

So there we were, Shoemaker versus Gots, as a backdrop for Bernstein versus employer. Somehow, the defense didn't even know about the prior mold report from the building. I didn't understand with all the legal maneuverings of discovery and whatever else they did, how the defense didn't know the building was a mold castle! Remember, all Marty wanted was her $1200. Now the company had brought in expensive experts whose reports cost many times more than $1200. Even worse for the defense, the employer's attitude about Marty was recorded somewhere and was going to come out in court.

All this legal action would be held in Marty's hometown and in my backyard. Somehow, I didn't think a fancy group of experts in expensive suits was going to impress a jury of local residents very much. Dr. Gots would be hard pressed to convince many people in this area that I was a physician *not to be believed* when so many from around here knew that when *I fought against* anyone who would *suppress* the truth about human health issues, *I had always been proven right*. And the defense didn't know about the long-standing fungal contamination in the building.

Marty didn't want just her $1200 now—she wanted that and a penalty. Punitive damages in mold cases can be incredibly high if bad faith can be shown on the part of the defendant. Marty had suffered mightily and paid plenty for health care. Her illness and costs and suffering were only due to one thing: mold in her workplace.

So the stage was set. Dr Gots would come to court in Salisbury; I would appear in court as well. Court costs aren't cheap. Attorneys aren't cheap. Dr. Gots isn't cheap and the defense would have to pay a whale of a lot of money *if* my opinion held over that espoused by Dr. Gots. Now how much money was the defense going to spend to settle that $1200 claim? The stakes were high and each side had met the ante.

• • •

Marty's lawyer okayed a settlement. No court case; no confrontation with Dr. Gots where "home cooking" in court proceedings is more than just an expression. Too bad. I'll argue with him again soon, I thought, but this case was such a slam dunk. I would have liked to see him try to make me look bad (it would be his job to try to do just that) on a case where we had such awesome data. At any rate, the insurance company's attempt to bully the little guy failed.

Marty won't tell me what she received for her $1200 case. It's really none of my business, and she wasn't allowed to talk about the case after the deal was done.

So what's the point?

Marty received her justice and her apology in the form of a check with plenty of zeroes divided by commas. Her illness flares occasionally, but she's learned to stay away from moldy buildings. The bad news: she's primed for mold illness for the rest of her life. The role of MMP9 as a marker for cytokine response in mold illness is well established. And using Actos, but *only* combined with the no-amylose diet, is a Godsend for those with high MMP9.

Back then, Dr. Gots tried to smear me as a wild-eyed physician and his attempt failed miserably. Back then, I felt his attack was personal and I was ready to take him personally to the mat. How dare he say such things about me with no basis?

But now, I'm not so sure. As much as it pains me to say this, "experts" like Dr. Gots may actually be *good* for people who have been harmed. Hire them! Give me a straw man punching bag hired by some blind insurance company as my medical opponent in a mold case, and my side will always win, if truth can win in courts. So, let's keep the Dr. Gots of the system healthy, testifying and wealthy as a result.

The current court system demands that the experts battle it out on health issues. What an absurdity: often, it seems to matter *more* to some courts what *credentials* an expert has compared to whether he actually knows anything. Pity the poor experts enriched by the bounty of the defense in mold cases. All their degrees and appointments and publications end up being reduced to one simple point: do you know how to treat mold patients? If not, go home. As a general rule, those non-knowers and non-treaters have to make the case to a jury that my ideas are worthless and I am not to be believed.

Good luck. I haven't yet found a jury gullible enough to swallow such nonsense from a medical "expert" who has no clue about mold.

Those experts are wrong, of course. The system that perpetuates the use of high-priced, back-stabbing experts is wrong. How many times have we seen low-budget movies where the fate of two fictitious warring factions is determined by a fight to the death of the solo gladiators each employs? Aren't both sides wrong? Can't we just get along?

No, we can't, not when a lot of money is involved. Marty's case is pretty typical—the forces of truth win by negotiation. The battle could have been "bloody," as the experts fought to the bitter end, but it wasn't. A far better solution would have been for the employer to say, "Gee, I'm sorry my workplace made you go blind for a while and caused you one year of constant pain. As a token of my good faith, here's a check for the time you lost from work." But we all know that won't happen.

What should a year of unnecessary blindness cost? How about if *your loved* one were the blinded one? Or if *your faulty* plumbing caused the mold growth that blinded the little girl from Massachusetts?

• • •

When I was young, watching Westerns on TV, we always knew that the guy in the white hat was the good guy; the black hat marked the bad guy. Life isn't that simple any more, especially in mold litigation.

Marty wears the white hat in her new workplace. She'll suffer from mold illness again, but she got out from her employer's Mold Castle with her eyesight and her health.

What about the people who still work in the building? Wouldn't it have made sense if the building had to be cleaned up as a condition of settlement?

CHAPTER 12

CHIMERA: Toxic Mold Invades The Halls of Congress

Growing silently in the dark, breeding invisibly, a sinister biotoxin threat has now reached the very heart of the federal government. As government officials worry about foreign terrorists bringing weapons of mass destruction to our population centers, undetected toxic mold organisms penetrated the heart of one of Washington's most revered public buildings by unknown means … causing a chronic debilitating disease blamed on everything except mold.

Yet one person is aware of these stealth biotoxin weapons. A very brave young woman confronts—and then goes public with—the source of the ailment that has devoured her life.

Suspicious Symptoms

The Setting: Rayburn Office Building, Independence Avenue and First Street, SW, Washington, D.C.

The Event: Attorney Melissa McDonald, Oversight Counsel to the House Judiciary Committee, discovers what may be the next generation of stealth bioterrorism agents in one of Washington's most closely guarded buildings. Despite losing her health due to mold illness caused by exposure to mold in her office, this Mold

Warrior refused to bow to unspoken pressures that would have her remain silent about the source of her illness.

• • •

It began harmlessly enough, back in the summer of 2002.

Melissa McDonald remembers looking at a flight of marble steps on that hazy morning in July, as she made her way away from the main entrance of the graceful old Rayburn Office Building to enter the side door.

For at least the hundredth time in her Washington career, the hard-charging lawyer from Missouri was passing by one of the giant sculptures that flanked the pink-marble façade of the complex, where 169 members of the U.S. Congress spend much of their working lives. Like every workday, the short walk from Lot 1, just 200 yards, could have been uplifting, almost spiritual. Here she was, poised for success in an incredibly demanding job that required brains, skill and political skills.

The sculpture, a 15-foot-high rendering of the "Chimera" from classical mythology, was enough to make you gasp with wonder. Somehow, the inspired artist had managed to tease from his inert pile of stone a winged and fire-snorting creature—part horse, part goat and part-eagle—that seemed gloriously but impossibly alive. The Chimera was a nightmare being; an unnatural fusion of living energies into a single grotesque form. Almost like her life, in a way. Success was hers already, if she only could enjoy it.

Chimera: The hybrid. A perverse distortion of nature, and yet completely real. Hurrying in the side door, Melissa stopped, gasping for breath. It's only five steps, for cripes sake ...

And then she stopped walking.
She'd just felt the symptoms again and felt the
Intense anger at her increasing disability.

I'm breathless.
I'm dizzy ... and all of my joints are on fire, they ache so bad.

I can't even move my hips without ice picks stabbing me relentlessly.

What is the matter with me? If I only had my energy. This is ridiculous, it's only a few steps!

Groaning with effort, she resumed her steps and finally managed to reach the top of the stairs. Lightheaded, short of breath she could barely pull the glass door to enter the storied office building, 40 years old, where she spent so many of her working hours.

Melissa McDonald—barely 30 years old and already a high-profile lawyer for one of the most powerful committees in Congress—was about to begin another day on the job. That is, if she could overcome the brain haze clouding her thoughts and somehow manage to keep her dead numb left leg from giving away on her as she walked to her basement office.

"Good morning, Ms. McDonald!"

"Hello, Chester," she nodded at the uniformed Park Police Sergeant who ran the Side Entrance Security Gate.

"Going to be pretty hot today, huh?"

"You're not kidding. Thank God for the air conditioning!"

"You bet—along with the taxpayers." He grinned as he waved her through the gateway and toward yet another set of stairs.

"Don't you work too hard!"

His words sounded hollow, almost echoing in the 50 feet of hallway that loomed in front of her.

If only he knew. Melissa wouldn't be putting in one of her 12-hour work-marathons today, and for good reason.

She didn't have the strength. And that was so pathetic! As a dominating athlete in high school and college, she'd once been able to rely on huge reserves of energy and endurance for the finishing kick of a victorious 100-meter dash or hours-long soccer games. But those days were long gone—replaced by the perpetual fog of her nameless ailment, both mental and physical, that seemed to have taken over her life in recent months.

The worst part was the fatigue. Increasingly, she felt as if she were carrying lead weights on her shoulders. Day after day, she would sit at her big executive desk in Room B351A and struggle for the courage to complete even the smallest assignment. With her joints aching and her dazed mind wandering from one vague topic to the next, she lacked the sharp mental focus that was the first requirement for legal work on the Hill.

McDonald knew she was sick—but no one had been able to diagnose her ailment, or to assemble the winning combination of drugs and therapy that might drive these enervating symptoms back into the dark place where they had somehow materialized.

Showtime. Dropping the briefcase on her mahogany desk, she settled into the padded swivel chair behind it. For at least the 20th time in recent months, she began scribbling on a yellow legal pad the long list of symptoms that she couldn't shake:

Constant muscle aches

Dry, non-productive cough

Mental confusion

Blurred vision at times

Abdominal cramps—not related to period

Short-term memory loss

Numbness in left foot

Occasional muscle spasms in legs

Severe pain in her left hip and low-back

Occasional weakness in left arm and hand, esp. after starchy foods, but really without much pattern.

• • •

A mega work-up by the neurologist—including an excruciatingly painful myelogram—had also failed to produce any answers. It just made her worse. The doctor just threw up his hands, like others before, wished her luck and sent her out the door with pain meds. They always found a way to say how good she looked, however, as if how one looks tells anyone anything about health and as if compliments would take away the implied stigma that she was making her symptoms up. If she heard, "all your standard lab tests results are normal," one more time, she really would scream.

Or cry. Or something. There was no place for solace, no medical help from anyone and no escape from her body.

Agony! How could she fix what was wrong if she didn't know what it was? Melissa knew she would never forget that ghastly moment during the myelogram—the moment when the neurologist had injected the dye into her spinal column and her left leg had suddenly jerked through an agonizing spasm ... a convulsive lurch that left it frozen stiff at a right angle in the air. The muscle contractions had been so intense that the anesthesiologist called in a squadron of "white coats," no one had ever seen such a thing and no one could help.

And what about all the prescription drugs she'd taken? What about all those months on the various antibiotics prescribed by three different physicians? And what about all the countless drugs with unpronounceable names? No help. How long was the illness going to last?

What the hell were the professionals all *missing*?

Again and again, they had quizzed her to the point of exhaustion.

Q. Did you change apartments, or are you still living in those upscale digs on the eighth floor of the D.C. apartment complex? ***No, I haven't moved.***

Q. Are you still in the same office in the basement of the Rayburn Building? *Sure, still there.*

Q. Have you experienced any depression in recent months? ***Oh, my God—don't even go there. I don't even know what the word means. I'm not depressed, I am sick. If anything, I am furious!***

Deep down, however, she knew she was terrified of what was happening. Constant pain, a lack of help from every professional. It wasn't like they didn't try; they just couldn't help.

Melissa was a fighter. Losing wasn't an option and living this way wasn't possible. Since the doctors couldn't fix her, she would take over. Traditional medicine? Alternative medicine? Her tradition of pain and loss wasn't an acceptable alternative!

All right, all right. If it wasn't mental, what about diet?

There was no denying that often she would "flare-up," after eating. But there was no obvious pattern to her symptoms. The elimination diets she tried didn't do much except convince her not to do them again.

Was there a link between what she was eating and how she felt? Hard to say.

A trip back to Missouri started her back to health. Dr. Charles Crist is renowned for his ability to heal the "incurable patients." He listened to Melissa a long time. His testing was thorough and complex. Diagnosis: Lyme disease. Antibiotics didn't help and neither did CSM after Actos.

Once back in Washington, while researching her symptoms again, she found Dr. Shoemaker, from Pocomoke, Md. He had a reputation for looking at Lyme disease differently than others, focusing on "neurotoxin-illnesses." Dr. Crist had in fact recommended Melissa to Dr. Shoemaker since Maryland was closer to Melissa than Dr. Crist in Missouri. Why not go see the new guy? All she had to lose was more time and more money.

Melissa doesn't think things happen for no reason. "Synchronicity," she calls it. Maybe there was a reason she and Dr. Crist had come up with the same possible next physician. Increasingly desperate, she had journeyed to Pocomoke in June of 2003 to consult with Dr. Shoemaker. He had listened to her at great length, asking only a few pointed questions. Shoemaker's picture was different than Lyme or Chronic Fatigue Syndrome.

From the Notes of the Treating Physician:

Typical biotoxin symptoms: fatigue, weakness, aching and unusual sharp, stabbing pains all over. She's also sensitive to bright light, with blurred vision, sinus congestion, shortness of breath after climbing just a few stairs (a trained athlete) and a hacking cough. Also fluctuating constipation and diarrhea, joint pains and morning stiffness. Cognitive deficits apparent: asked to divide 7 into 91, she laughs: "Get real!" Memory problems, inability to focus and concentrate, problems with word finding, reduced ability to assimilate new knowledge, and confusion. In addition, reports numbness and tingling. Mood swings worsen in the second half of her menstrual cycle—also appetite swings, night sweats and inability to regulate her body temperature. Constant shocks from static electricity. Refuses to touch a light switch with her fingers. Prefers taking "The zap" on her elbow!

• • •

After reviewing all of her symptoms with her, and running some preliminary tests that suggested that if Lyme had been present, it was long-gone now, Dr. Shoemaker turned in his chair: "Have you thought about the possibility that your workplace might be making you sick?"

Her response: a blank stare and then finally, a long-thought out, "No. You're kidding. The Rayburn Building? It's one of Washington's most famous buildings."

My God, she thought. Mold loves wet; the lower in the ground the better. I work in the basement!

Shoemaker was on a roll: "Is the Rayburn full of old heating ducts and unknown plumbing leaks in the countless private alcoves and offices? And how do you know with any certainty that being in the building is safe? We are looking at public policy being made by so many *who don't know* what mold can do to so many victims. Wouldn't it be ironic that the site of government most devoted to protecting us from mold was a source of mold illness?"

He was right. Just last summer, the maintenance crew re-worked all the sprinklers. And even now, they were remediating asbestos right outside my office door.

Melissa might still have looked terrific, but right now she was whirling. "I've been so preoccupied with terrorists attacks, like the anthrax scare, with its clean-up, I just didn't consider mold in my workplace."

Dr. Shoemaker added, "These days, Sick Building Syndrome is well on its way to becoming the next major environmental illness epidemic. And most of that SBS is linked directly to toxic mold."

• • •

The design of the building is a modified H plan with four stories above ground, two basements, and three levels of underground garage space. A white marble facade above a pink granite base covers a concrete and steel frame. One hundred sixty-nine Representatives were accommodated in three-room suites, with modern-for-the-time features such as toilets, kitchens, and built-in file cabinets; nine committees were also moved to this building. Amenities include a cafeteria, first aid room, Library of Congress book station, telephone and telegraph room, recording studio, post office, gymnasium, and facilities for press and television. A subway tunnel with two cars connects the building to the Capitol, and pedestrian tunnels join it to the Longworth Building.

• • •

The one thing Dr. Shoemaker did not tell his patient: Her symptoms could easily have been those of Sue Donahue, Carol Anderson, Craig Horsman, Ricky Jones, Marty Bernstein, Pat

Hayman, Martha Knight or Vickie Smith or any other mold victim. But was there any relationship between the Rayburn's anthrax and asbestos repairs to her illness?

As soon as he looked at her visual contrast scores, the Maryland physician understood the obvious: This was a biotoxin illness, all right. But McDonald had not improved after several weeks of a regular regimen on cholestyramine prescribed by Dr. Crist. Do three weeks again anyway; make sure the mold toxins are gone.

Still not much budge. And that could mean only one thing.

She was getting exposed to mold somewhere in her environment on a regular, continuing basis. The next step, while not as painful as a myelogram, was financially painful: paying for expert testing of her apartment and workplace. The bigger source of exposure? The graceful old Rayburn complex, one of the key headquarters buildings used by the United States Congress!

The Medical Evidence: An Overview

Her labs showed the expected. HLA was mold susceptible, MSH was low, MMP9 way too high and nasal culture showed the multiply antibiotic resistant coagulase negative Staph.

Here's where coincidence comes in. And where seeing countless mold patients enhances insights from seeing the first to the next Sick Building victim. Physicians who don't see patients like Melissa often would rarely have seen Carol Anderson before. It was Carol's suggestion not too long ago to look for myelin basic protein antibodies that had paid off big time.

The whole idea of molecular mimicry—where the body responds to a foreign antigen to make an antibody to the antigen that in turn mimics or cross-reacts with the tissues of the body (also called autoimmunity)—was high on my list of concerns. Just a few days before, I had read an article in the New England Journal of Medicine about celiac disease (a digestive illness made worse by eating foods with gluten and anything with the 18 amino acid peptide, gliadin) correlating with HLA DQ 2 and DQ 8. The authors had looked at 4000 people from Finland and showed convincingly that there was a genetic linkage in celiac disease.

The problem was that they weren't looking at the DQ as the second part of the HLA *triplet* I knew it was. So, it didn't take too much time to send off saved frozen samples of blood from patients with the full HLA triplets of 17-2-52A and 7-2-53, as well as 4-8-53, to check the hypothesis that extra gliadin autoantibodies occurred in this population. And sure enough it was true. But the illness and the antibodies to gliadin didn't appear in the groups of people who weren't sick. All they had was the genetic susceptibility, a potential if you will, but they didn't have exposure to fungal contamination. Those with the exposure usually had the autoantibodies, but not all of them did.

The antibodies were mostly found in the long-armed patients. These patients had wingspan greater than height, the reverse of normal. Wingspan is the length between outstretched fingertips. I remembered one of my early ciguatera cases, a woman with the ciguatera susceptible genotype, 4-8-53, who had been evaluated by all her neighbors for chronic diarrhea, cognitive problems and fatigue—the neighbors were mostly professors at Hopkins! No one could make a diagnosis. She turned out to have the dinoflagellate illness, as expected, but she was so flexible, almost like the "India Rubber Man," of local carnival fame, I never forgot her case. Could her body habitus, the one called Ehlers-Danlos III, translate to a general illness? I had other patients like the ciguatera woman and while I couldn't prove it, it seemed to me that body frames and environmental exposures might interact more than we had ever imagined.

And now Melissa is sitting cross-legged in my exam room, almost able to scratch her back with her right great toe! And she has the symptoms of celiac disease and she has the fair-skinned features and blonde hair of someone Nordic. Could she have *both* the Ehlers-Danlos and celiac, with mold exposure as the precipitating event? Her genes fit the scenario perfectly. Look at her flexibility: she can lean forward, clasp her hands in front of her and step through the handclasp, bringing her arms up beyond halfway up her back. She can fold one leg behind her neck too. Just like the ciguatera lady!

Sure enough, she had huge elevations of gliadin antibodies, as well as myelin basic protein antibodies off the charts. Melissa was telling me something. Loudly. She was genetically programmed to develop antibodies to gliadin based on body habitus and genotype, but that susceptibility was only unveiled following exposure to mold.

Comment by the Physician

"As I look back on that wild-assed scientific guess about body habitus and autoantibody formation, I think of what we know now and also of what we must look forward to. HLA linkages are waiting for us to be smart enough to find them.

"At the American Society of Microbiology meetings in Montreal (in April of 2004)—under the rubric of 'Integrating Metabolism and Genomics'—I presented 600 patients with HLA, biotoxin illness, antibodies and wingspan. Later, at another Microbiology conference, a researcher on autoimmunity from Louisiana State University confirmed my suspicions: 'We are just beginning to understand genetic linkages in human illnesses. The association of genetics and susceptibility to environmental illness is in its infancy. Keep up the observations! Even if we don't now understand the link from wingspan to genetic susceptibility to mold and autoimmunity, the process of inductive science begins with assembling what we do know.' So I had a whole pile of findings on the inductive scientist's bench, ready for the next step that would help me understand Melissa and the India Rubber Woman better.

• • •

"The findings of Melissa McDonald held true. And now we know that mold exposure is a major player in association with development of autoantibody formation in children. The association there is strong enough to become part of the case definition. And it all started with Melissa scratching her back!

"Her history suggested a *potential* for susceptibility to mold before she worked in the Rayburn Building. Her office in the Rayburn Building didn't look bad or smell bad. Anyone who lives or works in a basement ought to know if they are susceptible to mold!

"Indeed, it was a very familiar story. So many times, I'll hear about the potential for mold exposure, and especially when given the mold susceptible genotypes Melissa had. But there was an easy way to find out what environmental exposure was making her ill. Simply follow the five-step, repetitive exposure protocol!

"Melissa did the protocol, improving a little with CSM again, though she still had 14 symptoms to go, but down from her initial 34. Staying in her apartment didn't make her sicker off CSM and she was fine at every exposure other than the Rayburn Building. So off to work she went one fine autumn morning in the fall of 2003—without her usual dosages of CSM *to protect* her *from* the Halls of Congress.

"It didn't take long (less than 24 hours, in fact) for Melissa to suffer worsening fatigue, cough, shortness of breath and brain fog. But there was even more powerful evidence to come; soon an environmental report confirmed she was being exposed to toxigenic fungi, *Aspergillus* and *Penicillium,* in substantial numbers. In the absence of water intrusion in her work area, the tests indicated that the pathogens were flowing *from the HVAC* in her office!

"The fungal paradigm, based on the 5-step repetitive exposure protocol, gave us a giant arrow pointing to the perpetrator of Melissa's illness. It was looking more like the Rayburn Building.

"All at once, the outlines of a shocking truth came clear: Melissa McDonald might be poisoned by her workplace in the Halls of the U.S. Congress."

In short, all indications were that we had proved that she became ill after being exposed to her office environment. Just like the patients we had presented to the world's finest mold academics in Saratoga Springs and the hundreds more presented at other academic meetings, using Melissa as the canary in the mold mine told us that her illness came from only one source.

Environmental consultant Greg Weatherman had already thoroughly sampled her apartment—including her furniture, clothes and mattress—and had found less significant amounts of fungal spores and no evidence of amplified growth. But when he assayed Melissa's office in the Rayburn Building, the levels of toxigenic agents were high.

Where were the mold spores coming from? The subbasements were the most likely source. The further underground you go in living conditions, the better habitat is created for mold. What about the HVAC? Were there water leaks and mold growth somewhere else in the building, now being distributed by the universal contaminator, the heating and cooling system?

Could there be another explanation? The implications for national security of this medical finding were disturbing, to say the least.

If Melissa was sick, the scientific evidence showed that fully 24 percent of all exposed workers would *also* be affected by the toxic fungal spores floating throughout the Rayburn complex.

Did that percentage include members of Congress, along with their attorneys, legislative aides, staffers and secretaries? Of course it did!

Can you imagine trying to get all the people in the Rayburn building tested? Can you imagine *not testing* them all?

And imagine the reporters asking, "Congressman, did you have impairment of executive cognitive function when you voted for the Mars space project?" You could substitute any bill, any idea or any committee meeting decision, for example, for the space project. The horror of having neurotoxic government officials is enough to demand a comprehensive testing project, beginning NOW.

• • •

What followed in this saga of a brave young woman who dared to look her "toxic Chimera" in the eye was as predictable as it was tragic.

"You have got to get out of that office," said the doctor.

Melissa grimaced, and then shrugged. "Sure, I'll throw away my job. I'll scrap my career because the old offices on my floor happen to contain some mold!" as her terrified eyes pleaded with the physician on the other side of the room. "Get real."

But the doc's thoughts were already elsewhere, and he didn't like the direction in which they were headed.

"Melissa, do you remember, just after the anthrax scare in 2001, when all the offices in the Rayburn were vacated?"

"Sure. We talked about that."

"There could have been an HVAC intrusion of toxin-forming spores back then. The spores would have survived all the treatments they used. *Nothing* kills toxins.

"Melissa, do you know what was done after the anthrax to make the building safe? Think about it: instead of failed remediation of a moldy building like we usually see, did the anthrax remediators *change* the internal environment in the Rayburn building by use of their sterilization techniques to *let something* like an adaptable *Aspergillus grow freely*? Were all the competing, benign fungi and bacteria wiped out, leaving the resistant strains —the toxin-formers as we have seen—to grow without biocontrol? What happened to all the water that was used in the clean up?

Did it stay long enough to grow fungi that weren't there before the clean up?

"Let's face it: the Capitol Hill Office Buildings would be easy targets for terrorists. All they had to carry was a little bag of fungal spores to toss into a ventilation system under cover in a bathroom or vacated committee room!"

The two of them stared at each other.

• • •

So much work needs to be done to take the leap to assuming Congress has been attacked. Our data on mold illness has been accepted by my international peers and presented in Congress, itself. Neither Melissa nor I is a fear-monger, but based on finding one sick person, the next step is to look for a second and then a third. Maybe Melissa was getting sick from some unknown third source and the Rayburn is safe.

It is easy enough to find out. Let's do the testing. VCS takes about 5 minutes and symptom reporting is relatively fast as well. If the building is safe, who would be worried about a simple eye test?

"Dr. Shoemaker, are you suggesting that we could have been attacked more than twice? Are our enemies really so subtle and merciless and so vile that they would turn toxic mold into a weapon of war, right in the Halls of Congress?"

• • •

As of this writing, Melissa's questions remain unanswered. She left her job and will surface again in another high-level position in DC soon, I am sure. She still has so many health symptoms that if it weren't for her iron will to succeed, she might have given up. But, that isn't Melissa.

She cut out the starches to correct the gliadin problem, took her CSM and had her unusual biofilm-forming Staph eliminated from her nose.

But nothing worked. She is trying some new therapies designed to increase oxygen delivery to capillary networks in her muscles and lungs (see chapter 17). It is early yet to tell what the final results will be. I know that when I get approval to use MSH replacement from the FDA, Melissa will be getting a phone call!

• • •

I gave a talk at the 1st International Conference-Crossing Boundaries: Medical Biodefense and Civilian Medicine in November 2003. The heavyweight conference was sponsored by the National Center for Biodefense, headed by ex-Soviet Union anthrax guru, Dr. Ken Alibek and George Mason University. Attendees included medical experts from Walter Reed Army Institute of Research, representatives from USAMRIID from Fort Detrick, Md, the Russians, Israelis, Taiwanese and more uniforms from the US than I have ever seen. I was the only practicing community physician to give a talk on bioterrorism. The focus was primarily on smallpox, plague and anthrax.

I was nearly the last one to talk. The moderator, the head of infectious disease research from Fort Detrick wasn't familiar with my sponsoring organization, the Center for Research on Biotoxin Illnesses.

He might be now.

"Good afternoon, ladies and gentleman. It is getting late in the day, so let me make my point quickly: as a group, you are all missing the entire point of what biotoxins can do to us. We have a model in Nature for what biotoxins do to people, from molds to *Pfiesteria* to Lyme disease. The illnesses you fear are rapid killers. The possible results of attack with these agents are horrible to think about. I agree we need to know how they could be used against us.

"But you must know that an even greater threat will come from the Stealth toxins, especially those made by mold. Those toxins don't make their attack obvious by piles of corpses and skin lesions to alert us to a breach of our security.

"A much greater risk is what will happen when we begin to breathe small amounts of biotoxins. Your medical testing has no chance of detecting the illnesses, as you won't use the tests I know to be necessary. The threat from a foreign enemy to use those toxins isn't remote at all. You can buy quantities of mold toxins even now on the Internet. And the fungal toxins I refer to have already been used as biowarfare agents in Asian conflicts.

"A Stealth toxin would not kill us but would simply make us unable to function. Our sickened patients would tie up the resources of our health care systems and consume the time of loved ones as well. Instead of killing one person, the Stealth toxins would eliminate two or three from function. Those who would remain would likely have cognitive impairment, fatigue and would never be able to perform any activity requiring vigorous expenditure of energy. The toxin attack would disable us without killing us.

"Given the stunning lack of insight about fungal toxin illness nationally, especially from the CDC and 'authoritative' medical groups, if a Stealth toxin attack occurred, say in a Government Office Building, who would ever know that the aberrant behavior, fatigue and cognitive impairment of the workers in that building were the result of a silent attack?

"Which attack should we fear most? A Stealth attack or an obvious assault? If we are ill-equipped to manage an attack with aerosolized tularemia organisms, ones that can kill in 3 days, at least after that attack we could bury or dead and get on with the battle. With Stealth toxins our attacked would live a life of cognitive impairment, undercutting the ability of the rest of our population to carry on the fight."

• • •

Ken Alibek's group called almost immediately after the conference. What data did I have? What samples from anthrax victims could I share? How quickly could I go to their offices at George Mason to present another talk to his research staff?

His staff, largely Russian scientists who came to the US within the last 10 years, received my talk well in January, 2004.

No one has called from Fort Detrick yet to consult with me about a National plan for screening our population before the fungal toxin attack occurs. When we do find proof of a deliberate attack with biotoxins, will we hear on the Evening News that the intelligence agencies weren't aware of the threats posed by Stealth biotoxins, as if we hadn't learned from 9/11?

The attacks are ongoing in our schools, our homes and our workplaces.

What will be next? The Halls of Congress?

CHAPTER 13

Why's the Skinny Guy on a Diet? And No, It's Not Atkins!

By Marty Clarke

Some people just don't get it. They look at me and ask, "Why are *you* on a diet?" Here I am, six feet tall and lucky to break 154 pounds. I've got a nice career working as an automotive engineer and everyone says I've got everything going for me.

Except for my illness. Food makes me sick.

So I'm on a diet that doesn't include foods everyone else enjoys. Stop for a burger? Not me. How about a steaming plate of fresh manicotti? That would make me sick for a week. A hot dog with French Fries at the football game? Can't do it. I don't like to drink beer and rarely have a cold one, but even on that blue-moon day when I do, the commercial beers make me sick. How about a nice romantic dinner with a pleasant young lady, with steaks on the grill and a loaf of toasted garlic bread? Can't do the bread.

All this is because I have antigliadin antibodies. Autoantibodies, as they are called. If you already know that my mold-susceptible genotype sets me up for a dietary problem, you have learned well from Chapter 4. I don't know how long I've had this problem.

This illness takes gluten from my food and causes an inflammatory explosion in my gut. Or a "cytokine jolt" as Dr. Shoemaker calls it. Looking back on it, I must have had this for years.

Even now, I wonder about that little moldy place on Payne Avenue we lived in for awhile. Was that home my first real mold exposure? Was I starting to have gliadin-associated illness back then? I wonder—what would I have accomplished in my life without autoimmune gastrointestinal illness?

Make no mistake: I don't have celiac disease, though it might be easier for some people to accept my diet as medically necessary if I did.

If I eat gliadin, I die for a week. And I spent years that way. If only I knew, or Dr. Shoemaker knew way back when, that gliadin, the little protein that's the main toxic agent found in gluten, the gluten found in wheat, barley, oats and rye, could stunt my growth and make me skinny and tired. I don't know where I'd be now—perhaps well. But compared to the dark days I remember, at least I'm now more at ease both with my new life and what I eat, even though my diet is more restrictive than anyone else's at dinner time.

But some days, my situation gets the better of me. It's hard to go through life not experiencing excitement and other emotions that "normal" people do. As my life exposure to biotoxins and cytokine responses to foods took their toll on me, my emotional responses to major events dwindled.

I no longer wonder why my emotional responses have been diminished.

The havoc wreaked by biotoxins on my body chemistry is still with me like a slow-moving hurricane. Will the rain and wind ever stop? Sometimes my mind and my legs are like a couple with a bad marriage. At those times my mind says go and my legs just won't go, likely from the low-oxygen delivery caused by persistently low VEGF. My mind also wants that birthday cake, potluck neighborhood get-togethers and picnic food. But my body would beat me up in that fight.

I just wonder: how many others like me are there, out there? Lots, I'm sure.

And yet, I'm hopeful. *The rapid expansion of research on biotoxins and autoimmune diseases tells me to hang in there.* I'm young and there will be answers in the future. Dr. Shoemaker published several papers on MSH deficiency, so hopefully someone will be funding more research on MSH soon. Who knows, maybe next month I'll get some good news.

Life isn't as bad since I've bought some nice land. Just putting my hands into the dirt and on my good days getting an honest sweat going, clearing some brush and then checking out what my hard work has brought, brings meaning to me. That's a good example of how physical activity brings more relief to me than anything else. Just being outside gives me a deep sense of well being, too.

I have a desk job, but I am going to have to find fulfilling work outside.

Even at church, I have to fight against the mainstream; "Give us this day our daily bread." No, don't. I think I know a little bit about how that young girl with celiac disease must have felt when the Church said her communion didn't count since her wafer was made from rice. Deep down, I don't think God minds if those with my illness and hers take communion with rice wafers.

Yes, it can be frustrating to live in the world of the gliadin-unchallenged majority.

But I have to tell you; I really get frustrated sometimes when someone asks me to explain my diet. It's not the Atkins diet; it just sounds like it to the casual listener. And some of the comments about my physique still frost me: Thank you, I'm aware that I have to *gain* weight. I've given up explaining myself to those people—they don't want to hear about gliadin antibodies, not when the ball game is on, anyway.

Let me tell you what I've learned about gliadin, mold and fatigue. It might not help *you*, but the chances are excellent that you know someone who needs to know what I know, someone who's sick from mold or who cares about someone who's sick from mold. And don't be upset if I don't sit still for too long—sometimes just relaxing can be tough for me.

I grew up in Pocomoke, where it seems that everyone has some kind of allergy or another, so the fact that I was short and skinny was blamed on my allergies, too. My Dad has allergies and my Mom, too. I can remember asking my family doctor, Dr. Shoemaker, during a sports physical, you know the one where you have to be checked for a hernia, when I was going to grow. He just chuckled and said, "Any minute."

Well, I did grow, and I did well in high school. I was all set for a career in automotive engineering after I graduated from Virginia Tech. I interned with General Motors and before I knew it, I was helping set up a new plant in Mexico and then up to the big Motor City in Detroit, Michigan. And that's where my gliadin epiphany began.

• • •

I think that as a society, we're not accustomed to hearing about people with *low* body fat being on a diet. My body fat is somewhere around 4%. While I have to think ahead about my eating in order to be able to socialize, my choices are simple. I have to avoid gliadin.

I thank God for giving Dr. Shoemaker the gifts of knowledge he did, and for giving me the sense and tenacity to never give up hope. I thank God that I became one of Dr. Shoemaker's patients. Had those things not fallen into place, I have a hunch that for me, feeling better may have been delayed for years, if not indefinitely. When I think about some of my darker times, well, they were dark. I'll let it go at that.

Did I mention I'm feeling much better? I'll get to that, but first let me provide a little history.

I grew up not far from Dr. Shoemaker's office, and I was his patient throughout my childhood. As a child, I was very active and successful—in many ways, the kind of child every parent hopes to have. My life and family were not perfect, but nonetheless, I was blessed with many advantages. I participated in sports, but was never a star athlete. I was a Boy Scout and achieved Scouting's highest rank, Eagle Scout. I was also at the top of my high-school class and ran my own lawn business.

After high school, I left town to pursue higher education at a school a day's drive away. In college, I had a great time, learned a lot and made a few friendships that will never fade. The path to my engineering degree seemed like the road to hell at the time, but the memories that remain clearest are good ones.

College taught me a lot of valuable life lessons—not the least of which was what it feels like to be one of many worthy fish in a vast lake, instead of the biggest fish in a very small pond. Academically, I was able to craft my curriculum to allow me to get quite a bit of hands-on experience, in addition to reading somewhat nebulous theory found in textbooks.

Learning blue-collar skills—and getting callused hands to show for the efforts—and how to solve practical problems mark the most successful engineering students at Virginia Tech. That combination helped me land a co-operative education position with one of the world's largest corporations.

The corporate world provided me with opportunities for personal growth, including travel all over the world. Almost as a paradox, working for a big corporation has taught me much about myself. One of the things I learned is that I don't want to work for a big corporation! I now know that my heart is set on working for myself.

My life was good from birth through college, and the first few years after college would be considered very good by most people.

Two years after I left college, I felt emotionally flat. I'd realized that the engineering path I had taken wasn't the one I wanted to be on. I was living in a big unfriendly city too far away from my small-town home, and had gone through a tumultuous, bittersweet relationship. According to all the Internet surveys and psycho-fad books, I was suffering from depression and needed a modern miracle drug. I didn't like that idea and tried a myriad of ways to get myself feeling better, to no avail. After exhausting the options within my control, I went to a few doctors over a period of about a year. I didn't feel good and didn't feel good about myself, either.

Without my asking for them, and despite my lack of enthusiasm for taking drugs, they all prescribed psychoactive medications—mostly SSRI, selective serotonin reuptake inhibitors, a benzodiazepine or two, and maybe one or two other types. I followed the instructions, and after a few weeks there was no significant improvement. In fact, I often felt worse due to the emotional flattening effect of many of those drugs. To me, it was worse to feel nothing than to feel bad.

Let me feel what life brings, even if it's painful.

For understandable and all-too-common reasons, I lost faith in the doctors who'd been treating me. In most cases, they never spent more than 10 minutes with me. In all cases, they concluded that I needed some pills and everything would be fine. When everything wasn't fine, the next visit simply led to a different drug of the same class or a higher dose, resulting in the same effect—no improvement in the way I felt. Why weren't those doctors asking themselves, "What isn't right with this picture? Why does he appear to be depressed, yet he doesn't respond to standard anti-depressant therapy?"

My gut feelings told me that my initial instincts were correct—the depression and anxiety drugs that so many Americans use weren't *what I needed*. Those medications are wonderful for those who truly are depressed, but I know now that my illness was caused by biotoxins. But what *did* I need? I didn't know.

My exasperating medical experiences and fruitless search for answers brought me to seek and find a renewed interest in spirituality. It helped to get closer to God and gave me a positive focus. My mood improved enough that my critical thinking skills approached the level of a few years before. The timing of my illness had some lessons for me as well.

First, the timing of my "depression" never made sense to me. Probably because I hadn't followed standard engineering practice. Before trying to solve a problem, a good engineer clearly documents everything known about the problem. Then he or she makes some reasonable assumptions, followed by calculations. Then the theory he develops can be tested. Of course, feelings of depression and anxiety can't be solved using one of the various adaptations of Bernoulli's equation like a fluid dynamics problem.

The timing of my mood problem was too specific and that gave me some vindication for doubting that I needed an SSRI. I began to dig deeper into my feelings without reclining on a psychiatrist's couch.

My project at work sent me to another part of the country for a long-term plant start-up assignment. The plant was in a fairly small town—a place that felt a lot more like home, with people who welcomed me. I could relate to them. The improved comfort in my surroundings lifted my spirits a little, but I wasn't back to normal, not by a long shot. *After meals*, I still had strange physical sensations that could last all night and into the next morning. Breakfast brought more unusual sensations that lasted more hours, sometimes worsened by lunch.

Insight! The pattern held. Mood can be affected by diet, I'd heard that countless times. But the recommendations found in several books did only a little more than the prescription drugs did. I still had mood swings, pins-and-needles sensations in my face and body. Just what was happening, anyway? Well, I could always go home to take my mind off things.

I went back to my hometown for a welcome week off. I'd been putting in all kinds of hours; maybe I was just pushing myself to look for answers when I was too tired. The truth was, I felt like I'd exhausted all my options. And at age 26, that's not good.

I'm one of those people who truly enjoy physical labor. Give me a chainsaw or a shovel and I'm happy. While on that visit home, my younger brother Moose and I got into a log-moving project using a front-end loader. Neither of us had spent a lot of time using the machine, but my brother was more comfortable working with it than I was. I intended to be careful, but didn't intend to let Moose ace me.

Moose said I was working too slowly—move over and let him do it so we can get lunch before every restaurant on the East Coast closed!

I told him that taking over for me wouldn't be necessary. He wasn't even old enough to know what real work was, I said; I'd show him a thing or two! I began working faster and more carelessly, and oops, just like that, I was on the way to the hospital with some nasty lacerations. I still wonder how I have all my fingers. The next day, I visited Dr. Shoemaker's office for a second opinion on whether I needed hand surgery. While waiting to see him, I read some of the medical articles framed on the waiting room wall.

Most of the articles were about fish kills in our area a couple years prior and Dr. Shoemaker's involvement in solving the mystery behind them. That all took place while I was away. I was fascinated by those events and what Dr. Shoemaker was able to uncover despite a state government attempt to ignore or conceal the facts. When I got into the exam room, there were more articles to read and data from academic posters, too. Looked like the good doctor was saving on decoration costs by making the wall into a library!

A few years had passed since the fish kills and Dr. Shoemaker had plenty of time to do further scientific research on the toxins that had killed the fish and made people sick. I realized that he'd discovered a whole class of chronic illnesses caused by biologically produced neurotoxins.

I read a few excerpts from the book he'd written about his neurotoxin research and quickly found that the people in the case studies all shared the symptoms I'd been experiencing. They ranged from chronic fatigue to gastrointestinal problems to mood problems, particularly depression and irritability. There was another thing I shared with them: I always felt so terrible!

Even before I knew about Dr. Shoemaker's research, I'd wanted to talk to him about what I'd been going through. I knew him personally and trusted him more than any other doctor I'd ever been to. Every other doctor I'd seen knew me only as a face tied to a chart. They wouldn't know me in the grocery store, nor would they care. After reading all I'd just read, I got pretty excited about my prospects for feeling better. I was almost thankful my hand had been torn up and had brought me to the office that day.

Once my hand had been checked out, I brought up the other symptoms I'd been living with. We immediately did a neurotoxin exposure history and a visual contrast sensitivity (VCS) test. The results led to a blood draw for testing and prescriptions for a special three-week antibiotic followed by cholestyramine (CSM). The blood test results confirmed that I had hosted a Lyme-carrying tick at some point, and it had left behind the telltale signs of Lyme disease. With my immune response genes from Dad making me susceptible to Lyme toxins and the second set susceptible to mold (thanks, Mom), I was a set-up for what happens to chronic toxin-exposed patients.

They get to open Pandora's Box of autoimmune illness. And so did I.

• • •

I'd heard of Lyme disease, but was never concerned about it. I grew up running around in the fields and woods and had plenty of tick bites, but I just removed them and went on. I never had any rash or flu-like illness.

I didn't remember it during my visits with Dr. Shoemaker, but in preparing to write this article, I realized that one tick bite I had a few years ago was unlike any I'd had before. Even though I had carefully removed the tick, it left an inflamed area where I'd been bitten. The inflammation lasted for months—I can't remember exactly how long now.

I justified it by *assuming* that I had broken off part of the tick's head and that my body was just reacting to the embedded foreign material. I kept an eye on it—it didn't get any worse. I didn't realize I'd still be dealing with problems from that tick years later.

I finished the antibiotic; its purpose was to make sure that all the organisms passed along to me by the tick were dead. Then I started CSM and monitored the way I felt.

Frustratingly, I still didn't feel a lot better, but I didn't give up. The holidays came, affording me additional opportunities to see Dr. Shoemaker while visiting my family. We did a few more rounds of blood tests. I'll leave the details for the doctor to explain, but eventually, we'd done enough tests to conclude that I have an intestinal sensitivity to gluten, a condition very similar to and requiring the same treatment as celiac disease.

Very few people know what gluten is. It's a protein found in wheat, barley, rye and oats. In discovering that I had this sensitivity, it immediately made sense why I'd experienced symptoms that peaked just after meals and snacks. It also explained a lot of intestinal discomfort I'd had over the years. Did I have the gliadin problem before the tick illness? Having either one of the conditions meant I had no protection from the other.

Imagine my elation to finally know what was behind my symptoms, and my devastation to find out just what happened to patients with celiac disease, gluten intolerance and antigliadin antibodies. I had to totally eliminate all those ever-present grains from my diet, because just *one molecule* of gluten could set off the harmful immune responses when that large molecule is absorbed from the intestine.

It should be broken down and not absorbed intact. But in me, with my MSH deficiency, the gliadin sailed across the gut barriers into the bloodstream. The "big" molecule was targeted as a foreign invader by my immune system, and Bam! Just a little gliadin sent my inflammatory cytokine fighters into action, directed *against* me.

This information isn't new to people with gluten intolerance or celiac disease. I had to give up *almost every* food product on the market that's processed in any way. After my diagnosis, I researched which foods affected me for a few months, and confirmed by "challenges" what those foods did to my body. Occasionally, I forgot to be careful or didn't know I was crossing the gliadin line, which wasn't difficult, because I now estimate that about *85 percent* of the foods on the shelves of a standard grocery store either contain gluten or have ingredients that are potential hidden sources of gluten. However, because the FDA and USDA don't yet require it, almost none of them even mention gluten on the label.

Except in products that clearly contain wheat, barley, rye, and oats, gluten is almost always hidden in other ingredients with vague, seemingly innocuous names like "natural flavors." Other examples of ingredients that can contain gluten include caramel coloring, modified food starch, hydrolyzed vegetable protein, and the "natural juices" in meat products, among many others.

I've found that as a person requiring a gluten-free diet, I must do a lot of research in order to take care of myself while maintaining a diet that isn't the same every day.

The Internet is an excellent source of information on gluten, celiac disease and gluten intolerance. The people who post information on gluten-free and celiac disease online are generally dealing with the gluten-free lifestyle themselves. I'm certain they post the information because they care about and sympathize with the rest of us.

In researching the gluten-free lifestyle, I also discovered many other websites devoted to helping those people challenged by many different food allergies, intolerances and autoimmune diseases. Some good keywords to try in online searches are: gluten, gluten-free diet, special diet, celiac and celiac sprue. There are other keywords which will also yield helpful website links using any of the popular Internet search engines. There are food industry websites that demonstrate the need some people have for gluten-free or other special foods.

The diet and lifestyle change isn't easy. The first few weeks on the highly specialized gluten-free diet are very difficult, but eventually and with enough research, I found—and so can you—a reasonable complement of commercially available, acceptable foods.

With commercially produced foods, it's important to be aware that ingredients can change and to remember that gluten will most likely not be mentioned on the label. Perhaps more commonly, there may be a change in suppliers for one or more ingredients. A new supplier's manufacturing processes may derive the ingredient from a grain on the "no" list or may allow gluten contamination.

As an example of how gluten contamination can occur in the pre-packaged food industry, pretzels may be produced one day in a manufacturing plant, while the next day on the same line, potato chips are produced. Pretzel residue on the conveyor belts is now on the potato chips, but since "pretzel residue" is not required to be on the potato chip ingredient list, a new or unsuspecting celiac patient could be caught off-guard and harmed by eating pretzel residue when he thinks he's eating pure potato chips. Another example is that pure corn chips (corn does not contain gluten) may be dried after frying on a conveyor belt coated in wheat flour to prevent sticking.

If you have questions about the presence of gluten in a certain food, call the manufacturer. Tell them you are concerned about gluten. You'll be surprised to find they already know about your problem! Direct any gluten or other ingredient-related questions and comments to one of the company's dieticians; they'll have some familiarity with food intolerances.

Eating at restaurants and social activities can be particularly difficult and frustrating for people with food intolerances, especially gluten. Very few people who don't have celiac disease or gluten intolerance are fully aware of the cautions, which must be taken to prepare gluten-free dishes.

There are at least two restaurant chains I know of which offer gluten-free menus: Outback Steakhouse and P.F. Chiang's Chinese Bistro. Outback is pretty good if you like steak, chicken and shrimp and P.F. Chiang's is good, too. I just wish there was more than one Chiang's in my state! Another chain I've heard is gluten-friendly is B.D.'s Mongolian Barbecue. I haven't been to any of their locations since my diagnosis, but I went there frequently before moving. The food is delicious, and there are plenty of items that would be safe for people with intolerance of gluten and other ingredients.

At some restaurants and with some dishes, request that gluten-free foods be prepared on non-contaminated surfaces. Restaurant employees are generally willing to help, but don't expect them all to be familiar with special needs, even if their employers offer a special menu. Many waiters and food preparers have never personally had customers with special needs before. Eating at restaurants requires some patience with the staff and may require some educating, too. Or better yet, eat at home.

In other social settings where food will be served, such as a pot-luck dinner, it may be necessary for the "special diet" person to take one of his or her own favorite dishes and a side or two. Also consider convincing a friend who will also be in attendance into making a suitable dish or two. Perhaps the most difficult situation will be a banquet where all the food is provided by a third party. In this case, it may still be possible to take along a special meal. It just requires advance planning and communication.

Travel is another challenging situation for people on special diets. Again, research, communication and planning are critical. It's best to take along as much suitable food as possible. I've discovered that the gluten-free meals and other special meals offered by airlines are not necessarily "safe." I've traveled abroad by air and taken road trips since my diagnosis and found that I did best with the food I'd personally prepared and taken along.

For people with celiac disease or gluten intolerance, there's always something new to learn. Most likely, this is true for people with other food problems as well. Personally, my brand of autoimmune disease has led me to determine that I have adverse reactions to foods other than those containing gluten. Specifically, foods containing amylose bother me.

Amylose is a sugar, the most common complex carbohydrate. When I began the gluten-free diet, I was happy to learn that potatoes and rice were supposed to be safe for me to eat. I needed something to keep the weight on me! However, I noticed that although it wasn't as bad as the reaction to gluten, I did react to potatoes and rice, and to bananas. It turns out potatoes, rice and bananas all contain amylose. This crossover sensitivity is one of the differences between true celiac patients who avoid gliadin and those, like me, with biotoxin-associated autoimmune food illnesses, who avoid amylose *and* gliadin.

After eliminating both gluten and amylose, I'm left with a very simple diet that consists of meat, vegetables, fruit, corn and dairy products. Although I have a mild reaction to dairy and corn occasionally, they are still acceptable to me. I grill a lot, which is fun. Now that my diet is so simple, grocery shopping is somewhat less time-consuming and less expensive, so it doesn't break the bank to get good cuts of meat. I enjoy a really good steak one or two nights a week now.

There's no question that large numbers of gluten-intolerant people are also lactose-intolerant. Lactose is the sugar in cow's milk. Interestingly, it turns out that the lactose intolerance in those people is often caused by the untreated gluten-intolerance. When undiagnosed gluten-intolerant people eat gluten, the digestive absorption structures, called villi, in the small intestine become flattened or otherwise deformed. Villi are responsible for absorbing nutrients from food into the bloodstream, so that deformation causes improper absorption, especially of lactose.

This improper absorption can also lead to malnourishment or undernourishment, even for a person with a very healthy diet. The poor absorption keeps the body from getting all the good things it needs from food. Malabsorbing people usually don't feel real happy or healthy.

Fortunately, when I was diagnosed as a gluten-intolerant person and stopped eating gluten, the villi began to heal. Then dairy products were safe to eat again. Before I knew what was wrong, if I ate dairy products, I ended up like one of those people on TV who have a sensitive bladder and who have to find a bathroom right away. Only my urgent needs weren't from my bladder! Now, I can eat dairy without even having extra intestinal gas. And I don't take lactase enzyme pills anymore.

My life has improved as a result of starting this challenging gluten-free and amylose-free diet, so there's at least a thin silver lining to the frustrating diagnosis.

Although the diet is tough, and there's currently no treatment for me other than following the diet, I have hope of eating like a normal American again someday. I think Dr. Shoemaker's research on MSH replacement is groundbreaking and I'm glad to be a part of it. He thinks correcting my MSH deficiency will really help me, and after reading the literature on MSH and celiac disease, I'm convinced it will, too. The science makes sense to me personally, as a critical-thinking professional engineer. Don't forget, engineers have to learn a ton of biology and chemistry. We're waiting for the government to get on board and perhaps fund this research.

In the meantime, I highly encourage others not to give up on their illogical and unexplained ailments that frustrate them so much. Just because the doctors don't find out what's wrong at first, teach them to look for gliadin problems and don't let them assume that if the celiac test is negative, then there's no worry about gliadins. Don't settle for a fad diagnosis backed by poor science when instinct says there's more to it. We're all better off when we rely on our faith and on real science.

At least now, after all the doctors I mentioned and research I've done on my own, it's clear to me that I'm on the right track. I'm feeling tremendously better than I have in five years or more. The diet is working. I look forward to the day we get the go-ahead for MSH replacement so we can reverse what mold and Lyme disease have done to me. Hundreds of thousands, perhaps millions of other people will be cured, too!

CHAPTER 14

Mold, Multiple Sclerosis and MMP9

Mold Scars the Brain

"How many members are there in your church?" I asked the young Amish couple from Chester County, Pa. "And how many are ill?" They agreed to talk with me, provided that I did not use their names in the book.

The husband spoke, "We don't have a church; we have church in our house. There are at least 150 and many are ill. Some have been diagnosed with Lyme disease. Many are tired and hurt all the time. I honestly don't know why the children don't take to the work. The lady across the way from us is still so sick. She spent all that money for antibiotics for Lyme disease. She feels she should be well, but she isn't. I told her to come here, but she won't come; she has already spent all her money. And the woman with multiple sclerosis has her bedroom in the basement where it's much cooler. She was fine until she got married and moved into the basement. Her children are sick, too.

"I've been taking the powder ever since my wife took it and was given back to me. She can work all day now and always has a smile for us. When I was taking care of our five children and the farm and trying to help her, it was too much. Our two oldest daughters will still only take the powder in applesauce, but they know now when to ask for it. I need some more, please, many of us are much better now because of your medicine. We know we don't have to live feeling bad all the time."

"I have a question for you," I told my new friend, "Could I visit your community to do some testing? Couldn't we do for as many of the Amish there, as would let us, what we're doing for you here? We would draw blood for the HLA-DR gene test, do visual contrast testing and list symptoms on everyone. It might take a few days, but you would know if mold was making your church members and friends ill."

He looked over at his wife, who questioned with a shoulder shrug.

"We've talked to many already about your work," he said. "At first, they wouldn't hear of it, because you are a medical doctor, and if the remedy isn't herbal or organic, they won't take it. I know some would think about coming forward, but probably not many yet. You are outsiders. We'll talk to some others about what they think when we go home."

• • •

There was always something about the Amish that I couldn't let go. And I'm not talking about their clothes or horse-drawn carriages. Ever since I started questioning what life was all about, the Amish were a symbol of a different approach to life. As a culture, they have an integrity I don't see much any more. My introduction to the Amish was a result of chance more than design.

When we moved from Illinois to Carlisle, Pennsylvania, in 1966, there were many cultural adjustments for me. Real mountains, narrow winding country roads that were cow paths at one time, auctions and antiques. Limestone houses and basements that were root cellars. We lived on a bluff, just over the Conodoguinet Creek. The countryside was such a mix of cultures that were new to me. The local Civil War buffs still argued about strategy at the nearby battle of Gettysburg. Maybe a tactical change could have changed the course of history and kept the culture of the Old South alive.

That culture has died now; there's not much left of the Southern culture left in one-time strongholds of Charlotte or Atlanta.

• • •

The Amish would have been the same no matter who had won our Civil War. They didn't change just because everyone else did. Their "big cities" never changed.

To get to high school I drove past Amish farms every day. I could usually see people busy, building or working with livestock. The farms were always neat and tidy. At first, I thought they all had buried their electric lines, because I didn't see any. No, the lines weren't buried: they simply weren't there. And those carriages they rode around in, the ones with the black tops and highway warning signs being pulled by handsome horses—you could never get around them until the road came out of the woods. At first, I waved as I zoomed past their buggies in my Volkswagen bug, but the Amish simply looked at me without any change in expression. Later, I just smiled and waited until I could pass safely and respectfully.

They were different, that group. Most were from the Cumberland Valley, though some came up from Lancaster, leaving their farms when the taxes on their land got too high. I knew about a few of the teenagers who had gotten drunk when they first got a taste of freedom, but they didn't come back after that one time. Their culture remained intact, separated as they were from the rest of us.

• • •

The first Amish family I saw in my biotoxin practice had seven children. One of the twins was still sick after she'd had Lyme disease, and now two more had the same symptoms. They agreed to let me do HLA testing on the whole family if I paid for it, because they weren't interested in being studied. They *were* interested in becoming healthy again. All nine members of the family were 15-6-51 genotypes (Post-Lyme susceptible) and 17-2-52A (mold susceptible).

Their porous, stone-walled basement had sheets of *Stachybotrys* growing behind the shelves holding rows of jars of beans, corn, peaches, applesauce and tomatoes.

The three who were ill had suffered from Lyme and were successfully treated with Actos and cholestyramine after three weeks of antibiotics. The other six were healthy, even though they had mold-susceptible genotypes and exposure to the Stachy. It was almost as if there were something about Lyme that exposed its victims to a subsequent susceptibility to mold, "unveiling" it, if you will (*see Chapter 15, Libby Kyer*).

What a gold mine of information about genetic susceptibility! Here was an island population, with few contributions from the melting pot of human genes. What better way to look at the influence of genetics and environmental exposure on illness than to look at the Amish? If they would let me, of course.

As more Amish people came to Pocomoke for treatment of their illnesses, in cars driven by unknown others, my ignorance of the Amish culture gave way to a deep respect. I'd always be an outsider—I didn't even ask what the Pennsylvania Dutch word for someone like me was. When I asked one Amish elder why he allowed his family to travel to Pocomoke for treatment, despite me being "medical," he told me that his family was more important than avoiding a car or taking medication. I never heard one child complain about a blood test or anyone look at me crossly for being a few minutes late.

One man told me simply, "We won't complain about a doctor

who is trying to help us."

Although they're well known for their skills in carpentry and group collaborative effort, the Amish don't know much about mold. Their roofs dump water directly onto the foundations of their old homes. The building might survive to be more than 200 years old, but so would the potential for mold growth in their basements. Nothing is wasted in an Amish home, especially not storage and bedroom space in the basement. Wood shelves for food storage and beds on the floor are invitations for mold illness. Their genotypes are few, and so far, all the Amish patients I have seen have genes that give them mold-susceptibility.

And because the Amish rarely complain, I had to wonder, "How many are sick with a readily correctable illness?"

• • •

I call up experts in medicine and biochemistry a lot. I have real-world questions that the experts could help answer. So I regularly encounter experts reacting to a doctor from Pocomoke calling them with some weird question. Basically, there are two types of experts: the "Get lost" clan and the "Good question" clan. The age and specialty of the identified experts doesn't matter. Fortunately, the "Good question" clan has many members.

John Rees, Dr. Good Question

John Rees, MD is a one of the helpful experts. After years of reading his reports on MRI scans I had ordered, I started calling him with my questions. For example, "John, how can we image the ventral medial nucleus of the hypothalamus to determine if there is an infarct in the MSH production pathway?" He'd always try to answer. His answer to that one was, "You can't see the area on MRI; it's too small."

And when I asked him about memory problems, information poured forth on how we learn, recall what we learn and MRI imaging of the hippocampus. When Jim Frasca wanted actual MRI films on memory loss for his *Pfiesteria* show on Discovery Health, Dr. Rees made sure we had excellent and accurate MRI scans.

Dr. Rees aggressively seeks to discover tiny changes in the brain that appear before clinical disease appears. His thinking was just what I needed, since I wondered if mold biotoxins could literally scar the brain. I was struggling to understand the processes of memory-loss and executive cognitive deficits in my biotoxin patients. Whether it was only difficulty remembering what someone had just read or finding a word in conversation or the full-blown cognitive damage that eventually makes biotoxin illness so obvious (see the Epilogue for a recent, real world case), how did that damage occur? Cognitive impairment was perhaps the most frightening aspect of mold illness and, after fatigue, one of the most common.

Naturally, it didn't help that prestigious organizations, like the CDC and the Institute of Medicine consensus panel, didn't agree that mold even caused a multisymptom illness, much less profound cognitive impairment. And now, I'm trying to find markers for mold illness after it's too late, when there are already scars on the brain, as well as before there's damage, when the brain is being attacked, but is yet to be scarred.

I sure needed fewer "Get lost" docs and more like John Rees.

In trying to understand these unclear brain problems, I didn't want to be discouragned by mocking "Get lost" docs, or any ivory tower "experts" who denied mold illness. How they could turn their backs on the accounts of so many patients? Merely labeling patients with contrived psychiatric diagnoses after failing to ask routine questions in a medical history was outrageous. Thankfully, John Rees was in the right place at the right time to help me.

I was seeing lots of mold patients like Carol Anderson and a couple of her fellow victims from the Ritz-Carlton who had *markedly abnormal MRI scans*, high levels of antibodies to myelin basic protein (aMBP), and elevated MMP9. None had multiple sclerosis (MS), but their MRI scans had scars identical to those of MS patients. *These findings were a sign that mold illness could kill nerves in the brain.* There even were a few patients who the experts agreed had MS, who improved with biotoxin treatment, Actos and CSM. Actos knocked down the MMP9, TNF, PAI-1 and leptin; CSM knocked out the mold toxins and the MS symptoms cleared.

But the scars, the plaques in their brains, called "unidentified bright objects (UBOs)," stayed.

You can imagine my surprise when I read in June 2004 about researchers at the University of Chicago who were using Actos to treat MS! I wondered how many mold patients they had in their trials and if they knew to look at MMP9. No one from that institution answered my calls.

How could we tell if plaques in the brain were caused by mold exposure? I suppose we could do an MRI on every person going into a moldy environment, monitor their labs and then serially measure the changes in labs and look at MRI for diagnostic clues. Who could tell me what the fingerprint of mold toxin exposure was going to look like in the brain?

John Rees answered my call. He agreed to look at a group of scans done on mold patients and to do it for free. "This is important, Ritchie, I'm happy to help." He was confident that there was a relationship between exposure to biological neurotoxins and white matter disease, including illnesses like MS, which removed the myelin, the protective fatty coat on nerves.

So Dr. Rees looked at a series of MRI scans for me. Each patient had been told the brain MRI was abnormal, but the pattern of abnormality was not specific. After that, Dr. Rees would look at a group of scans for UBOs. He looked at the hippocampus and white matter for the earliest signs of demyelination. Unfortunately, what Dr. Rees found was not a distinctive marker. He found a wide variety of white matter abnormalities, but taken as a whole, there was not one location that was always damaged. There was no smoking gun, no consistent pattern of abnormalities on brain MRI that would link the diseases together.

• • •

The Biotoxin Pattern Predicts Brain Scars

BUT what we found in the blood tests was stunning. The patients with mold exposure, no HLA susceptibility, normal MMP and normal aMBP **never** had UBOs or plaques. But **all** the patients with HLA susceptibility, mold exposure and plaques **always** had high MMP9 and aMBP.

Dr. Rees wrote, *"There is evidence of unexplained white matter injury. Multiple sclerosis is still not fully understood and its MRI manifestations are protean. There may be a final common pathway for white matter disease that many different insults feed into. It's my belief that MS is many diseases and that mycotoxins-associated brain disease may share certain aspects of its pathogenesis."*

• • •

I read his report in May of 2003—a dead end in one way. We hadn't found a biomarker for mold in brain injury after the fact even though the blood tests said we should be able to find one. Well, I'll keep looking, I thought. We'll just have to look at patients with the unwanted combo of aMBP and high MMP9 and find out if there were a time course of exposure that resulted in the plaques. Good luck.

No one suggested we simply observe patients with aMBP and high MMP9 to see what their brain did **without** treatment. And yet such an unethical study might be necessary to convince the nay-sayers about mold and brain scarring.

So when Dr. Rees called in July 2003, I was surprised.

"Ritchie, what do you know about MR spectroscopy?" he asked me.

"Maybe how to spell it, John," I replied.

"I've been reviewing some of the papers from the recent meetings of the American Society of Neuroradiology," he said. "They can now find abnormalities in chemical and physiological parameters in the brain *before* plaque formation begins and *before* the damage occurs. I've been talking with Dr. Dan Nguyen at Georgetown, he's chief of neuroradiology there. He might be able to look at some biochemical markers for you. Call him. I've already spoken to him, so he'll know a little about what you want before you call."

What Drs. Rees and Nguyen were talking about was a way to develop a chemical spectrograph of white matter. Using special MR techniques, they could see particular peaks from particular chemicals in particular parts of the brain. It would be like taking a brain biopsy, analyzing its chemistry and putting it back in the brain, without cutting the scalp! The marker for nerve damage would be n-acetyl aspartate; lactate would indicate blood flow; and white matter markers would be sphingomyelin and choline.

This was amazing. I think I've seen all the old Star Treks—I can't tell you how may times Bones McKinney, the flight surgeon, could tell Captain Kirk all about what was happening inside someone's (or something's!) brain by analyzing chemicals at the bedside with a hand-held device.

And now Dr. Rees and Dr. Nguyen were going to do the same thing.

• • •

Bringing Diagnostic Aids to the Amish and Us

So what would I bring with me to the 20,000 Amish in and around Chester County, Pa.? I mean, besides my hammer and handsaw, as I hoped the Amish would let me work with them.

I would love to do MR spectroscopy on the precious children sleeping in 200-year-old basements to make sure they had no brain injury starting. Ideally, we use the Heidelberg Retinal Flowmeter to look at capillary blood flow in their retinas to see if there were signs of biotoxins. These tools would work great if we could just plug them into an electrical receptacle.

How about you? Would you consider a blood test for HLA-DR, MMP9 and aMBP? That would be easy enough. And we could do a history, too. That might be acceptable. VCS worked fine in Amish sunlight and anywhere else in the country.

But like all my patients, I needed an invitation to diagnose and treat the Amish. I'm certain there were people I could help who were living in front of kitchen cabinets with mold growing behind them. But no one asked me to come into their home. I wonder how many people could have been helped.

And no one has routinely asked me to analyze every person in a given building; with the data to be used as a baseline should there be a subsequent problem with water intrusion. It seems so obvious that we should be screening children and workers at risk, not to mention residents of high-rise apartments and condominiums, too. But I guess the initial cost of screening, low as it might be, coupled with the litigious nature of our society will prevent the mass screening we need from ever occurring.

It seems so tragic that we routinely screen children with eye tests at school; it would just take another five minutes to do VCS. We routinely have health histories taken by workplace health personnel; it would take just a few extra minutes to create a biotoxin history. And when someone comes to a physician for cholesterol screening, what extra time would be involved to measure MSH, MMP9, HLA and aMBP?

We could do so much good for so many, *if we're asked.*

CHAPTER 15

Libby Kyer

Thank You for Giving Me My Life Back

When did Libby start getting sick? She was born with mold-susceptible genes, although she didn't have the worst ones, the dreaded 12-3-52B or the 4-3-53; hers were the even rarer 12-7-52B. Having mold susceptible genes didn't automatically condemn her to having mold illness. She wasn't ill as an infant or a teenager. Something must have happened since then that made her vulnerable to illness from mold.

She was exposed to mold growing outdoors all her life, and surely there had to be a building with water damage where she lived or worked sometime. But mold never made her sick before. Oh, now she was sick all right, with the worst case of fibromyalgia anyone in the Colorado Front Range had ever seen. What had caused the illness to appear?

The other day she sent me a card agreeing to illustrate *Mold Warriors*. Her artwork is scientifically and aesthetically accurate; I was pleased she accepted my offer. The botanical artwork throughout this book is hers. While on her way to London for a juried art competition, stopping off in New York City for a show featuring her work, she wrote, telling me how terrific she felt. You'd never believe that she'd been essentially disabled for 14 years.

Looking back on her case, it's amazing she didn't just give up the fight. But her case shows how ridiculous the diagnosis of fibromyalgia really is, and how often even good physicians will agree that an empty, "fad" diagnosis is correct. The term fibromyalgia was organized into a socially acceptable diagnosis by a group of rheumatologists in 1988. For years, physicians had been baffled by patients whose chief complaint was pain, but who also had many other symptoms. What busy physician had time to listen to the long litany of complaints? No treatments worked, although antidepressants and chill pills were often prescribed.

It makes me cringe that a rheumatology consensus panel created this new illness. When policy is made by *consensus panels* of any kind, look out, there are shenanigans afoot! But with a formal "definition" of fibromyalgia, it was easier for baffled physicians to tell sufferers that there was a *real* illness present. As if the chaotic collection of markers for the illness—including trigger point pain, sleep disruption, cognitive problems—made a difference in therapy.

Fibromyalgia has no known cause and no diagnostic tests to confirm its existence, but most rheumatologists agreed that working through stress and adding some physical activity—maybe trying some yoga or Tai Chi—could help the patient have some kind of life. Helping patients *learn to live* with fibromyalgia was an important task for the caring primary physician.

Besides, now that fibromyalgia had become a recognized diagnosis eligible for insurance payments, there was a series of continuing education seminars available, usually featuring talks by the same speakers saying the same old things about fibromyalgia. Fibro became a self-perpetuating illness, for sure, complete with everything from expert panels to support groups on the Internet. I can't tell you how many of my Post-Lyme patients have been given that diagnosis, suggesting the main criteria for fibromyalgia that occurs after a tick bite is the absence of response to antibiotics.

If only fibromyalgia sufferers, especially the tick bite and mold victims, demanded a biotoxin workup, many would now have a life enriched with joy and hope like Libby does now.

• • •

Libby's Amazing Story

Libby was the administrative director in a neonatal ICU in Colorado. She was happily married, and since 1980 lived in a home that looked out over the mighty Rocky Mountains. The water came inside her home from time to time, as they lived in a run-out area of the foothills. There were streams around her house occasionally, with lots of other times when there would be standing surface water. They had seen mold indoors in the fall of 1987, but no one made a big deal about it.

February 1988 brought her a rapidly progressive illness. Before the tide turned, her survival was in question. It began with flu-like symptoms, followed by fever, aching and cough. Probably because of her work exposure and her role in a health-care facility, her docs worked her up intensively, confirming that her illness was caused by a virus called cytomegalovirus (CMV). This germ is a big player that often attacks babies born prematurely causing dire illness in the preemies, so maybe she picked up the illness on the job. At that time, there was nothing to do except provide fluids, maybe some acetaminophen and supportive care, hoping the patient wouldn't die. Adult patients usually survive, as the immune response kicks in and clears the illness. For most people with CMV, the illness is like mononucleosis—you'll feel bad for a few weeks and then bounce back.

For Libby though, it really was touch and go.

These days, we have great anti-viral medications for CMV, just like we have anti-virals that help treat mono quickly. But in 1988, Libby waited for weeks to improve. She showed modest improvement after a few weeks, enough to want to return to work. But she deteriorated when the spring rains and snow-melt began. Her illness was devastating for another eight months, and she was unable to work at all after the first three months. Another Christmas came and went. For two years her boss "let her work around her illness," going home to rest when needed. As time passed, the pattern to her illness was clear: her symptoms flared in spring, stabilized in summer, only to relapse with the impact of a sledgehammer when the October chill closed up their house.

Looking back on her case with what we know now, but no one knew then, the source of her illness is obvious. The ferocious cytokine storm created as part of her immune response to her CMV began the *unveiling* of her HLA susceptibility to mold. The seasonal mold exposure clawed at her, and finally, with her HLA-directed system for production of antibodies to mold toxins completely damaged, her illness progressed to complete disability.

Back then, physicians didn't have access to the information you now have in your hands. Unfortunately, many physicians still are unaware of the genetic basis of what a prior illness can do to change susceptibility to the next one that comes along. No one asked Libby if her massive cytokine response to CMV had initiated the process of priming her HLA-DR susceptibility to mold. No one asked Libby if she were exposed to mold that had never hurt her before. And no one treated the biotoxin source of her illness, either. No one knew.

For 14 years, Libby lived a life emptied of the professional satisfaction of doing her job well. Like a thief who'd stolen her life in the night, her illness was nameless. The joy of relationships, achieving personal goals, even just looking at the grandeur of the Rocky Mountains had changed from brilliant colors to faded grays and dark browns.

She couldn't work in her job any more, so she returned to her love of art. The illness hadn't stolen her talent, just her ability to work for more than an hour or two each day. Libby sought help everywhere, and was told she had Chronic Fatigue Syndrome, then picked up a diagnosis of the newly minted syndrome, fibromyalgia. The diagnosis stuck like it was branded onto her forehead.

If anyone had looked, they could have seen the predictable changes in her biochemical markers. From reading this book, you already know what they were. Her weight gain came out of nowhere, the tip-off that cytokine responses are nailing the receptor for leptin in the hypothalamus. High leptin will cause all of us to store fatty acids with incredible efficiency. Nothing gets the weight off a person with high leptin.

She had demanding fatigue—she never felt refreshed when she awoke from sleep. Her fatigue brought with it unusual pains that never went away, multiplied many times over if she tried to do some physical work on the rare days when she did have a sprinkling of energy. That was when her MSH started to fall. When her thirst started, with frequent urination and that curious susceptibility to static shocks appeared, that was when her posterior pituitary didn't make a sufficient amount of antidiuretic hormone.

Fibromyalgia? Hogwash. Fourteen years of chronic fatigue? Tragic! What a waste.

• • •

Her first visit to Pocomoke was October 10, 2002. She had a ton of symptoms; ones that we know now come from biotoxins. In case some Occupational Medicine doctor tries to tell you what mold illness *isn't*, here's what it *is*: Fatigue; weak, aching muscles; cramps; unusual, sharp stabbing pains that felt like an ice pick. Add to that blurred vision, sinus problems, shortness of breath, abdominal pain, diarrhea that interrupts sleep, and joint pain, especially in the small joints. Oh, and the cognitive problems: Trouble with memory, confusion, and word-finding difficulty. The neurological problems included numbness, tingling, metallic taste and vertigo. The hypothalamic system's difficulties included mood and appetite swings; sweats and trouble controlling body temperature; thirst and frequent urination; and constant electric shocks.

Having only half of these symptoms was a great day for Libby.

She started Actos, cholestyramine and replacement of her antidiuretic hormone. After the usual bumpy road of eradicating all those toxins in her body, lowering cytokines and correcting hormone abnormalities, she wrote me on December 30, 2002:

"My head cleared and the cotton fog was gone. Thinking was better. Memory was hugely better. My fibromyalgia seemed to be fading. I could LEAP out of bed in the morning, with no painful feet and joints. My energy was hugely improved.

"I had no GI problems—the 'GERD,' gastritis and spastic colon were all gone. I felt completely normal, HEALTHY even. I had full energy, better joint function, no roving joint swelling, and no fibromyalgia in legs or feet. I moved well and had a lot of energy and great spirits!

"I've lost about 12 pounds, despite making no changes in my eating habits. I don't eat a lot of junk anyway, so I never understood the weight gain in the first place.

"I don't have any 'fades' or 'wash out' feelings, which in the past have been associated with a strange, full feeling in my stomach. My completely dry, burning tongue gave way to a cool, comfy tongue. I could drink water and quench my thirst.

"I slept through the night, and I awoke refreshed. Even in dry weather, I'm not drawing sparks off of everyone and everything. I urinate less often—about four times each day versus eight to ten previously.

"Yow! I feel soooo muuuuucchhh better. The change in my life was just short of miraculous."

• • •

Her home is now cleared of mold, but Libby is primed for another mold illness. It is going to happen, no question about it. Whether it's mold in a trans-continental airliner or mold in the museum or mold in the big office buildings with windows that don't open, mold will make her sick again, some time, somewhere.

For now though, Libby feels that life has been given back to her. She told me in September, 2003, that she had walked all over England and France, with no pain. She had a new business, "Pigs Fly," and her art was being shown all over the world. She deserves everything good that's happening to her.

She somehow found a way to survive. She'd pushed for real treatment rather than a meaningless fluff diagnosis like fibromyalgia. She deserves her success. What an incredible story of courage! Hers is more than just a happy story; it's a symbol to so many that there's always hope.

• • •

And the answer to her question, "When did she become ill from mold?" isn't "when she was born." The answer is, once her HLA susceptibility was primed, she was *going to get it*. Is it always CMV that sets off the HLA? No. The list of other cytokine illnesses that result in new susceptibility to previous exposures includes: mold itself; Lyme (a major reason so many people who are sickened with Lyme once, think they have Lyme always, is linked to activation of HLA-DR and the progression of the Biotoxin Pathway), Coxsackie and ECHO viruses, Kawasaki's, anthrax (those who survived the 2001 attacks that I've seen), a swarm of beestings, mononucleosis, and anything that causes persistent activation of the alternative pathway of complement. But the truth is that we're still learning about the incredibly complex series of events that regulate the functional immune response.

Does the pattern of Libby's illness match that of one of your loved ones? Is the illness that won't go away simply the result of some *other* illness first opening the immune door to a biotoxin and saying, "Come right in?"

Unraveling the complexity of illnesses that lead to others will be the key to helping every person who doesn't respond to biotoxin therapies as well as Libby. But no one will accomplish that goal if we ever give up hope and resolve. Sure, it's hard to push for an effective treatment when you're receiving care from one physician or alternative provider after another for a non-entity like fibromyalgia, and they have little to offer you in the way of treatment or hope.

But just look at Libby. She never quit.

And we won't either.

CHAPTER 16

Turquoise Waters Blue-Green Toxins

Adventures with Pond Scum

The drought of 2002 was so bad that your sweat dried up before it hit the ground. And it was so bad that the turquoise water in our ponds told us we had new toxin intruders in our backyard. No other algae bloom looks that color except ones I fear.

Come to think of it, the drought was almost as bad as in the summer of 1997. That was the year of the first big *Pfiesteria* blooms in the Pocomoke River. It was almost funny now, looking back on the explanations from the state experts saying the *Pfiesteria* problem was caused by nutrient enriched run-off from farms. Bruce Nichols, District Conservationist for the Soil Conservation Service, was annoyed with the deliberate political spin and misrepresentation of the source of the *Pfiesteria* disaster. “Sure, I’ll believe that; there is always a lot of run-off when it hasn’t rained in weeks.”

In between moving hoses in the drought of ’02 to keep water on our trees, we had plenty of time to watch the new shore birds coming onto the newly created beaches of our ponds. Get out the bird books; I’ve never seen that sandpiper before. Bruce told me to enjoy the changes in our new habitat. But no, he said, running a water line from our well wouldn’t keep the ponds from drying down. All the well would do would be to pump more water vapor into the air, maybe change the water table for a few minutes and add unknown chemicals from our well into the ponds.

He was right, of course. Follow Nature, don't try to change it.

The ponds were changing color from the dry down. Every time I'd work below the dams now, I coughed and felt a little short of breath. I'd been working real hard; at long last, JoAnn and I had a few days without deadlines. There was so much to do outside, left waiting while indoor work and computer screens chained us. So the drought in Pocomoke kept me in the yard in July and August, doing imitations of the water bearer in the Revolutionary War, Dolly Pitcher. The ponds sure looked different these days, almost iridescent at dusk.

• • •

8000 acre Lake Griffin sits in the middle of the Ocklawaha watershed in Central Florida. The Chain of Lakes, from Apopka to Harris, Eustis, Yale, Dora and others were famous years ago for their clear water and great bass fishing, but not any more. In 1995, a newcomer, an "exotic" the algae specialists called it, entered these waters and stayed. *Cylindrospermopsis*, call it cylindro, is a blue-green algae. Cylindro is a major toxin-former. The algae species was not a stranger to those with concerns about world health. In China, it was causing liver cancer. In Australia, it was known to sicken bird populations. And in Brazil, it killed nearly 60 patients receiving dialysis with contaminated water. Who knew before the kidney patients died that the reservoir serving the clinic would be a source of silent death?

For Florida, cylindro was making people sick.

Before too much time passed, cylindro was sweeping all other algae from the water column in the nearby farm-chemical enriched lakes. From muck farms to nursery run-off, there were chemical inputs into the lakes that no one could imagine. Could the burst of cylindro growth be due to the agricultural chemicals? Of course. But the State blamed the explosive, invasive growth on enrichment in the water of nutrients! Does that blame surprise you? If the problem in natural disasters is nutrients, then all of us are to blame. If the problem is agricultural chemicals made by huge corporations, they are liable.

It wasn't much surprise to find that cylindro could grow in the Chain of Lakes like no other blue-green algae anyone had ever seen. It killed shorebirds and pelicans. It wiped out the brine shrimp populations and then the fish that lived on brine shrimp died out. You can guess what happened to the bass population and then the alligator population.

In *Desperation Medicine*, I talked a lot about the health threats from cylindro and the politics of environmental disasters. It turns out that the toxin-forming cylindro was resistant to fungicides and copper. Those compounds usually kill other organisms, but cylindro was unaffected.

Cylindro illustrates a key principle for indoor molds: chemicals can kill most, but look out for what emerges from the toxic chemical soup. We must beware what we leave behind. We can create Monsters in the Lake and Monsters Indoors by altering habitats.

If man-made toxic chemicals are agents of natural selection, we should be seeing toxin-forming algae and molds doing the same thing—altering their habitat. We have seen those habitat changes in air, earth and water. And Lake Griffin would be no different.

• • •

Who Would Be Stupid Enough to Do That?

The ponds at home were definitely growing blue-greens now. The water that was left in the five acres of ponds was a truly beautiful turquoise. Maybe aquamarine. I finally thought that maybe the reason I was coughing was due to some toxin made by the blue-greens.

I should have realized the health impact on me immediately. Perhaps the realization takes longer when the problem is from your home. Maybe it takes longer to admit the possibility. If the mold or algae biotoxins are far away, it is easier to face. Maybe the toxins were already damaging my thinking.

"This is my home!" I thought irritably. "These algae can't be here. No way am I going to admit they are making me sick." That thought helped me understand the irrational denials from Lake Griffin a bit better.

If I had toxic blue-greens, were they cylindro? Would they be with me forever? Good luck trying to sell this place if the ponds are toxic nightmares.

Some few months earlier, Dr. Wayne Carmichael had listened carefully to my ideas about biotoxins, including blue-green toxins, initiating cytokine responses following exposure. Wayne is widely acknowledged as being the world's foremost expert on blue-green algae and their toxins. I could speculate all I wanted: no one had any data that would shed any light on the question: "What are the cytokine and hormone responses seen in hyperacute exposure to toxigenic blue-green algae?" The answers to that question are no less important to mold patients and blue-green victims. We knew that the concepts of the Biotoxin Pathway were valid for mold, Lyme, recluse spiders and dinoflagellates. Did the toxin physiology also apply to pond scum?

Many experts say that the blue-greens were the first organisms to colonize the Earth with its chemically toxic atmosphere and water. How long ago did toxin-formers develop their ability to make toxins to survive? And what did 5 billion years of evolution bring to our ponds?

There was only one way to answer the burning questions about blue-greens and cytokines. Simple: Take a genetically susceptible individual and expose him to the toxic ponds. Measure levels of leptin and MMP9 before and after exposure. Monitor VCS. Once a patient has an illness, we won't know what he was like biochemically before he was ill.

Over the years of science, not to mention the old time sci-fi movies, there has always been someone who was willing to swallow a potion or expose themselves to learn and discover what it does. Marie Curie and her daughter Irène were scientists who both died from radiation poisoning as a result of their experiments. The physician who was sure about H. pylori causing ulcers ended up testing the hypothesis by infecting himself. And the guys who thought that willow bark would be good for pain relief and digitalis leaf would be good for dropsy had no guarantee that if they created illness by their experimentation that they would survive without damage.

I had plenty of evidence from treating cylindro patients from Florida that the illness was quickly corrected with cholestyramine. I could take some deep breaths in blue-green water areas and make myself sick, follow what was happening and then take the antidote for the poison with no harm done.

But what if my algae weren't cylindro? How could I be so sure that CSM would work?

Dr. Carmichael to the rescue. One of his colleagues, John Blakelock, was kind enough to analyze a water sample for me. I had toxin formers, for sure, but the main organism was a *Microcystis*, not cylindro! What had happened in Florida was that the appearance of cylindro coincided with subsequent blooms of *Microcystis*.

Was there something that cylindro did that predisposed to growth of the next toxin-forming algae?

• • •

Well, here goes nothing. Over Labor Day weekend, I had an extended time away from the office. I could get exposed, monitor my symptoms and labs if I got sick, and then take CSM, all in the time I would be out of the office. I had the multi-susceptible 14-5-52B genotype, an embryonic gift from my father, and since I had so many other toxin exposures previously, I was primed for illness. So there was no doubt I would get sick if I took a long stroll through toxin waters.

It never occurred to me what I would do if my theory were wrong. I mean, the guy who swallowed too much digitalis didn't live to tell his story. No chance of that happening to me. No way.

My baseline MMP9 was 220, leptin 17 and VCS perfect. I had one symptom, morning joint stiffness lasting more than 10 minutes from either Lyme bout II or Lyme bout III.

I had a nice slog in the once-deep pond. The blue-green colors were magnificent and they were all around me. After about 30 minutes of up close and personal with toxic algae, I started feeling it. The same distinctive hot taste on the tip of my tongue and back of my mouth, cough and shortness of breath told me I'd been hit. Back inside, a shower didn't help. Two hours later, my head began to hurt and the fluorescent lights in the office killed my eyes.

The next day, I was so achy in the morning. Probably because I didn't sleep very well once the diarrhea started. MMP9 shot to 440, leptin 20, VCS was positive. I had six symptoms now. This is acute biotoxin illness all right.

"We should be able to get some really good data out of this experiment with a cohort of one," I thought.

After 48 hours, I had trouble remembering what I read in the newspaper. My headache is a real blaster. I'm sucking on the albuterol inhaler because just walking up the steps to the second floor causes shortness of breath; it's too much. I've got to stop and rest. Then I couldn't remember where I put my deodorant. JoAnn was kind enough to show me the place in the bathroom cabinet where I have kept my Ban for the last 10 years. VCS continued to fall. My leptin is 27 and I've gained 13 pounds in 2 days. My MMP9 is nearly 650. I have 16 health symptoms.

Even now, writing this report two years later, I relive the memory and feel awful. I didn't forget everything: that memory of acute blue-green illness is burned into my soul.

Fortunately, CSM worked just as well for me as it had for all those Florida folks exposed to the Chain of Lakes blue-greens. You know, I had presented my data on the Florida cases at a CDC meeting in 2000. I even had lunch there one day with Dr. John Burns and the head of Florida's estuary toxin program, Dr. Alan Rowan. We talked all about the Florida data.

I guess they didn't believe me. And they would be wrong.

• • •

Wayne couldn't find the microcystin toxin in my blood, so I guess some scientist will be skeptical of my experience, especially if he is clueless about the time course of appearance of symptoms and markers for biotoxin illness. Unfortunately, isolating toxins from blood is a science in its infancy.

Absence of proof here really does mean that proof of absence has NOT been shown.

I hear the same arguments about mold cases from defense attorneys: "How do I know for sure what toxin caused the illness?"

And despite the obvious manipulative intent in the question, we can't yet reliably measure the minute amounts of biotoxins in blood that would tell us about the presence of biotoxins in the body. All we can do is follow the lessons of the Biotoxin Pathway to tell us what markers the toxins left behind.

• • •

Florida wasn't totally in denial about its blue-green problem. In 2003, I was asked to make a presentation to the Harris Chain of Lakes Commission, chaired by Skip Goerner. I had over two hours to tell them what I knew. The results of the meeting were predictable: the commission resolved to hear more testimony and called for more study. The head of the Lakes commission told the press that he and his children boat all the time on the lakes and they never have any problem. When the big CDC grant came to Florida to look at possible human health effects from exposure to blue-greens, none of the money reached Pocomoke. I have yet to see any publications on the Florida research.

After my talk in Florida, Commission member Charles Clark asked if I would like a boat tour of Lake Griffin. We could spend an hour on his boat first thing in the morning before I had to drive back to the Orlando airport.

Lake Griffin at sunrise is a spectacular sight. The rookeries on the far shore were in their morning raucous. The lake's shoreline was in full flower as well. The logs on the lake, the ones with two reptilian eyes and a long tail, weren't logs. Why anyone would want to water ski on the Lake with its big gators was beyond me. We boated down Haines Creek, looked at all the restoration attempts, including the flow-way. We went everywhere.

And nowhere did we escape the toxins. I guess I am supersensitive, because Dr. Clark didn't get hit. First was the hot burning taste in my mouth and on my tongue, then the fullness in my chest, queasy feeling and first inkling of a headache.

"Chuck, how about we head back to the dock. I really have to catch that plane," I said. Let those who aren't made ill by Lake Griffin play on its magnificent waters. I can't.

• • •

I don't know how many experiments in Nature have to be run to convince someone important that we must look at biotoxins in our environment as real problems. The politicians and government administrators need to understand the health risks as a threat to well-being and thinking and not just as an annoyance to their political agenda. We have toxins in our waters, toxins in our air and toxins in our schools, workplaces and homes. We can easily document illnesses caused by those toxins if someone in power says we should do the testing.

And we have easily accepted cures. Yet, no one seems to listen.

Just imagine what our lives would be like if biotoxin illness weren't silently, insidiously, eroding the basics of our health and our lives. And if we do nothing to prevent biotoxin illness, just imagine if our children are not only affected, but also primed for more biotoxin illness to come in their early years.

CHAPTER 17

Phil Ness Can't Breathe

The curse of the 14-5-52B genotype (and some others) is that once primed, you'll get every biotoxin illness that comes down the pike. Phil Ness retired to the banks of the Wicomico River on Maryland's Eastern Shore not knowing he bought five acres of Toxin Hell.

First was Lyme disease. Antibiotics alone from the other doctors didn't do the trick (they rarely do in the 14-5-52B victims). Phil came to Pocomoke for toxin-binding, cytokine-lowering treatment. His MMP9 and TNF, both off the charts, came down as his VCS went up and symptoms cleared.

"See you later, Phil. Careful with those ticks," I said as he left my office.

I *should* have said (but I didn't know back then), "Look out for the stairstep increase in illness severity with your second bout with Lyme. And watch out for the molds found indoors, because you're now *primed* for any kind of biotoxin illness."

Just like with Libby Kyer, once the susceptibility genie was unbottled, illnesses that Phil had breezed past all those years coaching basketball would now get him. How many moldy gyms and locker rooms had Phil been in all those years? Didn't bother him a bit. And as those basketball guys say, "No harm, no foul."

That's not true for him any more.

After Phil's Lyme was a recluse spider bite in 2001. This was an impressive lesion indeed! His blood tests showed what an awesome series of cytokine storms he survived. His photos documented the change in his bite with therapy.

I've heard several spider experts tell me that we don't have brown recluse spiders around here; our brown recluse is actually the Mediterranean recluse spider. Without seeing the spider, the bite itself could come from either species. The difference is immediately obvious to the expert following inspection of the spider's genitals. I had medical photos done on Phil to convince the experts that he *did* have an illness. Uncovering the thick skin of the blistered area, thereby exposing the dead and dying tissue, provided ready access to spider-toxin clinging to the tissues. The patient could slather on a thick layer of cholestyramine in a cream base, binding the toxins.

When we added cytokine-lowering agents and oral cholestyramine (my standard protocol for recluse bites of any species), Phil did well. The spider experts should see for themselves how well this protocol works.

But after his second biotoxin illness he was a little more tired than usual and his joints took a lot longer to warm up in the morning. The residual cost of the cytokine storm of his second illness was much greater than double the first illness.

He brought me a wonderful photo: There he was in his protective long pants tucked into his boots, gloves, thick coat and repellent-soaked hat, holding a can of Permanone. He could have been pitching the tick repellent for a TV commercial! Ah, yes, restful backyard living for the low-MSH patient in Tick and Spider City, Maryland.

• • •

Every fall, Phil makes sauerkraut; his family will tell you he's better at that than sinking foul shots and he once made 500 in a row of those. Into the office he comes, can't think, exhausted, joints flaring, coughing and more. His wife, Pat, had to bring him. No tick bites, no spiders, no leaks in the roof, but this is another obvious biotoxin illness. Phil is stairstepping now; it looks like he's on the stairway to Chronic Fatigue.

"Doctor, the only thing I've been doing before I became ill was making sauerkraut. I make it the same way I always have, working the vat of chopped cabbage until Mother's Mold grows on top. When that happens, it's getting ready," he said.

"Mother's Mold, what *kind of mold* is that, Phil?" I asked.

We don't have an outbreak of biotoxin-affected sauerkraut-eaters every year when the fresh sauerkraut is ready. At our local Oktoberfest in Ocean City, it's the *beer* and the *polka*, not the Mother's Mold, that makes people a little crazy. And lots of molds and yeasts are useful in cooking. Roquefort cheeses, anyone? And how about some yeasty hearth-baked bread with your beer? Food just doesn't cause a chronic mold illness in my experience, even though there might be some aflatoxin, another mold toxin, in the peanut butter.

"I don't know what kind it is," answered Phil. "It's Mother's Mold and we've always looked for it on our sauerkraut. I'll bring some in so you can have it analyzed."

He did and so, there it was, an unnamed *Aspergillus* growing on Phil's sauerkraut. Phil now gets sick from mold spores from his kitchen creations. Life really isn't fair.

He takes his CSM for his mold illness, but not Actos. His body fat is so low, and therefore his leptin is too, that if he takes Actos to lower his MMP9 and his TNF, and that would help him tremendously, he unfortunately drops his leptin at the same time. When leptin is too low to adequately drive production of MSH you will knock out MSH production. And in Phil, who doesn't have nearly enough MSH, you can make him immediately much worse. You have to be careful when you use Actos in skinny people!

This time Phil doesn't get better, at least not better the way he did before. His morning jog—forget it. He's lucky to climb a few steps without stopping for air. His brain isn't right anymore, either. At the family get together in New London, Conn., he's in a fog worthy of the raincoat. His pain is worse, with cough, abdominal pain and weird stabbing pains that feel like someone was sticking him with extra long, extra fine voodoo pins. And now Pat doesn't really trust him to drive back home.

• • •

What's happening to Phil? Simple stairstep worsening from another bout with Lyme? Is he worse from an undiagnosed third or fourth bout of a tick-borne illness?

No, Phil has suffered from biotoxins now turning on cytokine production one time too many. Now the cytokine response has damaged the ability of his body to ramp up production of VEGF. Remember good old vascular endothelial growth factor, VEGF, from chapter 4? This is the compound, a "growth factor," that can turn on production of new blood vessels to feed cancer cells (see 21st Century Medicine) *and* control blood flow in capillary beds. Without VEGF, there isn't the same blood flow to capillaries. Of all the factors that really scare me about chronic fatiguing illness, low VEGF that fails to rise after treatment is high on my list.

If my friend Phil isn't getting the correct blood flow into capillaries in his brain because his VEGF is dead low, I can understand his confusion and brain fog. The brain needs plenty of sugar and oxygen to work right. Low VEGF won't let enough of those essential nourishments get there. But why is he getting so short of breath? He acts like some sort of untrained athlete or someone with chronic illness from lung or heart disease. Like them, Phil just can't walk more than 200 feet on a level surface without stopping to gasp for air. And if he has a good day, one where he has just enough energy to dig in the garden for a few minutes or go with his wife to buy some one *else's* sauerkraut, he's wiped out for three days afterwards. I've heard plenty of experts in Chronic Fatigue Syndrome talk about "post-exertional malaise," and all they really mean is normal recovery from activity is delayed.

And now, Phil has it.

If his newly acquired low VEGF is the culprit causing his stairstep, or better yet, call it his headlong down-spiral into delayed recovery, then we should be able to show that he doesn't deliver blood to capillaries the right way. Measuring capillary blood flow is tough. We could dream about again using the fancy eye diagnostic device, the Heidelberg Retinal Flowmeter, a combination dual laser and Doppler. We already know that device can give you read-outs of flow rates in capillaries of the retina and neural rim of the optic nerve head. Without access to the laser, however, what we can use instead is a pulmonary stress test (PST), an exercise test that has long been used in assessing pulmonary reserve in those patients with chronic lung or heart disease. There even are standards for PST results used by disability examiners to assess ability to perform work by measuring uptake of oxygen with exercise.

If your use of oxygen, called VO2 max, is low, say under 15 ml/kg/minute, no one will argue you are is in trouble, and no one will argue you have a disability.

If low VEGF is causing the problem with capillary flow, then we should see profound reduction in the ability to deliver oxygen with exercise in low VEGF patients. Further, the PST measures just *when* the blood flow to muscles fails to keep up with demand for oxygen. We call that measurement the anaerobic threshold. When we correct the VEGF levels, we should see improvement in both VO2 max and the anaerobic threshold.

And that's exactly what we see, *if* VEGF responds to biotoxin treatment. For Phil, that "if" was a big one. His didn't.

• • •

Make no mistake, when I say anaerobic threshold, don't confuse anaerobics with aerobics. You know, aerobics are the vigorous exercises they do on TV videos and YMCAs. At first I thought aerobics meant a bunch of people doing funny-looking dances while dressed in Spandex. After they got sweaty enough, they would go have a healthy salad followed by two bowls of ice cream for dessert, since they'd been so healthy that day.

A person with a reduced anaerobic threshold usually has a low VEGF level. He will suffer like crazy for days after a brief aerobic workout. And if he also has a leptin problem (many with low VEGF will have both MSH deficiency and high leptin), eating just a little ice cream will cause him to gain and maintain weight. Tragically, this simple fact of physiology is lost on the aerobics people. They assume the low anaerobic threshold, high leptin, refractory weight gain patient is a gluttonous slob and they often will say it.

So how do many physicians treat a person like Phil? He goes to his doctor who weighs him, finds out he is short of breath and won't exercise because it makes him feel so bad. The treatment? The predictable suggestions that he cut down on the food he eats and start an exercise program. Maybe joining the aerobics program at the local YMCA would be a good idea. Well, you already know those suggestions are worthless in Phil's case, but I still hear them, especially from some of the docs who read the pulmonary stress tests!

"Eat less and exercise more" won't work for biotoxin patients.

• • •

Phil went for his PST. The results confirmed my worst fears. Here is this guy, 6' 3", who looks like he is made of solid muscle, who used to run all day and work the rest of the time, who has a VO2 max that is a perfect match to that of another guy who is in severe heart failure. The heart guy lives from bed to chair and we don't question why he's short of breath. The heart guy has a different reason he can't exercise: it's not low VEGF!

But a low VEGF is the cause of Phil's fatigue and low VO2 max. He can't live this way. So true, he can't. Thanks, VEGF. The dead low anaerobic threshold that's included at no extra charge is just a gravedigger's garnish for this illness.

What can be done to correct his anaerobic threshold and raise his VO2 max even if we must override his low VEGF? First we must recognize the illness! Then, find out what his genotype is, avoid biotoxin illnesses and treat them rapidly before the cytokine wars turn our immune responses against us. Finally, when the illness is not responding to anything, let's learn from Phil.

• • •

Enhancing oxygen delivery in capillaries: what a goal for the patient with diabetes and narrowed capillaries. And how about the athlete? Wouldn't he perform better if he could get more oxygen into his muscles? Would some athletes do anything, including taking performance enhancing drugs, to make up a fraction of a second or extend endurance? Of course they would; just look at all the drug testing at the Olympics and Tour de France trying to make sure that anyone who cheats is caught. Couldn't we use the lessons from the sports cheaters who enhance oxygen delivery to benefit those whose active, vigorous life has been stolen from them by biotoxin illness and oxygen delivery?

Let's not forget the lung patients who do so poorly as their illness progresses. What can we learn from medications we give to the unfortunate lung illness patients?

And finally, what can we learn from specialized athletic training performed by those who aren't elite athletes? They do more by doing less. What I mean by that is when someone like Phil has impaired oxygen delivery, they may exercise *only to the anaerobic threshold.* Once they go *beyond* the threshold, metabolic disasters arise quickly.

We designed a study, a treatment trial really, that would try to answer Phil's question; "What do I have to do to feel better?" I wanted to help Phil but there were so many biotoxin illness patients with low VEGF, shortness of breath, cough, unusual chest pains and the muscle symptoms following minimal activity that a clinical trial should be able to help them, too.

And along the way, we should be able to correct some of the misleading comments about respiratory symptoms that followed exposure to toxigenic fungi in buildings. Most of what I heard about "hypersensitivity pneumonitis" and other lung conditions was just plain wrong.

Telling physicians that they are wrong prompts immediate responses. The smart ones say, "Show me the data." No one will believe me, especially when I say "established opinion" about mold illness and shortness of breath is wrong, without data to prove my contentions.

We had a "control group" of thousands of "normal" patients who had taken PST before from all over the country. We knew what normal PST results were. Our long list of patients with low VEGF and/or chronic fatigue, and mold patients lead the way in both categories, were clearly different from unaffected, normal people. What happened when we intervened with meds or exercise?

Our first group of patients did nothing other than stay on their regular program. They are the control group of affected patients. We saw no change in their repeat PST when they did nothing to change VEGF or VO2 max. This means that the illness doesn't go away on its own.

When we added a lung medication, theophylline, one that improved diaphragm movement, making the lungs "stronger," we saw improvement of about 10 percent in three months. All other lung medications did nothing.

When we added creatine, yes, the same creatine found in high school gyms throughout the country, at hefty doses (0.3 grams/kg), we saw improvement of almost 10 percent per month for three months, leveling off to 33 percent improvement in 6 months. Even 50 percent improvement would only take Phil from near-dead to half-near dead. And he wouldn't take any meds that he didn't have to take. He wouldn't take creatine. For many people, especially those without an athletic background, creatine is a novelty item, and there isn't the same resistance to taking it that Phil had.

Creatine has been abused by so many young kids, however, that it has a bad name in some circles. If used at regular doses, monitoring fluids, taking creatine hasn't given my mold patients or the chronic fatigue patients a lot of problems. Perhaps its best use will be in combination with other treatments.

How about those diabetics? Actos raises VEGF if taken in combination with high dose Omega-3 fatty acids (the so-called fish oils). Actos doesn't change VO2 max when taken on its own *without the no-amylose diet*. With the no-amylose diet, we see 15–25 percent improvement in VO2 max with Actos. When the high dose Omega-3 is added, we can get another 10 percent improvement. High dose is 2400 mg of EPA and 1800 mg of DHA daily, taken in divided doses.

My favorite candidate to correct both low VEGF and low anaerobic threshold is erythropoietin (epo). This anti-cytokine has been recognized for years as being a stimulant for the bone marrow to ratchet up production of red blood cells. I see TV adds for it just about every night; the marketed compound, called Procrit, was cited as the 5th biggest selling drug in the US in one survey. Cancer patients receiving chemotherapy and dialysis patients with the "anemia of chronic disease," all know just how good Procrit makes them feel.

Could erythropoietin make a difference in VO2 max? You bet it does.

There is more good news about epo. Since the genes for VEGF and epo are both linked, that means that production of both is linked (see chapter 4). When we see simultaneously normal levels of epo in blood and low VEGF, there is a gene transcription problem. Why does this matter? For some patients, the use of epo as a supplement can override what is called, "transcription block," can boost VEGF as well. For these patients, a short course of epo "resets" the genes, defeats the block and there is no fall in VEGF after the epo pulse therapy is over. Those patients return to health and maintain that improvement.

Others with low VEGF aren't so fortunate. It looks like they will need longer duration of epo supplement. That treatment has all sorts of potential problems.

Obviously, we need a lot of work here, but with the development of the newly modified epo molecules that don't push red blood cell levels too high, look for many more uses for epo to come forth.

To finish the VO2 max, anaerobic threshold and VEGF trials, you've got to meet Fred first.

• • •

Fred Vanderveen could be a perfect match for his Harley. To tell you the truth, he would appear equally at home as the "Human Tree Trunk" in some professional wrestling event. When you look at him, all 315 pounds of muscle mass, perfect Santa Claus beard and nearly bald head (*not* shaven, thank you), you would never guess he is the National AAU director for Special Olympics. But that's Fred, too. Fred is the classic "Big tough guy with a soft heart."

For years, I've sent young athletes to Fred to help them develop strength, speed and conditioning in preparation for possible collegiate or professional athletic careers. Fred started his weight training academy nearly 20 years ago. At first it was only open nights and weekends after Fred's real job, teaching biology at Snow Hill High School. As time passed, he found others to help teach and train. During Fred's career, he's taken countless athletes, especially those with disabilities, to competitions and come home with all kinds of hardware—medals, trophies, plaques and more. The time he took his team of kids with Down's syndrome to Baltimore to join together to pull a jetliner down the BWI runway raised a lot of interest in Special Olympics.

"Fred, you're going to really get a kick out of Phil," I told him when I called. "He's hit hard with low VO2 max from a collection of biotoxin exposures, but you won't believe it to look at him. Hold him back, keep him under the anaerobic threshold and let's see what he can do. Nobody around the country has any protocol on how to do raise his VO2 max in the face of refractory low VEGF, *so invent one*. Just don't let him go anaerobic!"

"Well, what should I do? Bike, bands, just stretching?" he replied.

"Why don't you start with a combo, lots of warm-up, lots of stretching. Real gentle on any kind of isometrics. Keep him slowed down. Phil is like a racehorse with a healing cannon bone. Ease him in gently, very gently. You'll know when he went anaerobic because he won't recover well the next day. Keep a chart and make him slow down. Cramps, aching and stiffness the next day are all warning signs you went too far," I said.

"Ritchie, it looks like Phil will do best if we start him on the stationary bike for 5 minutes, keep his pulse rate under 90. We'll use set resistance. I'll give him stretch breaks every 2 minutes for 2 minutes. Then we'll do stretching of low-resistance bands, same stretching. Then he can work legs, more stretching, and add abs as he goes."

"Fred, I still think that's too much time on the bike, and don't let him increase the resistance or the workload. And no upper body yet!" I said.

• • •

In the end, Phil knew better about how to rehab his body than either Fred or me. Fred invented a terrific protocol and Phil re-invented it to match his VEGF response. He just followed his body. He kept his pulse low, monitored recovery, monitored fatigue in activity and kept detailed notes. He was doing three 90-minute sessions at Fred's gym every week. He started with bike, stretched, added the resistance bands, stretched, added legs, stretched, abs were next, stretched, more bike and repeat. After three months, he felt terrific, his brain was working, he had no pain and his fatigue was much reduced.

Repeat PST showed over 50 percent rise in VO2 max and anaerobic threshold. His overall work capacity increased nearly 75 percent. He was compensating for dead VEGF.

When he came for routine follow-up on April 22, 2004, I cheered. He did it! Just look at how well he's living now, despite low VEGF and despite low MSH. Muscle training improved his oxygen extraction despite his physiologic limitations. And now it was a joyous day. Just for completeness, I checked his C3a: a little high, but not bad, only 300 (normal should be less than 150).

• • •

I couldn't believe the phone call: Phil had another spider bite. The bite was on Sunday afternoon, April 25; I saw him 14 hours later. He was hit all right, and hit hard. VCS was already falling. Symptoms on the list were all re-appearing. Without MSH and VEGF, he had no protection from re-exposure. All the anaerobic threshold exercise in the world was just a house of cards—no match for the armies of cytokine Orcs shooting thick arrows through his chest armor.

In less than 15 hours, his C3a was over 5000. MMP9 normal still, as was interleukin-1-beta (IL-1B). In two days, IL-1B rocketed upwards and on day 3, it was MMP9's turn. Measuring changes in cytokines in a hyperacute exposure to biotoxins tells us the sequence of events in biotoxin illness and how the illness progresses.

Thanks to lots of Omega-3, cholestyramine, statins (see 21st Century Medicine), high fat diet and all the rest, Phil's case settled down. He was emotionally drained. Just like the long-suffering figure in Hades who keeps pushing a rock up a hill, only to see it fall down just before reaching the top, Phil was back at the bottom of the hill again. Time to begin from the bottom, back to Fred's place he goes. His exercise results were right back where he started.

All that work of three months, *gone in 10 days*. Cytokines will beat exercise any day.

We can't forget that Phil was physiologically impaired before the second bite: he had no MSH and no ability *to control* the sequence of over-response of inflammatory elements like MMP9, IL-1B and C3a. All his exercise had done was to compensate for his functional disability.

Back to the gym for Phil. More bands, bike and stretch. And more improvement too.

• • •

This should have been the end of the story. Not for Phil, the biotoxin magnet. Not too long ago, he called again. Another tick bite, another biotoxin illness, another surge in C3a, overnight. How many hits can his Blackjack hand take before he goes over 21?

Fred called, what was happening with Phil?

"Fred, Phil is the most unlucky guy I've seen as far as biotoxins. At the same time, he's showing us so much to help others and he wants his experiences shared," I said. "I remember when he first came in with his TNF and Lyme; he was more worried about his *daughter* than himself. Her psoriasis was world class, with so much joint problems, it couldn't be psoriasis alone. Sure enough, she had Lyme exposure; the lessons from Phil were instrumental for finding a way to attack the TNF basis for his daughter's illness. She was taking anti-TNF meds for her psoriasis long before anyone else suggested the idea. And here I thought it was only Mothers who sacrificed for their kids.

"What happens to Phil is a series of events in which cytokines limit blood flow to capillaries and then lower VEGF production. He remains well until the system is stressed, like walking upstairs, for example. Your workouts have enabled him to selectively enhance some of the fibers that are mainly anaerobic—the slow-twitch fibers, the physiologists call them. The fast-twitch fibers, the ones he needs for aerobic activity, are inaccessible to us because of his low VEGF. When he gets his global cytokine shutdown of flow, his fast-twitch fibers aren't a factor. They're cooked already. But the slow-twitch fibers can come back *as your training protocol develops*. Until we see the VEGF rise, he won't get fast-twitch back. To your wonderful credit, you've shown him how to push slow-twitch to carry him to functional life."

"You know, Ritchie, I've been thinking a lot about Phil and all the others you have sent me. It's almost like these Chronic Fatigue people don't have functioning mitochondria. They just don't deliver the energy they should when needed, they don't recover and they can't do much with acute increases in energy demands. The pulmonary stress tests show that in just about all of these patients."

"I think you're so close, Fred. Look at what goes on with these guys. They don't get the right blood flow to big time users of oxygen during exercise. Lung and muscle beds lead the list. So the poor muscles are doing what they are told to do during exercise, but without sufficient oxygen. And without delivery of the right amount of oxygen, the muscle won't be able to obtain energy from sugar efficiently.

"When the muscle needs fuel, energy in the form of ATP (call it the bodies' gasoline if you want, Fred is a biology teacher; he talks about ATP all the time) is produced, mostly from glucose. Glucose, call it sugar, has six carbons. It is the main source of fuel in muscles. Every cell in our body can break down sugar into two separate fragments, each three carbons long, releasing *2* ATP. But, here's the problem: if oxygen isn't present, the three carbon fragments, including lactic acid, just sit there, just causing trouble. If there is sufficient oxygen, those three carbon fragments are taken into the mitochondria and broken down into carbon dioxide and water, with a *by-product* of *36* ATP molecules. So Phil is burning sugar like crazy (about 5% efficiency!), getting 2 ATP at a time, just to make up the 38 ATP he needs. No wonder he runs out of fuel and feels so tired.

"Then he is wiped out for days because it takes so long to restore the sugar (sugar is stored in glycogen) inside the cell. It may take *several days* to replenish glycogen. If he tries to do something the next day, before his sugar levels in glycogen are back to normal, he is still wiped out."

"Well, that's what I was thinking. Instead of 38 total ATP, your patients only get 2 ATP. Man, if my Harley got that bad gas mileage, burning a tank of gas to go 20 miles instead of 380 miles, I'd have to sell it. I couldn't afford to pay for such inefficient use of gasoline.

"But back to my mitochondria idea, Ritchie. If Phil suddenly doesn't get oxygen for his mitochondria, he won't get the expected energy bonus from burning those fragments of sugar. So he ends up being tired without the bonus. And then he starts to build energy production from the slow-twitch fibers, and he gets hit with a biotoxin again and suffers more reduction of oxygen flow in capillaries. It's like someone is tantalizing him by giving him a functional mitochondria with enough oxygen to be normal, and then taking the mitochondria away. I can only imagine how some people must feel if they never get the mitochondria working again."

"Fred, it might help to imagine oxygen delivery to muscles as similar to a FedEx truck on the big highway. As long as the truck is on the main highway, you won't be getting any packages in your house. The truck has to leave the highway, go onto the State road, then the county road, to turn down your lane and then into your driveway. Now if there's a big snowstorm and the County snowplow pushes all the snow from your lane onto your driveway, the FedEx guy still won't bring anything to you. So, you can have normal levels of oxygen in your arteries, but if the entry of blood with the oxygen it carries into the capillaries is impossible due to low VEGF, your delivery system fails and you will not have the normal efficient consumption of sugar."

"And the patient, Ritchie, will feel as bad as Phil does."

• • •

"You know, Fred, cases like this one that get me thinking about evolution. Remember *Pfiesteria*? In one of the multiple stages of life forms that *Pfiesteria* can assume, it engulfs and eats creatures that have chloroplasts—the powerhouses that do for plants what mitochondria do for animals. Instead of eating the chloroplasts, the dinoflagellates let them continue to function, growing with them, using the plant chloroplasts as a walk-along energy source. And so, guess what? Those *Pfiesteria* grow to be huge. When the plant food isn't available, like when all those chemical poisons are in the water, then *Pfiesteria* has to resort to plan B and get small, mean, fast moving and toxin-forming to survive.

"Imagine if that simple act of engulfment of an energy generator happened to a different one-celled animal cell. Only this time, it sucked up mitochondria. And the mitochondria lived inside the cell, providing the energy needed to live in communities, make expensive (in energy terms) molecules like proteins, cholesterol and immunoglobulins and all the rest. That simple theft of energy would be the source of the rise of multi-celled organism from single celled organisms.

"That fundamental difference between bacteria and men was changed by engulfment. Engulfment is a really primitive method of defense too. It is part of what we call the innate immune response (see 21st Century Medicine). When we see it active in us we call it phagocytosis, we don't realize where the technique came from. We really are a combo of immune mechanisms aren't we!

"And just look at the cell membranes. Every little structure inside a cell has two membranes, just like the outer membrane of the cell. Not mitochondria. They have FOUR. Their outer two are the same as the rest of the other organelles in the cell, but the inner two are unique. The 'mother' *cell had to engulf the mitochondria* at some time, putting membranes around another organism (the mitochondria) that already had membranes. The DNA of the mitochondria is different too. And look at the replication schedule of mitochondria. They're so different from their host cells."

"That is such a cool idea, Ritchie. Hey, do you remember which of the animal germ cells have the mitochondria? It's not sperm. No way. *It's the eggs*. All your mitochondria come from some lady eons ago. I bet she needed mitochondria to deal with all the toxin-formers back then."

"Hey, what I think about sometimes is just how many toxins there were in the old days. Seriously, Fred, hear this out. Just look at where all our currently known toxin-formers come from. They all were really primitive organisms; they were here at the Dawn of Time. But don't be confused by primitive, because they are still here, aren't they? Dinos, (no, not dinosaurs, but dinoflagellates), fungi, spirochetes, blue-green algae: these guys have been around forever. Everyone agrees the fossil record shows that the blue-green algae were the first colonizers of our blue-green Earth. Can you imagine the biotoxin wars back then? No wonder the evolution of vertebrates has included immune defenses against toxins made by all these invertebrates. But those immune responses are ones that can be damaged by toxins too. Where did that capability come from in evolutionary terms?

"What kind of fungi did they have in the Pleistocene era anyway? And did they make toxins? What really changed aquatic habitats from growing trilobites to growing primitive sharks? Did some kind of reef-dwelling toxin-former like the ciguatera dinos of today come along to poison most of the predators to let some new life form flourish? And I still wonder how long it took for blood-sucking mosquitoes and ticks to develop following the development of blood forming organisms. And that means ticks and malaria type critters grew up hand-in-hand a long time ago."

"Ritchie, I love the idea. Evolution by natural selection, enforced by toxins. Do you think anyone else has come up with that idea?" Fred laughed. "But you're right. Energy-wise, Phil *becomes a single-celled organism*, without mitochondria, struggling to find enough sugar to use after his toxins get him. Do you think if he were still breeding he would compete for brides as well as a non-toxin guy? No way. If he doesn't breed, what happens to his genes? He would be a victim of natural selection. Without offspring, his genes would stop with him. And that's how evolution progresses."

• • •

For now, Phil is fine. He works out three times a week. He is happy as an unraked clam. Fred doesn't push him too hard and Phil doesn't push like crazy to do too much.

But I know I'll see Phil again sometime soon. Maybe when he gets back from his trip to Lake Griffin, just north of Orlando. He said he wanted to do some bass fishing.

No, Phil! Don't go to cylindro-land in the Chain-of Lakes!

CHAPTER 18

Joans of Arc

The scrupulous and the just, the noble, humane and devoted natures; the unselfish and the intelligent may begin a movement—but it passes away from them. They are not the leaders of a revolution. They are its victims.

Joseph Conrad, Under Western Eyes, *1911*

Joan of Arc remains a figure whose real story is shielded by the mists of time. What we know about her comes largely from the records of her trial before fire, courtesy of the English in 1431. I wish I knew the truth about the psychiatric make-up of this woman, because in our era, she'd likely be labeled as having schizophrenia, with auditory and visual hallucinations, possibly with paranoid ideation.

She heard voices that told her to rescue Paris from the English. In a vision she saw the disguised dauphin, leading to his ascension to a throne. Then, perceiving that the occupying English armies had to be driven from her beloved France, she organized a mob of peasants into a fighting force that fought, and angered, the mighty English.

And all this when she was a teenager!

Remember the image of Joan from your 6th grade history class, when garbed in gleaming white armor, she nearly single-handedly liberated the walled city of Orleans? Yet, despite her heroism and patriotism, she was sold for execution to the English by her trusted comrades, including the mighty Duke of Luxemburg, for 10,000 crowns.

She was sold-out by those she trusted. The English tried her, found her guilty, and ordered her burned at the stake. They didn't succeed with "death by fire" at first, but Joan had no hope for pardon. Back then, the popular way to deal with enemies of the state was to trump up charges and convict them of heresy and/or witchcraft.

We do the same things now in more subtle ways with those who offend the CDC, FDA and State Boards of Medicine. Accusations, some trumped up or politically correct just like Joan's witchcraft, include record-keeping errors, insufficient paperwork, deviations from standards (established by whom?), and billing errors. The English got Joan on heresy and witchcraft charges, finally burning her to a crisp in Rouen at age 19. In 1920, the Roman Catholic Church named Joan a Saint.

What really happened with Joan back then? How could this charismatic figure that mobilized the masses from their doldrums and won the big battles through the sheer force of her personality end up dead? The memory of her exploits is seared into my brain. She was greater than life; yet she ended up a martyr.

Imagine if Joan were alive today and she mobilized thousands of ordinary citizens to cry out against the oppression of the mold-opposing forces—insurance companies and corporations employing physicians to undermine the lives of ordinary people! She wouldn't ignore friends, relatives, neighbors, the weakest and the poor who are being trampled by those opposed to the truth about mold illness. Just imagine how Joan, sword drawn high, would lead the charge against the fat cats who continue to deny just compensation to mold victims.

She would be there in Louisiana at the big Naysayer Mold convention. In the streets, look, it's Joan and her crowd in New Orleans. They're carrying signs, singing songs and tearing down the artificial barriers of pseudo-science. They will not be ignored, stepped on or remain under anyone's boot any longer!

While these days Joan wouldn't be burned at the stake, she'd certainly be French fried in the media by the powers-that-be if she had any personal failings. Heaven forbid that she'd prefer to dress or look like a boy, or merely go out in unisex fashion. And if she dared venture out in funky suits made of metal, her dress would be the news story, much like Einstein might have his work summarized these days by a headline about his eccentric hair style. If anyone found out that she heard voices, well, that'd be it for poor Joan. I'm sure some serious threats and blackmail would try to intimidate her from appearing in her white mail a second time.

Think about it: what usually happens to anyone who swims *against the undertow?* They're swept out to sea.

• • •

Two modern women with the passion of Joan of Arc stand firm against the tide. Each noble woman is devoted to educating mold victims and getting them effective treatment. Vickie Smith is an accomplished triathlete who had the misfortune to teach in two separate moldy schools in the Central Dauphin School District, near Harrisburg, Pa. Each school had water intrusion and mold growth, and teaching in each made her sick. Because of the mold illnesses from the Dauphin's schools, she is now primed for mold illness for the rest of her life. Vickie can run now, but eventually, she won't be able to hide from mold spores in some building, at some time.

The second woman is Lorrie Taylor, long-time employee of Maryland's Prince Georges (PG) County Police Department. Argue with her at your own risk. Her mold illness, complete with plaques in the brain, was typical of what we read in Chapter 14. Lorrie has a high MMP9 level, high levels of antibodies to MBP, and exposure to massive levels of *Stachybotrys* in her workplace, the Clinton substation in PG County. With her findings, you know to expect to find plaques in her brain. But it isn't multiple sclerosis: it's mold.

Each of our heroines has been scarred by their mold illness and experiences. Like Joseph Conrad said, each revolution has a clear leader, someone who sees something that others do not yet see, and they're willing to suffer in order to effect change. Like Joan of Arc, our Joans looked beyond the concerns of daily living and each has sacrificed much. And like Joan of Arc, Vickie and Lorrie are *victims*—victims of the health terror that biotoxic mold can cause.

I hope their martyrdom won't go further than illness in its acute and chronic forms.

Vickie didn't ask for trouble from her school district. She was sick, not from her home and not from stress and not from depression and not from anything else, but the school district didn't agree. Here was a gifted athlete who was winded by *walking* up a flight of steps. She couldn't concentrate and couldn't function. She had mold illness.

On three separate occasions, she went through the repetitive exposure protocol. Her academically appointed pulmonologist documented the predictable changes in her oxygen consumption with exposure, treatment and re-exposure and re-treatment. Her symptoms, labs and VCS scores all matched the oxygen and breathing abnormalities. No court could ask for better data to prove what mold exposure had done to her physiologically.

She'd missed a lot of time from work due to her illness, so she filed a Worker's Compensation claim. She wasn't making huge demands for damages: she wanted her sick time back. She wanted her medical expenses and medication costs reimbursed. But more than that, Vickie wanted the school to be cleaned up. If *she* were ill, especially given her incredibly high level of physical fitness, what would happen to the students?

One reason Vickie is a Joan of Arc is that she was willing to pierce through lies and deception to bring justice. The district wrongly denied responsibility for ignoring the water intrusion and mold growth. During her painful journey, more than a few hurtful things had been said about her. If her claim had been filed in a high-profile mold state, like California, where mold awareness more widespread, her award likely would have been substantial.

Pennsylvania isn't California.

The Worker's Compensation insurance carrier contested Vickie's case. She saw their doctor, a professor at the University of Pennsylvania School of Medicine. Somehow, the academic skills that won this physician his multiple appointments simply evaporated when he wrote a report on Vickie's illness. Vickie found enough errors in his report to fill *three* pages of corrections. The essential importance of her exposure trials was ignored. The professor didn't even know what MMP9 was or how to use it in evaluation of mold illness in the 5-step protocol. And of course, he didn't bother to investigate them. Why learn, when you already know everything?

So when the judge (not having any medical knowledge of his own) agreed with the professor and not the two treating physicians, and then and denied her claim, Vickie was shocked.

"I don't care how far I have to take this. Oh, we'll appeal, you better believe that. We'll go all the way to the Supreme Court if we have to!" she said angrily. "What a joke. What an insult. How can an Independent Medical Examiner expect his opinion to be credible when there are 44 stated errors in his report? How does the opinion of this same medical professional, who saw me at most for 90 minutes, nearly two years after the illness began, carry more clout than two treating physicians who saw me repeatedly for more than 15 months? I thought the Court had the responsibility to reach an unbiased decision based on *all* the evidence presented! I don't know how they can sleep at night."

• • •

Vickie had to stop teaching before she could begin to feel better. She used CSM and all my other therapies to improve progressively. She isn't 100 percent yet, but her performance in a recent triathlon supports her claim to be healthy as a horse.

Vickie lights up a room when she enters—strong, upbeat, not afraid to say nice things to people she barely knows, she's as personable as anyone you would ever want to meet. But she's a Warrior, all right. Whether she's referring patients for treatment or participating in writing the legal briefs to bring justice to mold-injured people, Vickie won't ever back down from a challenge. If hard work will win justice, then Vickie will emerge victorious. Just look what triathletes do.

Now, whenever Vickie Smith refers a patient, I know they'll already understand mold illness and the Biotoxin Pathway. They'll be prepared for the complexities in treatment required to eliminate complications.

But I'm worried about Vickie as she returns to teaching in a brand new school this fall. Mold likes new HVAC systems that aren't designed properly, just as it likes old systems with air leaks from moldy crawlspaces. Vickie will be off CSM while at her new job, so if there's a mold problem, we'll know by her relapse.

What we didn't know about mold illness when Vickie first came to Pocomoke for treatment is the importance of elevated C3a. Hers is still sky high. We know now that in a 5-year follow-up study of patients completely healed after treatment of their *Pfiesteria* illness, some don't stay well. Compared to our groups of unexposed and unaffected people, *Pfiesteria* patients have a tremendous increase in death, chronic fatigue and multiple symptoms. And for those who do poorly, and are still alive, elevated C3a stands out like a bulls eye.

C3a silently and slowly causes severe body damage. C3a never stops recruiting so many other harmful sources of metabolic damage. It's like the creature in the movie, *Alien*, relentlessly digesting its host, with the first indication of its presence coming when it's too late.

Can we fix the elevated C3a in Vickie? Can we prevent the chronic stealth illness it causes? Vickie won't rest until C3a is wrestled to the mat and defeated. Then she'll make sure the mold community has their C3a levels measured.

She'll stay on track with her activities, her interests and all she does for other mold victims. In her zeal, I just hope she'll remember to eat and sleep!

• • •

Lorrie is like the British "lorry" or truck. I think that her coworkers would agree that if you stand in Lorrie's way when she is right, she'll run you over. She's been working for Prince Georges County for 20 years. Never has there been a more dedicated employee.

"I don't care how much trouble it is to do things right, they must be done right," she says.

When I met her several years ago working at a mold illness evaluation clinic, she stayed all day. The county Fellowship of Police was worried enough about mold that they brought my entire staff in to do testing. Many of her fellow officers were exposed to one or more contaminated police stations within the county. The worst one, Oxon Hill, had been closed. Based on the officers' statements, the other stations had water intrusion problems as well.

Staffers working at Clinton station, Lorrie's workplace, would routinely put out buckets to catch the water when it rained. The crawlspace was full of mold, and don't even think about looking inside the air ducts running from "down there" into the HVAC system. Her ceiling tile sample, pulled from the trash and given to her (you have to be careful about removing samples from the trash, someone might say you're stealing!), had massive *Stachybotrys* contamination. Once again, based on the knowledge that *sick patients* signify sick buildings and *not* building inspections, the other stations were likely to be just as bad.

There was no shortage of sick officers, but the county wasn't doing anything about the problem.

Lorrie was severely affected. She had 34 symptoms, mold-susceptible genotypes, antibodies to gliadin and myelin basic protein, low MSH, high MMP9 and the typical screwed-up levels of pituitary hormones we see in low-MSH patients. Lorrie's brain was full of plaque—everywhere the radiologist looked, there was another "non-specific" plaque.

I tried the standard approach: "You must leave the workplace *now*. You can't afford any more brain damage. The job isn't worth the illness. Let's get you treated and out into a new job or you'll destroy more brain tissue."

"Look, Dr. Shoemaker, you don't understand," she said. "Those officers in those buildings are long-time friends. This is my family. I can't just walk out and leave them to fight this alone. I don't care what it takes to prove the building is making people ill. If you need me to go back in there to prove I get sick again, you got it. How many times do I need to do it? But I'm telling you; I won't stop working, not as long as my fellow officers are still getting sick. I know it's too late for me. And even if your treatment lowers my MBP—and I'm glad it has so far—those plaques aren't going away. I'll be damned if I'm going to sit back and let what's happened to me happen to anyone else. You got it now, doc?

"If I need to drive the three hours to see you every week and have all the blood tests and everything else done, that's what I'll do. What the Hell, I'll drag the entire station down to you with me. I'm not going to let this disaster happen to anyone else!"

"Lorrie, I hear you and believe me, I'm not going to argue with you. I respect you and your loyalty," I told her. "But you should work here with me for a week or two and just look at the illnesses my patients have. Once a person goes over the MSH line, it's like jumping off a cliff—the point of no return is passed. You're on that edge right now. What good arc you to the other officers if you can't spell cat correctly three times in a row?

"We've talked about what happened in 2002 when I petitioned the FDA for emergency, compassionate clearance for me to use MSH replacement in cases like yours. I understand that agency has a duty to protect the population as a whole, but for cripes sake; they won't let any MSH treatment happen unless it's done by their rules. Maybe the FDA will let me use MSH sometime in the future, but not yet. Many of your buddies down at the station won't show much improvement without MSH replacement regardless what I do for treatment.

"Every time you or your fellow officers go back into the building, even with CSM to bind toxin out of bile, you suffer from what toxins do as they make their way into the liver. Sure, the toxins are eventually pumped into bile and sure, CSM can then bind and remove them. But what happens on the way to the liver? Before CSM even has a chance to protect you, the toxins have turned on fat cells to make cytokines to cause more damage. And the cytokines will turn on MMP9 and cause more myelin basic protein antibodies and destroy more of your brain.

"Lorrie, just look at your MRI," I begged her. "It's like Swiss cheese—it's a sieve with holes arranged all through the white matter. You only get one brain! You can't afford to bind toxins *after they've run amok in you*. The damage from the over-stimulated immune response to biotoxins, whether they are from mold or something else, is being done as those little molecules of poison make their way out of you.

"You can still lead. And what happened to you this afternoon in my tiny parking lot? You couldn't find where you left your car. Just look at the tremor you have and all those funny muscle twitches that the neurologists haven't explained. That's mold."

But Lorrie stood her ground.

"Dr. Shoemaker, thank you for telling me what I already know. But you still don't get it. I know I'm damaged goods, and the brain damage isn't going to go away. But I'm fighting this even if it kills me. Besides, I just read some nonsense report about mold, what the Institute of Medicine or some group paid by the CDC wrote? It's bologna, from page 1 to the end and I know it. Here is mold illness, caused by cytokines in genetically susceptible people. We know that. But look at the report—it barely mentions cytokines and doesn't have a clue about the genetics. And they are supposed to be up on the science of mold illness? Not in my book.

"If all I do is prove those types of people wrong, that's fine with me. Just let one of those CDC big shots *work at my desk* under the moldy tiles for a while. Just let them suck some spores into their lungs from our HVAC for a few weeks. That will put an end to their BS about mold 'maybe' hurting people. What did they say? 'A dearth of evidence?' I've got news for them; we got your dearth right here, buddy."

Lorrie laughed for a couple of seconds, but soon started coughing again.

"Can you just see those eggheads begging for some cholestyramine? I would give a day's vacation just to see that."

For a brief moment the bravado and anger cleared. She looked at me with eyes that have never lied: "I've got holes in my brain from that mold. And they're trying to hurt my friends."

• • •

Like I said, don't argue with Lorrie Taylor—she's right. She's been permanently injured by the mold in her workplace, and nothing I can do will correct that. So what does she have to lose?

Fortunately, with the protection of high-dose CSM, Lorrie's illness hasn't progressed even though she stays at work. For her, however, there will be a time that the preventive use of CSM no longer protects her. We're following her MMP9, C3a, leptin and VEGF too. When those numbers bump upwards, it's time for Lorrie to leave.

• • •

I'll never forget how Dr. George Demas was killed by his repetitive exposures to *Pfiesteria*. He died doing what he wanted, sampling the soils in submerged sediments in estuaries along the Chesapeake Bay. His death was at first called pneumonia by the State, but I have the autopsy slides and the original autopsy report. *No one* can cover-up that fact from me.

What I don't know is how many others will die from biotoxins. And even if they aren't killed, how many people are living empty lives, with no spark, no spirit and no hope? How close is Lorrie to such a life? Or Vickie? They function well despite their illness and they're not going to rest until they help others.

As they help others, and as they help fuel a movement of public outcry, they *will* disclose the false and misleading reports from government agencies and they *will* expose insurance company-funded consensus statements. False statements and misleading pronouncements from experts we should be able to trust can ruin quality of life and eventually kill people. But our complaints fall on deaf ears when both politics and money lead science and medicine by the nose-ring.

How long will it take to bring truth to the mold illness arguments? I fear that Vickie and Lorrie will become victims of the movement, just as Joseph Conrad told us, before too long.

Maybe before then, the FDA will stop unnecessarily obstructing patient care and we'll have MSH replacement and erythropoietin too to offer Vickie and Lorrie. But until that time arrives we have to keep the Joans of Arc away from spores bearing toxins so they can keep from becoming modern-day martyrs.

CHAPTER 19

Lawyers, Lawyers And More Lawyers

At one time, it was the church that played a central role in life's main events—birth, marriage and death. Now it's the lawyers.

Many physicians dislike lawyers because of medical malpractice lawsuits; but like everyone else we rely on attorneys for our daily practice. Need to form a corporation? Want to create a partnership agreement? You need an attorney. Want to patent something? You need an attorney. The list of paper work lawyers do is a long one, but still, most physicians don't have anything nice to say about lawyers. Given that medical malpractice insurance costs have gone through the roof, some physicians now won't provide routine care to malpractice lawyers and their families. All of these things tell me that doctor/lawyer antagonism is likely to grow.

What's behind those physicians' attitudes? Powerful feelings—physicians' anger at being sued and a pervasive belief that attorneys will do anything for money.

Somehow lawyers have ended up with a bad reputation with almost everyone. You've read the surveys that show Americans don't think much of lawyers. And who hasn't chuckled over the jokes about defense lawyers: "When the boatload of tourists sank in the Coral Sea, why did the sharks eat everyone except the defense lawyers? Professional courtesy: they don't eat their own kind." The plaintiff attorneys get targeted too. "It's 99% of them that give the rest of the profession a bad name."

Part of the reason for attorneys' bad reputation is many of them make a comfortable living by suing someone or something periodically, even if the suit doesn't make sense. When there's a dispute and a lawyer is involved, you'd better believe that there's going to be some money changing hands. It used to be true that one person taking money from another person using threats was a crime like armed robbery or extortion, but now it's just another day in Court.

We have a malpractice crisis in medicine, with the cost of malpractice insurance exceeding annual income for some medical specialists in more than a few locations in the U.S. What are the sources of the problem? Do we suffer from an epidemic of surgeons cutting off the wrong leg or obstetricians maiming babies with bad delivery techniques? No—but those tragic events have happened.

Then who's to blame? Everyone blames someone else.

In medical circles, we hear that trial lawyers are the reason for all the lawsuits, and it's thought that greedy lawyers remain the biggest stumbling block to malpractice reform. They bring frivolous lawsuits and take a hefty percentage of any settlement they win. They certainly block any kind of malpractice reform measures in many states—after all, why would the lawyers in State legislatures change what's working so well for their profession?

Conversely, trial lawyers say that if a physician doesn't want to get sued, he should simply practice good medicine. They claim trial lawyers are simply fighting for a patient's right to recover lost income following an injury caused by the "egregious acts" of negligent physicians. Trial lawyers will quickly remind all those within listening range that there's no malpractice payout without malpractice occurring.

When the trial lawyers say a few "bad apples" in medicine are causing the high dollar payouts and that if physician groups would only discipline those errant physicians, they miss the obvious reality in malpractice claims. It will always be true that a few physicians, usually those in the highest risk specialties, end up being saddled with the biggest pay-outs for malpractice. That doesn't mean that a few bad apples are responsible for the malpractice crisis!

Any physician who's been wrongly accused of malpractice might feel lawyerly comments such as "if a physician doesn't want to get sued, he should simply practice good medicine," is self serving, empty verbiage. Yet any patient who's suffered loss as a result of medical negligence would agree with those comments.

Let's look closer at the medical malpractice situation, as it's strikingly similar to what's happening with mold lawsuits. One effect of the rising costs of malpractice insurance in some places is that the public is losing access to care from high-risk surgical specialists, like trauma surgeons, and obstetricians and neurosurgery specialists. Insurance costs for emergency room physicians are rising, too. The highly taxed and regulated physicians don't have many choices: they can quit practicing; move to a different state with cheaper malpractice insurance costs; offer you less services so they're not doing anything that might get them in trouble, denying patients necessary methods or operations or alternatives; or go without insurance, a dangerous decision in today's litigious environment.

The parallel in mold circles is that insurance policies are now hard to find at any cost. If you want insurance coverage for mold, you can move away from a non-covered home, change your policy (and probably pay more) or forget it. Just the other day, I received documents from our insurance company that listed mold exclusions for both my office practice and my home. I'm now responsible for costs arising from mold damage to my property—before I received these documents, the insurance company was responsible. Yet insurers also claim that mold isn't harmful.

The attorneys involved in mold cases have taken note of the legal climate in medicine. Recent statistics from Maryland suggest that doctors aren't being sued more often, but average jury awards per case are going up at a dizzying rate. A lecturer from our insurance company tells us juries are sending a message to doctors: we're being called to task for poor empathy, arrogance and a failure to listen. These juries are saying, "Even if you don't cut off the wrong leg, you should act like a human being about it."

From the insurance companies' perspective, abusive, arrogant doctors are one of the causes of the excess awards underlying the malpractice crisis.

Building owners and supervisors should pay attention, since the mold lawsuit Grim Reaper is coming for them soon, too. If someone can prove an owner or manager showed arrogant disinterest or disregarded known existing damage or dismissed health symptoms called to his attention due to mold, it's likely to result in aggressive actions by juries. With insurance companies both denying the danger of mold and also running away from liability for mold claims, there are going to be arrogant building owners and "denying" physicians held personally responsible for health effects due to mold exposure.

The lawyers know that in mold claims, juries will award more now that we can prove the health effects from mold exposure.

Insurance companies often pay out and consume in overheads more than what they take in from premiums, and they live on the interest (called the "float") money generated from investing the premiums paid. Since the float is reduced if they make bad investments and/or the stock market falls, as it did in the last few years, then premiums have to go up to make up the loss. Given that insurance companies took big investment losses and payouts are higher now, then this viewpoint says the malpractice crisis is the insurance companies' fault. Trial lawyers never fail to play this trump card.

Trial lawyers might take on a medical malpractice case because there's a possible big payoff for them if the case settles before trial. In court, defenders of doctors win more than 70 percent of cases against them, so settling before trial is often a good option for the malpractice plaintiff. Early mold cases used to be typically won by the defense and settlement became a win for the plaintiff.

If the trial lawyer only got paid for his time and materials in a lawsuit, the way we pay a plumber to fix a broken pipe, I suspect we wouldn't have some lawyers looking at some medical malpractice cases as a lottery. The incentive for an attorney is that if he wins, he can expect a good return on his investment. The lawyer is protected in some states; he receives payment for legal expenses, regardless of the jury award. If the jury wants to punish the defendants by making them pay a massive sum, the plaintiff's attorney can make millions. The bigger the award, the better the return. From this perspective, trial lawyers who chase after indoor hygienists look like the source of the problem.

Maybe all three—insurance companies, doctors and lawyers—are at fault.

Let's put aside the war clubs and look for a solution. Bad things are going to happen in medicine, so patients injured or killed out of neglect should have legal recourse. Wrongful death is a physician's nightmare and an unscrupulous attorney's dream.

Back to our parallel to mold cases. Despite our best efforts, everything made by man that's designed to keep out water will eventually fail. Mold will cause human health effects, guaranteed for at least the 24 percent of the population who are genetically susceptible to mold.

The victim's compensation fund for the World Trade Center (WTC) disaster established a formula for deciding what a human life was worth. In cases of wrongful death caused by malpractice, should juries follow the WTC commission guidelines for the value of a human life? Should we compensate someone who lives with a permanent disability more money than someone who's dead, since it costs more to maintain a life in a compromised state? As machine-like as it may sound, there must be a formula to use to decide damages fairly and uniformly for mold and malpractice, rather than let huge jury awards control entire industries. We already have financial formulas in law and government, such as divorce and estate formulas, not to mention IRS tax tables.

What if the attorney who lost a malpractice case were forced to *pay* for the costs of the defense and of the defendant? Sounds like a winning idea until you realize that clients will bear the brunt of extra legal costs. "Loser pays all" might stop some frivolous lawsuits. And how about the pain and suffering that wrongfully accused doctors feel? Shouldn't they receive a check for that from the losing plaintiff attorney?

The problem for the analogy here is that I've seen very sick mold patients lose in court due to bad lawyering, biased judges and poorly informed juries. Making those plaintiffs's pay would add insult to injury, especially if mold has cost them their health and their most precious investment, their home.

And what if all malpractice cases were required to go through a board of experienced, dedicated experts from medicine and law to determine if the malpractice claim had enough merit that the "loser pays for all" proviso would be waived? A reasonable board, balanced between consumer and professional, makes a lot of sense, provided the board is given adequate power. Maryland set up an arbitration board for malpractice claims that's routinely bypassed now by plaintiff attorneys, because they say, "The board is toothless, who cares what the board says?"

I think we need to really look carefully at arbitration boards: they could be the answer to expensive trials and out-of-control jury awards for malpractice and mold claims. So if a physician had numerous verified malpractice payouts, a board with license authority could clarify if the cases are due to the doctor treating difficult cases or due to the malpractice. The board could keep costs of bringing a case low and they could calm the out-of-control jury awards. The loser of a board decision could always appeal in a regular court. Just like the bad doctor should fear malpractice boards, so too should the building or school owner who refuses to make timely repairs and thus makes people ill from biotoxic mold. Sure, the builder should have a right to quickly "cure" a water intrusion problem before litigation can begin. But when he reneges on his obligation, especially when health issues are added, then the penalties for his failure to act should be increased as well.

I wish I could tell you that a reasonable system for mold litigation might be in place next week, but it won't be. It could happen someday, but until we have some decent understanding of mold cases by both the plaintiff and defense, we're going to see legal fights get increasingly rough. Let's face it: Willie Sutton robbed banks because that's where the money was. Now that insurance companies have eliminated most mold coverage, effectively removing the property damage issue, there isn't much money in mold property damage claims. Health issues—that is where the money is going to be in mold cases.

Sutton's Law of Mold: Mold makes people sick and we can prove it.

Shouldn't we be able to present a health case to a "mold board," determine the health problems, treat the condition and resolve the issues? Mold injury isn't rocket science if the involved parties are sincere and the board has reasonable guidelines to follow. We know the symptoms of mold illness, and we can treat the illness, reducing needless suffering. Chronic injury is more complex; defense attorneys, plaintiffs and politicians need to see the clearly defined basics of mold outlined in this book and stop playing expensive litigation games!

Don't forget, the guidelines need to be developed by persons, including physicians, indoor hygienists and engineers who *actually practice* their trade. We don't need any more appointed "expert panels" of experts that don't *treat* patients!

Maybe a board would work, but don't forget lawyers are amazing talkers, skilled debaters and there are so many of them! Any solution to the malpractice issue and mold accountability will only come when the elected attorneys agree on a solution.

The truth about lawsuits in our country is aptly summarized by Judge R. Patrick Hayman, who wryly observed some years ago, "Just look at the litigious nature of our society. At the same time that there is evidence of failure in schools with national trends showing a fall-off in test scores in math and science, we are arguing who is at fault when someone spills a cup of hot coffee in their lap. We have our priorities wrong: Japan trains three times more engineers than lawyers and the U.S. trains three times more lawyers than engineers."

Turn on late-night TV and there will be the one-stop shopping law firms that will handle personal injuries, drug reactions, malpractice claims, challenges to wills, asbestos (got mesothelioma?) and more. Now, the "*more*" often means *mold* claims. That's right, mold, the new asbestos. It's not a stretch to imagine some TV lawyer solemnly intoning over the photos of mold growing everywhere, "Just a few spores can make you sick. Give us a call on our 800 number."

Whether we believe the courtroom is the proper forum for a fight about truth, the reality is that the legal profession guides societal change. Elected lawyers make the laws, they safeguard the laws and they challenge the laws. Just look at the attorney who sues a big company successfully, obtains a verdict big enough to make the mega-corporation wince and force societal change. Don't forget, a big company will look at a $100,000 loss in a legal case as "chump change."

So if an attorney wins big against Big Tobacco, after years of fighting, how much should he be paid? And wouldn't it be nice if the mold attorney, instead of buying a palace in the Hamptons, could be like Robin Hood, taking his percentage of the award and voluntarily giving some back to a public fund that would support mold treatment, school mold removal or the like. The tobacco fund has supported many worthy projects, why not mold cases?

• • •

In this chapter, we're going to look at the legal battles over mold through the eyes of plaintiff attorneys. The next battles in the Mold Wars are being fought in the courts. Societal change in attitudes about mold and public health issues relating to mold illness will only happen if it's forced, regardless of how many academic papers we publish proving that mold makes people sick. When it costs too much to let construction defects hurt people, we'll begin to see modernization of construction techniques. The construction industry is going to wince when juries say they used pro-mold inferior building techniques. With a financial incentive, construction industries will develop the needed technological changes. The engineers who developed the technology that put men on the moon could easily make the modest changes in materials and design required to make buildings that won't be biotoxin greenhouses.

The mold problems we face now are ones that have readily available solutions. Everyone involved would have to give up something in order for all to benefit. Unfortunately, what that means in America, is "forget it." Here come the lawyers, the lobbyists and the politicians, ready to protect their special interests!

When the jury awards for *failure to diagnose and treat* mold illnesses begin, an up-coming version of medical malpractice, doctors will be forced to learn what readers of *Mold Warriors* already know: the illness is as obvious as a steamroller running over your foot. And then we'll see a new phase of malpractice education from our insurance carriers—suddenly, they'll be offering doctors courses on mold illnesses. Similarly, when it costs too much for water intrusion to remain undetected or human illness to be ignored by a big building owner, then we'll see changes in monitoring and repair. Note who's making society and professions change: the attorneys!

Until then, we'll continue to see plaintiff attorneys leading the way in mold cases. Don't forget, if we're involved with anything covered by our laws, we need a lawyer to guide us through the maze of statutes and regulations other lawyers agreed was a good idea. And just like the medical language of the Biotoxin Pathway requires some translation, so too does the legal language.

Expert Witnesses

If you or a loved one is a mold patient who has been injured physically and financially by water damage in a building, what can be done? You can ask the building owner to fix the problem and ask an insurance company to cover your property damages, but only if the damage falls within the terms of your policy.

If the injury were just to property, we wouldn't have the controversy in mold litigation that exists today. The health issues are where the big bucks are in mold cases now. Due to the work of dedicated practitioners and researchers, the medical and legal establishments are slowly accepting the facts that mold causes illness. They have moved from vague, unhelpful medical thinking on mold illness that said "maybe mold causes illness," to the concession that, "chronic respiratory illness is caused by mold." And soon, because of the vast array of powerful data on human illness caused by mold emerging into scientific literature, even the most diehard naysayer will concede that mold causes other symptoms, especially chronic fatigue, joint pain and brain fog.

The change in public awareness of health effects from mold is driven by what medical and mold experts know, and by what they're *allowed* to reveal about what they know in court. Without expert opinion, the controversy over mold boils down to a variation of the playground argument, "Did so," versus "Did not."

Plaintiff: "The mold made me sick. And it's your fault."

Defense: "You aren't sick from mold. And even if the hypothetical illness existed, and we believe it doesn't, it's not our fault unless you can prove it, which we will insist you can't and we don't want the jury to hear your evidence."

As you might expect, the law has a mechanism to accept new theories. And, as you also might expect, there are all kinds of rules attorneys must follow to introduce new theories as valid scientific evidence for the jury to use in deciding a case. The process usually begins with an expert saying that the cause of the adverse event can only be Process A, followed by an opposing expert saying Process A not only couldn't explain the event, but there may be many alternative plausible explanations not yet considered. Furthermore, Process A is unproven and shouldn't be allowed into evidence for consideration by the jury.

Which expert will you believe? The answer is skewed by the conservative, backward-looking nature of our courts. Rules regarding what can be said in court about something new arose long ago in the days when technology made slow, incremental advances. Certainly if new information isn't based on sound science, it shouldn't be given the credence that's given to the designation of "expert" testimony. And it's also certain that if the new testimony concerns health effects from toxic mold, it's likely the defense won't want that testimony allowed into evidence. The traditional fail-safe of the law has been the role of cross-examination.

Columbus Meets the Insurance Company

For an inside look at how the law handles new theories of medicine, like mold illnesses and the health effects caused by biotoxins, let's turn our perspective from a plaintiff's side to the defense. Understanding this may help you in court someday, no matter what side you are on.

Imagine a lawsuit brought in 1493 against a boat builder

because he failed to install restraining ropes to keep the boat from falling off the flat world. All the sailors are dead and can't be found. Clearly, they've fallen off the world; an event that wouldn't have occurred if the boat had restraining ropes, according to accepted beliefs. Now the families want compensation. Although the money won't bring back their loved ones, the law will try to make the families whole. The insurance company will be asked to cover the loss with a "green poultice."

A simple set of defenses might be that the boat didn't fall off the flat world because the world is *round*. There's no proof that the men are dead (maybe they just decided to leave and enjoy the good life in the New World). There's no causal link between the men's disappearance and the absence of restraining ropes. Hence, the lack of restraining ropes has nothing to do with the lamentable loss of life.

The court could refuse to admit any evidence regarding the theory that the world is round because the theory is novel—not generally accepted. The theory might be true, but there are no proponents of that theory other than one sailor, Mr. Columbus. All *accepted* opinions, including those from the prestigious Organization of Exceptional Minds (OEM) and the national opinion leader, the Kingdom-funded Conference of Denial and Consensus Purveyance (CDC), agree that the world is flat. There are no published studies showing the World is round. While Mr. Columbus may have experienced what he *thought* or *believed* was a round world, he is unable to establish a causal relationship between his navigational skills and a circular world. Moreover, he hasn't presented any proof of his theory since he hasn't published his findings or had them reviewed by his peers.

Even worse, his findings, even if published, haven't been validated by anyone else, because his view is tainted by bias—Mr. Columbus is a well-known expatriate who will do anything to get his aberrant views accepted because he has a patent on the "round world" idea, and thus has much to gain by its acceptance.

The plaintiff's expert, who has a list of professional degrees in astronomy, cartography, shipbuilding, and who is also a forensic pathologist specializing in fall-off-the-globe injuries, will quickly point out the clear conflicts of interest. The lawyers for the families will attack Mr. Columbus personally, using any innuendo they can. Didn't Mr. Columbus bring some sort of hallucinogenic leaf back with him on his last journey, something called tobacco? And what about Mr. Columbus' prior treatment with mercury for the sailor's disease? Mr. Columbus clearly relies on anecdotal testimony. His findings are not grounded in sound legal or scientific data. He has not tested his theories in the lab using a control group of flat versus round Worlds. In short, he's a quack and he's not to be believed.

The court can't accept both sides. Will you believe the "discredited" expert with the *novel idea* that the world is round or the well-established expert whose work on the flatness of the Earth is widely recognized by his peers (who are also his sherry drinking buddies, each paid handsome retainers by the same insurance company who has hired them all so many times previously that they wear color coordinated suits to the courthouse)? If the court won't agree with the one astronomy expert, why, there are seven more out in the hall, each waiting to attack Mr. Columbus, too. If the well-respected expert had never been out to sea, would that make any difference? And since Mr. Columbus only cites reports from ancient mariners to bolster his viewpoint, the other side can quickly discredit those reports variously as having been oral and not written; and cast innuendo that the oral reports were made under the influence of mind-altering spirits.

If the issue is simply "who do you believe?" it's for the jury to decide. If the issue is "we aren't even going to let Mr. Columbus appear in court," then we're looking at an expert witness issue. Underlying the argument to exclude Mr. Columbus is the requirement that a *judge* make a decision about the worthiness of Mr. Columbus' science.

Based on what omnipotent power invested in his gavel? How can a judge know enough about all science to make such a call? The law provides a structure for the judge to use, as we'll see.

Now, How Do You Think My Claim That Mold Makes People Sick Plays Out in Courts in 2004?

Columbus was able to prove his point over time, in part because he had the backing of powerful kings and queens who thought that they could make a lot of money using Columbus' idea. (They'd already figured out how to get around his patent —the world was not round, it was oval). His point was eventually proven years later because *he was right*. The views of the opposing experts with no hands-on experience became irrelevant and discredited over time. The length of time required for Columbus' view to become generally accepted, however, points out a flaw in our current legal system regarding expert testimony.

The law went through similar growing pains regarding radiation effects on soldiers who endured above-ground A bomb testing in Nevada; the validity of DNA testing; the health effects of smoking; and the toxin-associated effects of Agent Orange. Maybe Justice really is blind to what's going on in science.

We verified water damage and mold growth and that certain molds make toxins. We verified that toxins can make people sick. But mold growth's ability to cause people in water-damaged buildings to become sick wasn't "accepted" as verified until our research came along.

The economic consequences underlying the link between mold growth and human illness are gigantic. Unfortunately, even though many studies show sick people in sick buildings, and many esteemed professional organizations say it isn't a good idea to inhabit moldy buildings, where's the *proof* that mold exposure *causes* fatigue and cognitive effects and autoimmune disease? Organizations like the Centers for Disease Control and Prevention (CDC) and the prestigious Institute of Medicine (IOM), which is partially funded by the CDC, at least now finally acknowledge the respiratory effects of mold exposure, with sinus congestion and lung problems leading the way. Yet they pull away from saying mold can *cause* much more, citing a lack of published evidence. But then their conclusions are that water damage should be corrected promptly and that living in water-damaged buildings should be avoided.

That statement *sounds* like they're concerned with more than a runny nose, doesn't it?

So why do all the prestigious organizations say it's not a good idea for people to be in moldy buildings? Because they get sick? Then why can't they *say* that being in a moldy building is associated with illness? Association doesn't equal "cause." To show "cause," what's needed is a methodologically sound group of scientific studies to confirm *prospective illness acquisition* in people exposed to water-damaged buildings, with no other explanation for that illness, other than exposure to the indoor air in those buildings.

Fortunately, that's exactly what our studies show repeatedly: mold makes people sick.

• • •

I began gathering answers about mold by treating patients living or working in water-damaged buildings in 1998. All I had then were a few cases of illness that looked like *Pfiesteria*, but weren't. A few experts, like Dr. Ken Hudnell and Dr. Arthur Raines, knew what I was talking about, but not enough to get past the legal label, "novel, unproven," and therefore, not admissible. By June 2002 a few cases had become 103 patients from 43 different buildings, and experts began to take notice.

My friend, Dr. Hudnell, presenting our studies at the International Conference on Neurotoxicology held in Brescia, Italy, made mold a direct cause of extensive illness. And by September,

Hudnell and I were on a *serious* roll. My paper on 156 people from 150 water-damaged buildings was an academic rumble. As the first shot, I presented 11 different elements from the Biotoxin Pathway at the 5th International Conference on Bioaerosols, Fungi, Bacteria, Mycotoxins and Human Health held in Saratoga Springs, NY. Dr. Hudnell presented a second paper on 19 patients at the same conference. Both had to pass expert peer review to be presented. Each paper was published in the winter of 2005, after being approved by still other reviewers. The science of toxic mold illness was strong.

At the Saratoga Springs meeting, I had to stand by the poster that presented my talk. Much to my encouragement, the CDC's Dr. Stephen Redd was checking out the data like a 15-year-old boy looks at a pretty girl. He spent a long time looking at that poster, especially the Biotoxin Pathway material. I wondered what was going on behind his impassive face.

Finally, he said, "There's more data on human health and mold per square inch on this poster than I've ever seen before."

I happily gave him my business card and asked if the CDC would be interested in a presentation. He suggested that he would wait for the Institute of Medicine (IOM) report on mold and human health. But that meant the IOM report wouldn't include my data. Incredible! It was *Pfiesteria* all over again. If you're going to write an all-inclusive report on the world being round, wouldn't you want to talk to those who sailed to the East and returned from the West? And if the report is about human health, wouldn't you talk to patients and physicians alike who had experience with the illness? To hide behind the academic line, "If it hasn't been published, it doesn't exist," denies the time-honored importance of the *clinical practice of medicine*.

Wait for the IOM, why? What would they know *first hand*? They don't treat patients! And who paid that consensus panel for the time they took to say what they wrote? Even scholars who choose for their own narrow reasons to ignore the validity of actual cases of sick patients, can only do that for so long. But with academic papers rolling off the presses, proving that mold illness slams multiple body systems and causes multiple symptoms, including chronic fatigue, cognitive deficits, joint pains, neurologic symptoms and more, pretty soon the opposing opinions will become untenable.

Dr. Shoemaker, Defend Yourself

Can you just hear the defense lawyers relentlessly attacking my work? They can not successfully attack the validity of my treatment results. If they can't attack me, however, the plaintiffs are going to win easily. So here are their wind-up toy, standard attack questions:

"Where are your studies published?"

"We don't think much of the peer review at that Big World Conference."

"And you are *just* a Family Practitioner?"

"Please, in simple layman's terms, could you tell us what biotoxins are?"

Or the really common question, "Have you measured the biotoxin levels in Mrs. Jones? And since you can't measure biotoxin levels, and we have only your word for it, how can you explain to the court why the seven esteemed scientists hired by the insurance company arrayed in their color coordinated suits in the hallway don't agree with you?"

And then the same defense medical experts chime in with the same chorus they have used for years (though with different fill-in-the-blank words now), "We've never heard of the Biotoxin Pathway. MSH? Never heard of it and never used it, either. HLA? Harrumph, harrumph. Mere speculation. And who else has published anything on the subject using your protocols?"

Until the causation data was published, it "didn't exist." Now, with publication and peer-review available, they attack those who did the peer review, saying the journal isn't rigorous enough.

Defense lawyers are creative people. They'll find something else, like the IOM report, to hide behind for a while. And then they'll move on to something else.

Dodd Fisher

One attorney who's followed my biotoxin ideas from practice to publication to acceptance is Dodd Fisher, from Grosse Point, Mich. As early as July 2002, Dodd was talking to me about mold and human health. As legal firms go, solo practitioners like Dodd rarely make a big splash in legal circles. Sure, they might win an occasional big verdict, but as a rule, the investment of time and money needed to bring a mold case is just too great for small legal firms to handle.

Lawyers often end up making a difference in mold cases because they're the last resort for the injured victim. When health issues, especially chronic fatigue and neurologic symptoms develop, all the defendant has to do is say, "Prove it." Without proof of causation of the injury, accepted by other scientists, there's no case. As Dodd said in early 2003:

> "I actually fell into my first mold case by chance. It was a homeowner's issue in Warren (a Detroit suburb). A lady had a tenant there, and we started off by just trying to get the insurance company to pay for the property repairs. The owners were out of town, and they came back and the house had flooded. The water bill showed that probably 12,000 gallons had flooded in during a short period of time. They had a water-based furnace system, and the pilot light went out or something. It was December, the pipes had frozen in the furnace and it had flooded upstairs and downstairs, through the ceilings, and even the ceiling fixtures were full of water.
>
> "Well, the insurance company didn't want to pay for the damages; they tried to treat it as a cosmetic issue. They didn't want to pay for things like moldy mattresses or living expenses, and they didn't want to repair the ceilings. The

homeowner actually repaired one of the ceilings, herself. So they were forced to move back into the house while the repairs were being done. And that place was full of *Stachybotrys*, *Aspergillus*, and *Penicillium*. She went to a doctor, and to the industrial hygienist who worked for the insurance company. He said that the place was safe in December 2000. The industrial hygienist didn't do any air sampling to determine if the place was in fact safe. He merely stated that there was no acceptable standard of safety and that on visual inspection, the home appeared 'safe.'

"The case was filed in January of 2001, and the hygienist kept saying the place was safe, although independent tests showed it was full of black mold, *Stachybotrys*. This was my first homeowner case—and I ran up against a pretty well-known attorney in this field. We had an environmental company go in there and do some testing, and the house was in really bad shape. This particular client was forced to move into a mobile home. The majority of these people don't have the money it takes to survive during the litigation. You know, there's nobody out there to help them. So I try to help them, but I have to identify experts who can bolster the case.

"That's what I like about Dr. Shoemaker in this regard. He's got the wheels turning, and he's actually moving things along. And we're seeing some actual progress. But there are a lot of places where there are wheels turning, and it doesn't seem to be changing things.

"In Michigan, the statutes basically say a landlord has to keep the premises free from unsanitary conditions. But what does that mean, to keep it healthy? Must you repair defects that you know or should have known about? Many defense attorneys in these cases point out that mold is often invisible in walls and ceilings and ubiquitous, and ask how landlords can be responsible for something they can't see, using undefined standards, even though it may be everywhere in a defective building.

"And my response is always that if they're aware of an unsanitary condition—such as a leaky pipe, or something leaking in the bathroom, and they know about it, or there's something wrong with the roof, and they know about it—then they're obligated to remove that hazard. The first line of defense I usually get is that they didn't see it, or they didn't know it was a problem. And the second line of defense is: 'Well, nobody knows what problems mold *does* cause!'

"They usually argue that there are questions about what mold can do to people's health—and how can a landlord be responsible for something that the medical community doesn't even agree about?"

Dodd doesn't hesitate to tell us how he thinks the law is going to evolve.

"I think the next step in Michigan will be a seller's/landlord's disclosure liability. If there's water-damage or mold problems, as landlord you can't say you don't know about it. In California, for instance, it's an actual criminal misdemeanor if you don't disclose a mold issue. In Canada, covering up environmental dangers is a criminal offence. You can cover up mold by putting up paneling or new drywall or paint, but that shows *intent* to hide the problem and that's going to become pretty serious.

"But that's a tough issue—trying to protect yourself from something you're not aware of. That's why you've got to have some accountability from the seller and the landlord. And you can do that by establishing some building codes and developing some law there. Put some teeth in the law, so that it says, 'Okay, Mr. Owner, if you're aware of some problem and you don't disclose it, you can be sanctioned—either by fines or a misdemeanor.' In some states, though, the way the law reads, even with disclosure, there isn't any obligation to repair.

"What we need are the kinds of local mold laws that will require some accountability on the landlord's or sellers' part. But if you're asking, how can a buyer protect himself, I have to wonder: Who can afford to pay $2,000 to do environmental testing before you buy a house or rent an office or take a new job? But it may come to that. I'll tell you this much: If I were going to buy a house today, I'd pay someone to go in and do testing to make sure the place was safe.

"I learned this the hard way. I went to one house that was contaminated half a dozen times, and every time I went in there, I got really sick, and I stayed sick for a couple of days after that. So I don't go to these places anymore, I let the environmental testing companies go there. I don't want to take any chances at this point.

"And I've learned the defense lawyers and the insurance companies won't take you seriously if you haven't lined up some good experts. You have to rely on the inspector—and a lot of the time, you get an inspector who's been referred by a seller's agent. And this is where the inspectors get away with a lot of things, because the inspector has a contract with the purchaser that says their liability is limited to the amount paid for the inspection. Say you've paid $200 for an 'inspection,' about the cost to *drive* to a location in metro Detroit; it's no surprise they don't see much! They don't pick up on the mold intrusion problem in the roof, or a leaking problem in the basement. Or maybe there's some kind of defect causing a water problem. But the inspector's off the hook because of that little 'hold harmless' provision.

"I think I'm very creative in my pursuit of justice, when dealing with water problems. Let's say it's a water-repair issue and it involves an insurance company, and the insurance company brings in their own contractor and cleaning company. At that point, everybody's on the hook, and so I often use an 'agency argument,' by pointing out that the insurance company brought in these people, so they're acting as the company's agent.

"So that's how I pursue these cases. I can line up two or three multiple defendants, and help these claimants as a result—because the more players the merrier, and it means there's a better chance of getting some settlements along the way. You focus on the primary target, which is often the insurance company.

"But to answer the question about how do you protect yourself when buying new property, I think it's very difficult. And I do think we need building codes that will better address these issues. You know, it's nice to have low utility bills in the winter because your building is tightly sealed, but if you have inadequate airflow and inadequate ventilation, all it takes is one moisture issue, and you're going to have a mold problem in the house.

"That's the major problem with new construction—the drywall they're using often allows mold growth—and you're going to see some lawsuits against the drywall companies in the future. So you need to find out what kind of drywall was used on the property you're thinking about buying, and whether or not it included any mold-retardant products in the materials involved. There's a pretty steep learning curve on this.

"But one way you can protect yourself as a purchaser is to make sure somebody inspects the building and makes sure there's no water problem, and that there's adequate ventilation, and that the builder didn't have any prior water problems before you got there. You should also make sure that the builder has been out there for a while, putting up buildings. And if it's a brand-new builder, I'd be a little nervous—because they're not necessarily going to know anything about mold. And they're less likely to understand the complex structural problems that are involved, when it comes to mold.

"So you want a builder who's been around for a while, and who has a good reputation. Watch out for these 'overnight' builders—they probably don't even have insurance for structural defects.

"But I do think this is going to be an increasingly important area of law, starting in the near future. I've already dealt with condo issues, and apartment issues and management company issues. But it's a very creative area of law, as well, and I like challenges of this kind."

The incredible thing is that Fisher has won settlements in legal cases, despite the legal climate in Michigan, which is chilly to mold cases.

"I'd say the key point is for homeowners and renters to 'be aware!' These health issues related to mold are serious. I recently had the first Social Security disability case in the State of Michigan in March of 2002 and it involved neurological damage as the result of exposure to toxic mold. Our application for permanent disability in the State of Michigan was approved.

"This issue is not going to go away. It's a real serious problem, and people should know that doctors like Dr. Shoemaker can use new methods and techniques, combined with traditional processes in medicine based on a differential diagnosis, to determine whether or not an individual is susceptible to mold. We should also remember that the elderly and the children are most susceptible.

"I tell my clients to remain hopeful, because the medical world is learning more about this issue every day, and the legal community is starting to catch up. I think you're going to see a proliferation of cases involving toxic mold and personal injury—to say nothing of property loss—all around the country.

"I had a case where the plaintiff wanted the court to issue a citation against a management company—but the court had to dismiss the citation against them because there are no standards, no codes to deal with the mold issue. So some landlords are getting away with it. In Michigan, they don't have to pay for these damages because of the lack of codes. And until we get those, you're only going to have a few legal settlements here and there, and it won't amount to more than bits and pieces.

"The tragedy is that a lot of mold victims are getting delayed justice—or no justice at all—and it's happening at the cost of their health. It's a real problem, and the lawyer tends to become their savior, because family members don't understand the mold, and the people who complain are often looked at as being crazy. And except for a couple of physicians in Michigan who know how to deal with the problem, the doctors here don't know how to deal with this issue, either. They don't learn about it in medical school so there's a real void. When those doctors *have* to learn about mold, and they will when Dr. Shoemaker's data is in every front page article on mold, then we will see medical change. But until then, the change will come from the courts first.

"I suspect that if this becomes a new field of medical specialization, Dr. Shoemaker will be in the forefront. But that's something down the road. His work is still considered new, even though it's based on established science. What *is* new is that he's helping many mold patients regain their health. Unfortunately, not all. Maybe that's why Melinda Ballard (President of Policyholders of America) said, 'He's still considered a fringe scientist.' But that was in 2002. Still, legal people are arguing about his work. We've got a great deal of self-education to do in this country, when it comes to toxic illnesses caused by mold and water damage, and I'm just glad that Dr. Shoemaker and a few other researchers are out there leading the way."

• • •

Daubert and Frye

As Dodd says, "I need to present medical evidence that will survive the challenge from the defense that the testimony shouldn't be allowed." The courts have tried to set up standards for experts, but there is no law that says one standard applies to all states. The Federal standard is derived from the US Supreme Court ruling in a landmark case, Daubert versus Merrell Dow. States have a collection of standards, including Daubert, and some states use the older Frye standard. Each time one of the standards is used, someone is going to ask, "Does the result make good sense?" Even if the answer is no, which it often is, the lack of logic doesn't change the standard. One reason the law changes very slowly is the "flat world concept"; whereas modern technology means that what we know to be true is changing more rapidly than ever before.

Michigan uses the standard, "As long as the basic methodology and principles employed by an expert to reach a conclusion are sound and create a trustworthy foundation for the conclusion reached, the expert testimony is admitted no matter how novel." In other words, does the process matter? It does, and it should.

So how can a court discern if an "expert opinion" is really expert based on the best knowledge of today or if it is no more substantial than smoke rings? There is much discussion of "Rules or Tests or Standards" that have been set down by the Supreme Court and by the States to determine if the opinion is going to be accepted. Dodd Fisher writes often on the subject of expert testimony and whether or not the expert will be allowed to testify in front of a jury. In one of his papers (note: for simplicity, I have deleted the references from Dodd's academic paper) he states:

"Michigan rules of evidence (MRE) 702 and Federal Rules of Evidence (FRE) 702 provide rules to determine the admissibility of expert opinion testimony. The rules are as follows:

"MRE 702: If the court determines that **recognized** scientific, technical, or other specialized knowledge will **assist the trier of fact** to understand the evidence or to determine a fact in issue, a witness qualified as an expert by knowledge, skill, experience, training, or education, may testify thereto in the form of an opinion or otherwise.

"FRE 702: If scientific, technical, or other specialized knowledge will **assist the trier of fact** to understand the evidence or to determine a fact in issue, a witness qualified as an expert by knowledge, skill, or experience, training, or education, may testify thereto in the form of an opinion or otherwise, if (1) the testimony is based upon **sufficient facts or data,** (2) the testimony is the product of **reliable principles and methods**, and (3) the witness has **applied the principles and methods reliably** to the facts of the case.

"Generally, both the Michigan and the Federal Rule permits a qualified expert to testify if the expert's testimony will help the trier of fact understand the evidence or determine a fact at issue. The purpose of the testimony is to aid the jury in understanding and interpreting complex evidence so that it can be useful to the average lay person.

"For an expert to be qualified, the court, when determining the qualifications of an expert witness must at a minimum evaluate the following:

a. the educational and professional training

b. the area of specialization of the expert

c. the length of time the expert has been engaged in the profession and specialty

d. the relevancy of the expert's testimony

"The Federal courts provide essentially the exact type of qualification for expert testimony.

"Both jurisdictions also consider the following:

a. whether the expert's technique or theory can be or has been tested.

b. whether the technique or theory has been subject to peer review and publication.

c. whether the rate of error of the technique or theory is known when applied.

d. whether there are the existence and maintenance of standards and controls.

e. Whether the technique or theory has been generally accepted in the scientific community.

f. Whether experts are proposing to testify about matters growing naturally and directly out of research conducted independent of the litigation, or whether they have developed their opinions expressly for testifying.

g. Whether the expert has unjustifiably extrapolated from an accepted premise to an unfounded conclusion.

h. Whether the expert has adequately accounted for obvious alternative explanations.

i. Whether the expert is being as careful as he would be in his professional work outside of paid consulting for litigation.

j. Whether the field of expertise claimed by the expert is known to reach reliable results for the type of opinion the expert would give."

A court may admit novel methodologies or forms of scientific evidence only if their proponent establishes that they have achieved general scientific acceptance among impartial and disinterested experts in the field.

The subject matter of an expert's testimony may be scientific, technical, or other specialized knowledge in a recognized area of expertise. If expert testimony is elicited on the basis of a particular scientific technique or theory, that technique or theory must be sufficiently established to have gained general acceptance in its field. This standard is the *Davis/Frye test.* However, after many years of use in the federal courts, *Frye* was rejected in 1993 as the standard for admission of novel scientific evidence in federal cases. In *Daubert v. Merrell Dow Pharmaceuticals,* 509 U.S. 579 (1993), the U.S. Supreme Court ruled that FRE 702 supersedes the strict *Frye* standard. *Daubert* applies a relevancy and reliability approach and employs a balancing of a number of factors listed above.

"The Rules" are straightforward, although you need a directory to see which states do what and why because they differ in how they apply the rules. Some follow Daubert and some follow Frye. Others follow a more insightful or perhaps subjective standard, such as Schreck in Colorado and Haverton in North Carolina, that allow the judge to determine if the process is reliable. After the expert is admitted, let him withstand the time-honored process of cross examination. For the expert that means learning what defense attorneys like to do. For the attorneys are not under oath as the expert is. They need not apply truth and only truth.

To complicate things, within each state the standards for applying whichever doctrine the state espouses are subject to the judge's discretion. Never lose sight of the fact that what happens in a Frye hearing in Florida in a District Court has nothing to do with what happens in a Florida appeals court or in Colorado in a Circuit Court.

And then we have the "Judge is King" theory of standards where little that the jurist says seems to make sense based on any logic except that the Judge is the King of his Court.

There is one standard for a physician to be admitted as a treating doctor and another standard for the physician to be an "expert." But what if the physician is both? The Judge has all the power in making these decisions. If the Judge has a reputation for being conservative, the plaintiff's attorney might try to find a more liberal venue if his case depended on a liberal interpretation of a given statute or opinion on expert testimony. While we would like to think that Justice is blind, the reality is that scientific knowledge is constantly changing and the law seeks to keep up, not always successfully. That isn't the "blindness" we want. What may be unacceptable in court today is likely to be different tomorrow if new information comes forward. As the science of mold advances, there are academic papers being prepared, reviewed, published and being presented at professional gatherings. Justice demands that the courts are not left behind.

To me, Dodd Fisher defines what a mold attorney should be. He knows the issues, he writes articles to show colleagues in his profession about the problems in toxic mold cases, and he helps his clients salvage something from their mold-shattered lives.

Dan Bryson

At the other end of the legal spectrum from the solo practitioner, Dan Bryson is a partner in the North Carolina-based law firm, Lewis and Roberts. In the bigger cases, including class action suits, the "little guy," like Dodd Fisher, either has to pass up the case or try to participate as co-counsel with a larger firm. The paperwork on 350 clients from one apartment complex would bury a smaller firm. Mold cases are slow to develop, take a tremendous amount of organization and staff time, not to mention the staggering costs associated with all the different aspects of a case. Bryson's firm can handle the big cases that Fisher must pass up.

The first time I spoke with Dan was in April 2003. He had the kind of voice that made me feel I could trust him right away, and that's rare for me. Dan was disarmingly pleasant, punctuating his responses with "right, right." As I have subsequently found out, like the TV ad, when Dan talks, others listen. He lectures frequently to other professionals on the "structure of a mold case" and construction defects.

Dan talked about the Daubert issues and standards for expert witness testimony. He was polite, but frank, saying that with all due respect for our group, until we had published papers and established our work in the medical community as a whole, that he probably couldn't use our work in trial. I didn't think I'd hear much from his firm after that conversation, at least until we were better known.

Dodd Fisher's report that I was a "scientist on the fringe of knowledge" was still fresh in my memory. Now Dan Bryson was gently saying something similar.

And his point was well taken—it's one thing to know how to diagnose and treat mold illness and quite another for those in the medical and legal communities to know what you know. Even after publication, there has to be repetition of any new idea, with confirmation of the findings. That's part of the standard. I took some comfort that there were clear precedents in legal cases for new data. If I followed generally accepted methodologies (I did) and applied them to new information in a scientifically acceptable way (I did), then they would assist the "trier of fact" (the jury), and then my material would pass the expert witness test. As the North Carolina Supreme Court (Haverton) said in June 2004, if the expert is "reliable," then don't *make the judge* decide validity of science, let the established mechanism of cross-examination begin. Trust the jury with the intelligence to decide which weight they will apply to conflicting testimony of paid experts. Don't trust the judge with all-seeing omnipotence, especially when his last scientific discussion was in his college biology class.

Dan's firm is swamped with calls from prospective clients. While the cases all might be legitimate, on average, even a big firm like Lewis and Roberts can handle at most one in twenty. Dan suggests that before anyone calls an attorney for help in a mold case, the client be sure that they have seen a doctor, have had substantive environmental testing done and that the statute of limitations hasn't expired. In some states, the patient can't sue their employer for personal injury due to mold, so be prepared and find out first if you can. For example, if you are Pat Romanosky, you might not be able to sue all the parties that have harmed you. Government treats itself well, so if the at-fault organization is government, then it might well have statutory immunity from your claims. This is Bureaucracy as King.

The legal battle must be won before the Mold Wars to be won.

One attorney told me confidentially that I should carefully choose cases, because if I had either poor lawyering or judge, it could effect whether my data was admitted in later cases. Once an adverse precedent has been established in court opinion, succeeding defense attorneys and courts can cite the opinion as part of their reason for keeping evidence from the jury.

This time, we'll win. We have a valid case definition and approach and tons of validated data. We follow the process of science, building on what was known and applying those facts to the unknown using scientific logic. If the methods are sound, the evidence should survive challenges. The jury will decide whether to accept the testimony after hearing all the evidence.

And in that battle our weapons will include: valid case definition; elimination of any other alternative causes; valid methodological approach; and lots of validated data. Given the entrenched opinions of the CDC and the IOM, as well as old, negative comments from the insurance industry about the reality of toxic mold illness, the ultimate victory will come from treating patients successfully, presenting and publishing the findings, and documenting increasingly widespread use of my protocols.

As Dodd and Dan both later told me, the health issue was one they considered part of their duty in a mold case. They would rather have their clients healthy first; considerations of a lawsuit were a distant second.

Frye in Florida, May 2004

Leading the way in patient care and research doesn't equate to winning the legal wars. Sure, as a treating physician, I'd accumulated mountains of data on the mold illnesses of nearly 2000 patients. Would that make me a treating physician or an expert in the eyes of the courts? The question was important in ways that I didn't understand.

If I was allowed to give my opinion as a treating physician using standard procedures in medicine, such as differential diagnosis, then I would not be subject to the Frye challenges in Florida. (Don't forget, each state is different!) I would be allowed to give an opinion to the jury on how the mold caused an illness. If, on the other hand, I were considered an expert, because I had relied on scientific data and not established medical procedure to form my opinion, then I would be subject to Frye and its "generally accepted" test.

The more unusual the case or the more unusual the kind of allegation of mechanism of injury, the more likely the plaintiff will run into the "gatekeeper" of the expert opinion system (aka, the judge). We had just such an experience in Florida in May 2004. The case appeared straightforward.

Mrs. D'Angelos* lived alone in an apartment complex. The ventilation system caused water to condense in her closets, creating the same water intrusion a leaky pipe in the area would cause. Environmental reports documented amplified growth of indoor resident toxigenic fungi: *Stachybotrys, Aspergillus, Chaetomium* and *Penicillium.* She had a progressive deterioration of her health, including cognitive problems. Her series of HMO physicians, each seeing her infrequently, diagnosed Alzheimer's, even though the medical record clearly showed multiple symptoms from multiple body systems well before her cognitive decline began. If you are elderly and have cognitive problems in Florida, the Alzheimer's label is quickly attached, like a Scarlet Letter in indelible ink. No one even considered the cognitive effects of mold exposure—no such notation appeared in her medical records.

Gordon Koegler, one of the solo practitioners partnering with Scott Mager, a principal in his own large firm, asked me to consult on her case with her Florida attending physician to see if she might have an illness caused by mold exposure. Sure enough,

---

*Just before Mold Warriors went to press, the judge in this case recused himself following a motion for recusal alleging bias made by the plaintiff attorneys. The names of the plaintiff and the judge have been changed.

after conducting a differential diagnosis, I concluded that she met the necessary criteria for a case definition of a mold illness. Lab tests, including HLA, MSH, MMP9 and other assays were done, adding strength to my previously made diagnosis. The attending physician decided not to treat her with cholestyramine because she had a problem with constipation (which is often made worse by this toxin-binding drug). Mrs. D'Angelos remained ill. That's tragic, because the constipation from cholestyramine is common and usually responds with simple over-the-counter preparations.

The defense team was familiar with mold cases, having been on the *losing* side in a $14 million verdict (call it Centex) in Martin County, Fla. They weren't interested in being big losers in another big mold case. After two days and eight grueling hours of deposition, the defense decided to challenge my testimony. If I were allowed to testify, they would have an uphill battle on their hands. They challenged the testimony of Dr. Hudnell as well.

If you're like me, when I hear that some guy, guilty as sin, gets off from a conviction because of some court room technicalities, I get ticked. And now, some big bucks attorney is going to keep me from telling the jury *the truth*, based on some technicality? This is totally unacceptable.

• • •

Judge Joshua Jerico had been sitting on the Court bench for 16 years. He would hear the Frye challenge. The bailiff spoke in glowing terms regarding Judge Jerico: "He's at work every day at 8. Always takes the stairs to his office on the 10th floor. Works out everyday from 12-1 and is back on the bench at 1:30. Takes his work real seriously, and he hates to be reversed on appeal," he said. "He's tough, but fair."

The court proceedings looked just like what you would see on TV. Both sets of attorneys made their statements, postured, interrupted and objected like two sets of Greco-Roman wrestlers struggling to stay upright. But it was the judge who had the power in this arena. He wanted the challenge to focus on two issues: 1) is there a link, accepted by the scientific community, between exposure to toxigenic mold and human health? 2) Did the doctor use an accepted method to apply a differential diagnosis in this case?

Judge Jerico considered that the medical testimony in Centex had passed a Frye hearing, but even so, wouldn't just let Dr. Hudnell and me testify solely based on that precedent, holding that the Frye standard was always applied on a case-by-case basis.

And to no one's surprise, the defense disputed whether Mrs. D'Angelos' adverse health effects resulted from exposure to toxigenic fungi. In his testimony to the challenge to his appearance, Dr. Hudnell began the war of dueling medical papers, complete with references from the EPA and multiple additional sources. The piles of papers on each side of the courtroom prompted Judge Jerico to comment, "This case has single-handedly contributed to global warming." Under my breath, I added, "Based on the hot air from the defense."

It seemed clear at the end of the morning session that Dr. Hudnell had held up well under an aggressive attack on him and his use of the Visual Contrast Sensitivity (VCS) test. All we needed was a discussion of differential diagnosis and we would have met the standards outlined under Frye; Scott Mager was upbeat. "I think we got the Judge to grasp the facts. He came around a lot."

I had the afternoon shift in the witness box. Gordon went through the basis for my work, showed the scientific process. The defense tried to say my causation studies weren't peer reviewed (they were, on two occasions, as you've read previously). Then they tried to say my methods were so new, so "novel" that they couldn't be allowed. No, the Judge didn't buy that, either. It's the method, not the application and opinion, that's subject to

Frye. I use traditional methods to collect my data. The data is reproducible and others use the protocols. Gordon's questioning brought out all those points.

We talked in detail about the differential diagnosis. The Judge was adamant that no new tests should have been used. No problem; I was using standard tests that used for years, applied to show new disease after they'd helped to rule out other illnesses.

"The tests I use are all available in a standard laboratory. The tests aren't novel," I said. "The opinion I have in this case is based on my experience as a treating physician. The additional testing *adds expert opinion* to medical certainty as to the cause of the illness, but her history, exposures, absence of confounding diagnoses and multisystem, multisymptom illness gives me reasonable medical certainty. It is the elements of the case definition —the same elements used by the CDC in *Pfiesteria* cases—that are the basis for my opinion."

Gordon went back to the differential diagnosis. Over and over again, he asked how I had ruled out other diagnoses. Are reproducible scientific data parts of a differential diagnosis? Of course.

The events of the next few minutes are unclear to me. Abruptly, the Judge asked me to leave the courtroom. I could see Scott, aggressively gesturing, holding a copy of a recent Florida Appellant Court decision, State Farm versus Johnson, showing that opinion testimony of a treating physician wasn't subject to Frye. I couldn't hear anything, but it didn't *look* good for Mager and Koegler. I was asked to return to the witness box to again review the differential diagnosis I had performed. With all due respect, the work-up of an elderly woman with selective cognitive changes, without across-the-board deterioration over time and improvement with removal from the affected apartment, is not typical of Alzheimer's and could reasonably be discarded as a possible diagnosis. My differential was detailed and thorough. The HMO diagnosis of Alzheimer's was based on superficial medical data, not enough to satisfy a concerned family member from Pocomoke, for sure! Then, out of the blue, I was asked to leave the courtroom again.

What was going on?

More gesturing, more gnashing-of-teeth and stressed expressions. Then it was over. The Judge ruled I hadn't done a differential diagnosis, but that I had applied my novel theory up front. Ken and I were both struck, meaning we could not testify.

Unbelievable. The Florida opinion in Johnson was right there, in front of Judge Jerico but he seemed to be following a Frye hearing the previous month, where a different judge had ruled I hadn't done a differential diagnosis. Just what does, "ruled out all possible confounding diagnoses," actually mean to a judge? Was I excluded because I had relied on reasonable scientific data (as an expert) whereas a "treating physician" had a lower standard of expertise or opinion and I wasn't Mrs. D'Angelos' primary care physician?

Dan Bryson was understandably concerned about D'Angelos. "We need to find a way to cure the apparent judicial error," he said.

Gordon and Scott were livid.

"At least this gives us a solid basis for appeal," said Gordon. "Johnson clearly applies here. We gave him Johnson, referred to Johnson, he can't ignore Johnson. In my opinion, the judge made a reversible error. We need you to do a proffer. Basically, that's a process in which you will be allowed to put into testimony what you would have said had the judge not struck you and Dr. Hudnell. It will be part of the basis for our appeal."

So five days later, the proffer began. Only this time, the Judge said he had a chance to read Johnson from another source and he felt that the Johnson ruling applied to the D'Angelos case and that he wanted a re-hearing on whether I would be allowed to testify, as he concluded the two cases were "indistinguishable."

The defense jumped up and down, feeling that victory had

been snatched from them. A re-hearing was scheduled to argue the points of law.

Hadn't somebody forgotten Mrs. D'Angelos in all these legal proceedings? Where were the patient advocates, saying she hadn't been given a chance to have her cognitive function returned to her? Wouldn't it be easy to show if improvement occurred with cholestyramine that Alzheimer's wasn't the correct diagnosis and that mold was? As it turned out, delaying the court hearing after all the legal maneuvering had a significant benefit. Mrs. D'Angelos was finally treated with cholestyramine and the result was wonderful improvement in her symptoms and abnormal labs. In the end, who cares about the arguments and the lawyers when sick people aren't healed? Mold Warriors care about people.

If my victory in the Mold Wars wasn't going to come from a Frye hearing, this is just the beginning, as legal cases can go on for years. The greater win would be to benefit sickened patients. After all, I am a treating physician. Let the Fishers and Brysons of the world argue over the fine points of law. Let Koegler and Mager deal with the judges and juries and getting expert testimony admitted into Court.

After these few mold cases, I have a different opinion of attorneys. Yes, I agree they make business decisions based on money. So do I; and so does the manager of the local supermarket and the owner of the local hardware store. But money is not the only consideration at work here. There are issues of principle, fairness and justice that motivate attorneys to bring change that forces Congress to enact new laws or to force new interpretation of existing laws.

The mold arguments continue at the cutting edge of law. The forces that will compel Congress to write new laws on mold are analogous to discovering the new aspects of physiology in mold illness—and the consequences—that we'll see in 21st Century Medicine. There's no greater intellectual and personal challenge than to ask new questions and to find answers in what we already know.

Penicillin chrysogenum

CHAPTER 20

Moldy Buildings: It's a Jungle in There

Why Are There So Many Moldy Buildings?

Because there's too much moisture indoors.

That's right, good old H2O. Too much water, more than 60 percent, makes us sick by letting molds grow around us. And the reverse, too little water in our air, less than 30 percent, makes us sick by letting viruses and bacteria grow too easily on or in us. We need moisture for health like Goldilocks needed her porridge: for us, "just right" means between 30–60 percent.

When indoor moisture is excessive and there are no competitors, mold growth is guaranteed. Under amplified mold-growth conditions, people genetically susceptible to mold toxins will become ill and those not susceptible can become ill if exposure is massive or long lasting. Without amplified mold growth indoors, there *won't* be mold-induced human illness. No unnecessary moisture equals no illness.

However, raise the indoor moisture levels to 70–80 percent relative humidity and just watch the Stachy and *Aspergillus* come from nowhere within a few days to grow, as if they were fed a miracle growth fertilizer.

The humidity needs of indoor fungi are well defined and are

supported by a solid scientific body of knowledge. Although much of the research is complex, one basic theme recurs: Indoor toxigenic fungi are capable of manufacturing many different poisons that pose a health threat to those who share living space in water-damaged buildings.

But, they can only do it in the presence of excessive moisture.

The Critical Meaning of Humidity: The Life Blood of Mold

The relationship between indoor humidity and growth of toxic molds is also related to temperature. When we measure the amount of moisture in air, or the humidity, we're referring to the amount of water in the air that can be held in air at a given temperature. Cold air doesn't hold much water, but warm air holds a lot.

To illustrate, just think about the humidity in a typical home in Vermont in January. Cold air outside means not much moisture in that air. When we take the dry outdoor air indoors and heat it, we expand the *capacity* of air to hold water. If that Vermont homeowner doesn't work really hard to add water vapor indoors, the warmed air in his living room in winter will be drier than the Sahara desert.

On the other side of the moisture concern is a home in Florida in July. The air conditioner runs constantly just to keep the temperature at 74 degrees. You can be sure that the AC is working overtime to take moisture out of the air as it's cooled.

There's no doubt that managing these complicated temperature/humidity dynamics requires some engineering expertise. In the U.S., that expertise usually comes in the form of a Heating, Ventilation, and Air Conditioning (HVAC) system built to regulate temperature and the moisture in the indoor air. If the HVAC is designed properly, mold won't be a problem. But where you find mold, you'll often be able to pinpoint a defective HVAC system that failed to control indoor moisture.

In theory, if the HVAC and the building envelope—the walls, windows, doors, roof and floors—are each designed properly, you could have a sieve for a roof, ponds forming in the basement and windows constantly dumping rain indoors, and yet mold *would not grow*. When it works, HVAC protects people from mold illness; but when it doesn't, it's a facilitator for mold. This simple fact tells us something about legal liability and mold: when illness occurs, it's usually someone's fault.

A simple example of HVAC ignorance causing liability can be found when school administrators who don't understand the mold health threat try to save a few dollars by shutting down their school buildings for the summer. September brings an unwelcome surprise. The school is now a lush lawn of colorful molds growing on books and carpets, in air ducts and even under the windows of the principal's office. *Welcome back, kids, but don't worry, a little mold never hurt anyone. We'll clean it up this weekend.*

But the cleanup won't work if mold spores have already migrated throughout the building. And that's exactly what *will* happen, the moment the maintenance folks turn on the AC again. Why? Because an HVAC system is also the most efficient "mold spore distribution" system anyone could ever hope to design!

Indoor temperatures in residences, workplaces and schools-in-session will be around 70 degrees. HVAC is the key controller here. If there are no sources of water intrusion and the HVAC is designed well and working properly, indoor moisture will remain under control and any localized mold outbreaks won't spread. But if the HVAC isn't working correctly, it becomes your worst enemy. Once spores bearing toxins are released into the air as "bioaerosols," airborne bits of biologically active material, the HVAC is no longer the controller; it becomes the distributor.

Miss Kelly Asks for a HEPA Filter at Her School

Miss Kelly was angry. As a long-time public relations staffer at her college, Salisbury University (SU), in Salisbury, Maryland, she was a veteran at making friends and influencing people. But now she was sick, *very* sick from mold exposure, and no one, especially her immediate supervisor, seemed to care. As a matter of fact, no one at the entire university chose to contradict the Department of Human Resources, who continued week after week to insist that mold wasn't a problem. They sent in a worker from the Environmental Safety Department to look around her office. He spent about 30 seconds and left.

Nope, not one employee was sick from mold, the administrators said.

So when Kelly continued to complain about her illness, the visible mold and the musty smells, she became *The Troublemaker.* Like administrators everywhere, troublemakers are pigeonholed into three groups: one person is a "complainer," two are a "conspiracy" and three are a "class action lawsuit."

But Miss Kelly refused to back down.

"They've got the AC running, but it doesn't work. It actually *increases* the humidity!" she told everyone who would listen. "The humidity in Holloway Hall stays at about 80 percent all summer long until the heat comes on, and then it bottoms out at 15 percent. We get the mold from that soggy basement blown up here in the summer, and then in winter we just get the dried-out mold from leftover summer blooms!

"All I want is a humidifier for the dry season and a *de*humidifier in the rainy season, with a HEPA (high-efficiency particulate air) filter to keep me from getting sick from whatever is going on in this building. And Human Resources won't do a damn thing. I don't know why our Environmental Safety department is allowed to leave any mention of 'Safe' in their name. Hey, if I wanted to work with molds, I would have majored in ceramics!"

Well, Why Don't They Fix the HVAC?

There's no good answer. You and I both know the problem is probably going to be on the deadly sin list: pride, greed or sloth; you decide. Some administrators are so intellectually rigid that for them to consider anything other than their bottom line (or with liability imminent, their bottom) is impossible. Costs are more important than people, say the bean counters by their actions. Other bureaucrats are so defensive that they'll argue against any employee who dares to say their workplace makes them ill. But wait. Those administrators will have to live with their ego-driven decisions when they find loyal workers being forced out of employment with heart ailments, brain scars, chronic fatigue and serious cognitive problems. Perhaps one day the bureaucrats will also have to live with legal judgments against them personally.

Why Not Give Her a HEPA Filter Or Let Her Bring in Her Own?

Nope, Human Resources and Salisbury University refused to help Kelly breathe any easier, literally or figuratively. They said no to any use of any filtration device. They absolutely forbid her to bring in her own HEPA filter. Like we saw in Marty Bernstein's case, Kelly had no choice but to seek legal assistance. So she did.

Her only option was to file a claim with Worker's Compensation for the right to bring in her *own* HEPA filter. What a ridiculous situation!

Her hearing was one of a long list of items on the WC docket, but there seemed to be more members of the Salisbury University administration attending than the sum total of lawyers on both sides.

Somebody at Salisbury University was worried.

The outcome was incredible. Kelly's attorney hadn't asked for a formal medical report, so the 5-step repetitive exposure protocol that proved causation of her illness wasn't introduced as evidence. Without proof of her illness, the Commissioner decided there wasn't proof she needed a HEPA filter, and since the school said she didn't need any filter or air protection, he wasn't going to allow her claim.

"We're not going to put up with this silliness," Kelly said. "This opinion stinks. Oh, by the way, Dr. Shoe, my attorney told me that the Worker's Comp lawyer told him that you, Dr. Shoemaker, are some kind of quack, anyway. This isn't right. So now, I don't care how far we have to go, this isn't right. I'll take it to the end."

Kelly just enlisted as a Mold Warrior. When defense attorneys call me names, my position becomes stronger.

Kelly was a member of the double-blinded, placebo-controlled clinical trial demonstrating the validity of the 5-step repetitive exposure protocol. When a judge hears her case with a jury to decide whether or not she could use a HEPA filter at work, she'll have overwhelming scientific data to prove her point. When she wins, she'll get her HEPA filter. Think about it: what reasonable employer is threatened by an employee using a HEPA filter? If she had instead asked for the right to open her egg on the little end, like the Lilliputians in Gulliver's Travels, the case would have been equally absurd.

Part of the reason we have so many moldy buildings is due to owners/administrators who will fight the truth tooth and nail when their pocketbooks and egos are at stake.

Moisture and Mold: Inside the Toxic Jungle

As any good microbiologist will tell you, the battle for survival among indoor resident species of fungi, bacteria and other germs is fascinating to watch. Especially interesting, say the experts, is the way so many of these organisms make biotoxins in order to fight for dominance in their particular environments. Remember, these indoor organisms don't have to face much competition for their place in the food line. In most indoor environments, they have plenty of energy available (beyond the requirements of living and breeding) to make expensive chemical and biological weapons, such as biotoxins. Essentially, toxins are part of the weaponry employed by microorganisms to compete against other microbes; it's a jungle *in* there.

When we find sick patients in a place like Salisbury U's Holloway Hall, we can be sure of one thing: There are toxigenic fungi present, due to amplified growth, due to excessive moisture occurring somewhere in the distribution range of the HVAC. Sick patients are the markers for Sick Buildings.

Another certainty—and this was especially true at Holloway Hall—is that the administrators responsible for the outbreak will insist that *their* testing shows no toxic mold effects and therefore, no one could possibly be sick. Here, we smile wryly, knowing that the repeated battles for truth and justice that you've read so far in *Mold Warriors* will once again be waged against denial and deception. The new venue, of course, will be Holloway Hall.

Wouldn't it make more sense to simply admit there's a moisture problem here and fix it, instead of bringing a legal battle to the front pages of the Daily Times of Salisbury?

Let's face it: Now that medical doctors can precisely identify cases of mold illnesses with screening, then diagnose and treat them successfully, patients need not be denied either justice *or* the possibility of a return to good health. Why don't we simply identify the problem when it occurs, fix it and tell the ignorant, obstructionist bureaucrats to get out of the way!

Here's a simple prescription to ease Kelly's pain:

1. Find the mold problem using a company that's not golfing with the Boss
2. First fix the moisture problem; then pay the remediation company enough to remove the mold—all of it.
3. Find the sick patients; heal them
4. Go back to work

No one at Salisbury University did anything right for Kelly.

While we're at Salisbury University, we'll ask our independent building inspectors to actually do the job correctly instead of "sham testing." Meaning, they will sample not only the boss-designated areas, but also in the *unaffected* areas. Why? Testing for mold means you must look everywhere; mold lurks unseen often.

Of course, finding mold in obvious sites is easy for companies with integrity. Cross the street to sniff around in the Alumni

House, for example. Even in winter, the indoor humidity is over 80 percent, the windows sweat and I'm told the drapes smell like closed-up locker rooms.

Oh, yes, the records at my office clearly show there are sick workers there, too. When a graduate of '48 comes by to visit Alumni House to walk down Memory Lane, why should he or she end up carrying away a bucketful of mold spores as a memento of "good old SU?" And why should the workers in the house remain at risk for mold illness?

OPEC Caused the Moldy Buildings

The most significant danger from energy-saving construction techniques is toxic mold. You can seal a building tighter than a sardine can and save a lot of money on heating and cooling costs. But there is a reason we are so fond of fresh air. Fresh air is as important to health as fresh food. We wouldn't buy moldy oranges in the grocery stores, why would we willingly work in a moldy building? Often the mold problem in buildings can be prevented by simply letting in the fresh air!

Do you remember the energy crisis of the1970s? Back then, the Middle Eastern sheiks in the Organization of Petroleum Exporting Countries (OPEC) made the decision to cut their crude oil production to increase demand and boost prices. Americans were squeezed for fuel as a result. Long lines and long waits for high-priced gasoline prompted a surge of "energy conservation" and "environmentalism" among politicians who were suddenly demanding we develop alternative sources of energy.

Of course, the energy crisis did produce a lot of lip service about solar energy and energy-efficient automobiles, with a few plugs for hydrogen fuel cells and wind energy. In the end, nothing of substance was done then to develop alternative fuel sources. Alternatives, like sealed, moldy, gas-trapped box buildings, were hardly the answer.

Today, we are still dependent on Saudi oil, (it's *déjà vu*, all over again) but what have we learned? For one thing, we've learned how to make toxic buildings.

One lasting feature of our national resolve in the 1970s to solve our dependence on foreign oil was to make buildings frugal, conserving energy in every way possible. But it's now clear that strategy has had an unintended consequence that's been a real killer: indoor amplified mold growth. Mold toxins hurt us daily in ways that silently and stealthily erode our quality of life. Even worse, the damage usually takes place in the dark, away from public scrutiny, at least until *Mold Warriors* came along.

Tragically, the so-called energy-efficient buildings built since the 1970's, and that we *continue* to build almost always have oversized insulation capability, windows that are sealed shut, vapor/air barriers of all sorts and *very* limited air turnover—structural characteristics that are prescriptions now being filled for a public health disaster.

All too often, these structures work to provide a pleasant, unthreatening environment for *fungi*. Without predators, such as other fungi and many bacteria, toxic molds have an ecological situation ripe for explosive expansion of growth. No enemies, ample food, and wonderful growth conditions in a greenhouse called a high rise: we've got our indoor nightmare. We've got a new creature emerging from our polluted environment.

Pollution, after all, is too much of the wrong thing in the wrong place at the wrong time. Put a spoonful of Hershey's chocolate syrup on vanilla ice cream and everyone's happy. Load barrels of the same syrup on a barge, accidentally dump them on a wetland and it's a front-page screamer about the new pollutant.

The Solution to Pollution? Dilution!

Molds growing in buildings are too much of the wrong thing in the wrong place at the wrong time, but the problem here is often a lack of fresh outdoor air to dilute and flush out indoor toxins.

Your Aerosolized Contributions to Indoor Microbes

Although the primary cause of SBS is simply a lack of fresh air that could dilute and flush out indoor biotoxins, we need to remember other negative consequences produced by sealing ourselves inside offices and homes.

We've talked about moisture, the *numero uno* cause of SBS. Take a shower, dry your hair, steam some vegetables or merely exhale, and right away, the water content of the indoor air begins to rise. As you recall, we prefer that indoor humidity remain between 30 and 60 percent. Take too much water out of the air, and the mucus membranes of our airways will dry out and crack, triggering inevitable increases in sinus and lung congestion. Let the moisture in the air rise to 60 percent and watch the fungal spores wake up.

Of course, closed air circulation can lead to the build-up of *other* bioaerosol—health threats. For example, dander from pets, house dust, fumes from the new carpet or the newspaper and computer, all add their volatile components to what we breathe. And some of the worst gases are ones that are odorless.

Even such a simple act as turning on the kitchen exhaust fan while cooking has a surprising drama. As you direct the kitchen air outside, air has to come from somewhere to fill the vacuum. The exhaust fan creates pressure that pushes inside air to the outside; guaranteeing that air from the bowels of the basement and whatever is in it will be sucked upwards. Exhaust fans have a motto: Excelsior! Ever upwards, for the dank components of the fungal-enriched basement air rise up the stairs, like dark ghosts, to travel into your breathing passageways.

Will Attic Exhausts Solve Stale Air Problems?

Some indoor engineers remove stale air by exhausting it from the attic. Yes, install a whole-house fan that exhausts all the hot air into the attic and out into the cool night. Does that make sense?

Unfortunately, these powerful fans are worse than the small kitchen exhaust fans. These fans can lift spores from any resting place, turning them into bioaerosols howitzers. The result: a hellish marriage of spores, volatile organic compounds, dust particles and bioaerosols from your worst nightmare, and that spreads throughout your home. That flood in the basement last spring that gave a mold colony a chance to amplify and manufacture spores, covered with toxins, just contaminated the $40,000 heirloom rug, Grandma's soft sofa and all your clothes. Horrible? You bet.

But it happens every day.

So forget the exhaust fan. What about the ever-present HVAC? Given water intrusion and fungal growth, eventually some spores will settle in the ducts of the HVAC. Most ducts provide a ready supply of food and moisture. It's no surprise that the fungal growth and spore-formation process will continue daily. So instead of the spores coming solely from an area of water intrusion, the spores that poison us can come right out of our air ducts.

These basic biological mechanisms, repeated endlessly throughout our energy-efficient buildings every day, are the agents that cause us to suffer illness caused by exposure to indoor toxigenic fungi.

Fungi are well adapted to grow in the range of temperatures in our homes and workplaces. If a building is more than two stories tall, we can add the "stack effect" to the mold equation. Heat rises, as do air currents from the sub-basement and the basement. So as the air in stacked floors rises, the suspended particles, including mold spores, from the sub-floors and basements are distributed upward. The typical office setting will also likely have the added burden of volatile materials from copy machines, faxes, and the heated components from computers and other office machines.

But the contamination doesn't end when the workday is over. Soon the janitors arrive to add their air-polluting load of cleaning chemicals, everything from quaternary ammonium compounds (quats) to bleaches, disinfectants and oxidizing agents. Indeed, the air people breathe in a typical commercial building has more than enough entries to start an organic chemistry lab!

The bottom line is clear: we've polluted our indoor environment with compounds and organisms that can make us sick in exchange for saving dollars on the heating and air conditioning bill. Irrational? Wasteful? Injurious to health? Yes.

We Created the Toxin Formers and Asked Them to Stay for Supper

Have we created a new race of indoor-dwelling, toxin-making fungi that make us sick? Are we the guilty parties changing the pace of evolution?

The challenge presented to public health officials by the surging SBS epidemic is further complicated by the ugly fact that some of the invader-microbes are entirely new on the scene. Increasingly, the evidence shows that chemical pollution and our modern lifestyles are combining to create genetically altered strains of indoor-dwelling, toxin-forming fungi, and these new life forms are making us sick. Of course, it wouldn't be the first time that humans have contributed heavily to the emergence of disease agents which then emerge from the muck to occupy newly created "niches" in our ecosystem.

Some examples:

Pfiesteria: This algae-like dinoflagellate, nicknamed the "Cell from Hell," made the Baltimore and Washington, DC Six O'clock News throughout the autumn of 1997. The coverage began after blooms of *Pfiesteria* swarmed into many estuaries of the Chesapeake Bay, especially the Pocomoke (Md.) River, killing millions of fish and making hundreds of people sick.

As you might expect, the first response of politicians in Maryland and Virginia was to deny the problem. Later, when forced to acknowledge the incontrovertible truth that these new toxin-formers caused illness in people, the actual cause of the blooms in the water was misidentified as "excessive levels of nutrients from chicken manure, washed in from local farms."

Ludicrous? I'll say. But it wasn't until 2002 that the actual cause —environmental chemicals, and not just nutrients—emerged when the Environmental Protection Agency (EPA) reported that *Pfiesteria* wasn't killed by copper (added to chicken feed to prevent spoilage). We helped those dinoflagellates grow by adding dithiocarbamates (DTC) fungicides to the water, too. DTC helps copper kill the predators of the toxin-forming stage of *Pfiesteria* so they can *grow unchecked* (for more on this microbiological mystery story, see my previous books, *Pfiesteria: Crossing Dark Water*, and *Desperation Medicine* for an in-depth review of the *Pfiesteria* disaster). When the predators of *Pfiesteria* were wiped out, *Pfiesteria* growth was amplified.

What happens to the rabbit population when the foxes are killed? What happens to the deer population when wolves are killed or hunting isn't allowed?

Cylindrospermopsis: These algae didn't appear in the U.S. until 1995, when it was discovered in Florida. These blue-green algae were well known to health officials in Australia and Brazil, where it was already a public health menace. Toxins from a *cylindrospermopsis* bloom killed nearly 60 patients at a dialysis center in Brazil, after the reservoir supplying water to the facility treated the water with copper sulfate, killing the algae but releasing a flood of the algae's lethal toxins into the water.

In Florida, the invading algae rapidly took over the lakes in the St. Johns River watershed, including Lake Griffin and Lake Apopka. Why? Because it was no longer killed by the chemicals we'd used in the past. "Cylindro" spawned out of control, without being slowed down by the three factors that usually limit algae blooms: nitrogen, phosphorus and sunlight. Even more ominously, the Cell from Hell II was resistant to copper and to benomyl, a fungicide widely used in nearby fields and nurseries.

Aphanomycetes: This mild-mannered fungus never harmed a single fish—not until it mutated in the early 1990s. Almost overnight, the fungus became resistant to copper and who knows what else, killing fish in estuaries from Florida to Virginia. In order to study the fungus, U.S. Geological Survey researchers had to add copper to the lab aquaria, since the other "wild" fungi in water would normally out-compete *Aphanomycetes*. Copper killed the wild-type fungi, but not the *Aphanomycetes*.

Have We Altered Populations of Indoor Fungi the Same Way?

To answer that question, let's take a moment to look at a gallon or two of household paint armed with the latest "mildew-cide" in it. At Home Depot, they have it as a paint option. And while we're at it, let's also take a closer look at the new wall coverings that are bacteria and mildew resistant.

At first glance, these advances in wall coverings will save us money on maintenance costs, right? Think again. These new chemical-laced products actually create *additional* work for those who clean office buildings, condominiums or homes. Toxic molds adapt to the new environment by transforming themselves into resistant strains no longer killed by the fungicide in the wall covering.

What makes these commercial products so low-maintenance? Agents of natural selection: fungicides, biocides and bactericides. But they won't be low maintenance for long! When the mold starts *growing on the fungicide-laced vinyl wallpaper*, is it a new strain that's resistant to the chemical that used to kill it? You bet it is.

Benomyl, like other fungicides, was added to commercial paint beginning in the 1970s. Did the use of fungicides by the paint industry create a monster race of resistant organisms, the ones that usually are toxin-formers? Or was it the use of fungicides in sunscreens, furniture fabrics and the like that selected for our small group of indoor toxin-formers? Fungi are resilient creatures, having survived countless millennia. It should come as no surprise that every time we attack fungi, whether with anti-fungal medications in a leukemia patient, or simply with plant disease-controlling chemicals, some resistant strain of fungus is going to emerge. Maybe the answer to the problem of toxic in-door fungi is to bring in some shovelsful of dirt to use the micro-bial battle for survival to roust out the reigning fungi.

Nothing we've done so far to eradicate toxin-formers from our buildings has worked, so maybe this idea isn't so far-fetched after all.

For some perspective, let's go back to the *Pfiesteria* outbreak for a moment. What did it teach us about toxin-formers? When we see toxin-formers emerging from chemical-induced selective processes, what's usually included is a group of organisms that can laugh at what mankind throws at them, all the while using their toxin weapons to keep the competition at bay.

Example: Dr. Chin Yang has already noticed some fungi resis-tant to copper in his labs at P and K Microbiology. Other re-searchers have identified fungi that now bloom wildly in cleaning solutions *containing* quaternary ammonium compounds, the same chemicals used to disinfect moldy surfaces!

Perhaps the best example of people becoming ill from chemi-cally altered organisms comes from a group sickened following exposure to *Aspergillus fumigatus* in a Maryland chicken hatchery.

For years, this hatchery had relied on a fungicide, Clinifarm (enilconazole), specifically approved and labeled by the US De-partment of Agriculture (USDA) for use in hatcheries to control this *Aspergillus* species. Left unchecked, *Aspergillus fumigatus* kills newly hatched chicks.

The enilconazole worked for a while. But then some workers at the hatchery began to come to Pocomoke for medical treatment of a newly contracted biotoxin-induced illness. One savvy patient even brought in a series of Petri dishes, each growing multiple colonies of a black, filamentous fungus. P and K Microbiology definitively identified the wooly Petri beast as *Aspergillus fumigatus*, and there it was, growing *on* the Clinifarm, using the fungicide as a source of food! Amazingly, it was living off the chemical meant to kill it. Imagine a human child growing up and flourishing living on a diet of cyanide, arsenic and hemlock.

In short: The fungi had adapted to repetitive use of the same fungicide by mutating and becoming resistant to it. Among a billion organisms are the very rare fungi that have the genes to survive the killing effects of the fungicide. They survive, breed and give rise to a new race of resistant fungi. They quickly take over the turf vacated by the fungi killed by the selection agent.

This is the same situation we see in bacteria: if you want to select for the growth of a new strain of a germ, *Staphylococcus aureus*, a common germ found on our skin, for example, keep treating it with the same kind of antibiotic. It won't be too long before the only staphylococci still growing will be ones *not* killed by the antibiotic. They breed happily, and then spread, until everyone is asking, "Where did the resistant strain of Staph *come* from?"

Another critical example of how health-threatening organisms survive our chemical attacks comes from the history of benomyl, another member of the same family of fungicides as enilconazole. The product has a long history of use in fungal research labs. It was used commercially as a systemic fungicide worldwide, but was eventually pulled from the market after allegations that it caused crop damage. Even before the fungicide was blamed for killing plants, however, resistance of fungi to the fungicide was seen.

Benomyl prevents cell division by interfering with the cell division process in fungi in which replicating chromosomes are separated, providing a full copy of the chromosomes of the fungus to each of the new "daughter" cells (the same way some drugs widely used to effectively treat breast cancer work). The plant "feeds" a dose of benomyl to any fungus trying to digest the plant.

During later cell division, the resulting distortions created by the fungicide poisons deprive the daughter cells of the right amount of chromosomal material and they can't live. But occasionally, one of the offspring gets *less than a lethal dose of deformed chromosomes,* ultimately giving rise to new generations of mutant, benomyl-resistant daughter cells. Put them into an environment with benomyl (and that's incredibly common), and the growth of the mutated strain will take off.

Natural selection moves quickly when mutations begin.

So what does this have to do with human beings? I can tell you that the emergence of "selected-for" mutations from cultures treated with this product is so certain that in academic centers, including the National Institutes of Health (NIH), benomyl is widely used as an agent to cause fungal mutations. Just the other day, in fact, two researchers from the NIH, each ill from mold growing in their condominium in Washington, DC, visited my office for an SBS evaluation. Both said they used benomyl in their genomic research and both confirmed first hand what I had read.

The mutagenic effects of benomyl on fungi are easy to observe. Just pour some of the pesticide into a test-broth with fungi and watch the new fungi "come out of the mud." Labs use benomyl at concentrations of 5–10 parts per billion (ppb) to demonstrate the mutagenic effects, the same levels seen in wall coverings.

Someone needs to fund studies that would conclusively demonstrate the cause-and-effect relationships between the chemical biocides in our wall preps and the subsequent human illness due to the biocide-resistant strains of the fungi. I don't think the wall-covering industry or the paint industry will be asking to do those studies, but I am certain that industry already knows what a biocide in wall covering does to fungi. I still ask, "How can we ignore the lessons of biocide use *outdoors,* creating changes in populations of fungi, when they can logically be applied to changes in populations of *indoor* fungi?"

I believe we've created mutated species of indoor fungi as a direct result of our shortsighted efforts to reduce building costs. Those fungi loom as a major public health problem.

Destroying the Argument That Ubiquitous Molds Cause Illness

During many years of treating SBS patients, I've become quite familiar with the arguments used by defense lawyers. Their job seems hopeless at first: the defendant didn't clean up rotting walls or didn't even have the human decency to permit use of a HEPA filter. And no honest person could ignore the evidence that mold makes people sick. But defense attorneys are resourceful. They love to mimic a magician—you know; make you look away from the reality of what actually happened. I call them smokescreen artists.

I try to be patient, but if they suggest that mold *anywhere* could cause the illness, I struggle to keep from popping a brain vein. They are using a deliberate distortion. Yes, smokescreen is wrong.

So here are a couple of the standard junk science attorney arguments used to keep buildings and schools sick.

Bogus argument No. 1: Molds are ubiquitous. They are everywhere, so how can a mold—just a little mold growing in a building—be responsible for chronic, debilitating illness and cognitive impairments?

Bogus argument No. 2: Molds in *outdoor* environments don't hurt anyone, do they? Some of them are similar to those found *indoors*, but you don't hear about gardeners, lumber workers and mulch workers getting sick. Why aren't those workers sick, the same way that indoor workers are? If mold is everywhere, therefore, the building alone couldn't be the culprit.

To put these arguments to rest, let's talk "ecology," the branch of science that studies the interactions of organisms in a given environment. There are thousands of genera of fungi and we are constantly being exposed to many of them.

But inside a water-damaged building, we find there are *relatively few* types of mold because the environment will only provide ideal growth conditions for a few genera and species in that habitat. Toxigenic molds like their indoor weather balmy, with afternoon showers and plenty of food. Not all outdoor organisms flourish in those conditions. The indoor organisms lack the same competition for food, moisture, cover and reproductive space. There are few predators indoors also. With abundant resources, limited competition and ideal conditions for breeding, the indoor mold populations explode. We call that growth "amplification." With amplification comes manufacture of metabolically "expensive" molecules, like toxins, used to safeguard the young spores.

Outdoors, though, the species diversity is huge. The blunt reality is that it's total war between species with only the biggest, baddest, most bodacious or fastest-reproducing, surviving. The greater the diversity of species that are present in a given ecosystem, the more controls are exerted on any runaway species. In fact, some species often release compounds, called "fungistats," into the environment, slowing growth of fungi.

With such battling just to survive the day, molds rarely are able to spend extra energy to make toxins. Amplified growth, therefore, is controlled by species diversity in an ecosystem. Without amplification, toxin production is markedly reduced. All resources are devoted to survival.

With this simple principle in mind, let's go back the two bogus arguments from the defense attorney. Are the molds "growing everywhere" growing in an amplified manner? *No.* Is mold growth in soils amplified in the face of thousands of nematodes, bacteria, other fungi and other organisms in each gram of soil? *No.* Are molds growing indoors in an amplified manner? *Yes.* So, the argument about the ubiquitous fungi causing illness, proven to be false by our repetitive exposure protocols, *is not related* to the amplified conditions inside water-damaged buildings, with relatively few species of fungi ever routinely found and reported in environmental samples.

The diversity of wild populations of microorganisms is good. Restricting fungi to only a few species, as is the case indoors, guarantees the potential for toxin formation by a few dominant organisms. Why do we rarely see other genera in indoor samples, other than *Alternaria, Cladosporium, Stachybotrys, Penicillium, Aspergillus, Acremonium, Trichoderma* and *Chaetomium*? Why don't we see thousands of species growing indoors? *These few are the ones able to survive in the unique habitats called homes, work and schools.*

Construction Defects = Mold-Spawning Moisture

Now that we've established that toxic mold can—and does—make people sick, let's look at the vulnerabilities in the buildings in your life and the lives of your loved ones. Most patients who have been injured by mold have faced the same culprit: water intrusion.

And the cause of the intrusion? In most cases, the mold-spawning moisture is a result of construction defects or even design. Sloppy construction, usually fueled by the need to build as quickly as possible, increasing profit-per-unit time, leads the list of sources of indoor moisture.

New house starts continue to drive the U.S. economy, whether the hurriedly mass-produced structures wind up crammed into every buildable square inch in Southern California; Atlanta, Georgia; the coasts of Florida; or in once-bucolic locales such as tiny Chestertown, Maryland. Land values drive sales of new ground for construction. Every time you drive past an old farm, surrounded by a string of new homes, you know that the farm was subdivided and sold, lot-by-lot.

Land that can be used for new construction is valuable, indeed.

Any real estate expert will tell you that "tear down" of many outdated buildings is the established practice in older communities, as the land is worth far more than the existing structure. It's easier and cheaper for the developers to tear down the old structure and put up a new one, recycling the land, than it is to find unbuilt land.

The availability of land for that dream home is far more limited than demand. On Barnegat Bay, N.J., for example, the demand for construction sites remains extraordinary. A postage stamp-sized lot can easily command $400,000 and more. And why not, when the market is jammed with buyers eager to escape metropolitan New York each weekend, year round?

You don't need a realtor's license to understand that when demand for new housing is high, *the time to complete construction* becomes the most important step controlling profits for developers. "Fast construction" usually means shoddy construction, and that often means water and mold.

The second most important factor in the profit equation is labor costs. As the demand for skilled labor grows, it gets harder to find enough skilled tradesmen. The pool of skilled workers in any given area is limited. Given the massive demand for masons, roofers, carpenters, plumbers and electricians, among many other tradesmen, hourly wages naturally go up.

Given this reality, is it any wonder that "saving money on labor costs" becomes a builder's goal? An unskilled laborer who can do plumbing and roofing, not to mention planting some bushes around the front of the house, for example, is going to get a lot of jobs because he is cheaper than an experienced plumber or roofer.

Guess what? Construction defects occur everywhere, regardless of whether the building is an inexpensive subdivision or a $200 million Miami Beach high-rise. These mistakes are occurring with increasing frequency.

Whether you let the water in via an incorrectly installed picture window on the St. Martin's River in Maryland or by defective HVAC coils in that expensive high rise in Los Angeles, that one defect can cause the whole structure to become contaminated.

Judging by the patients I see, it's simple construction mistakes —loose plumbing connections, inadequate flashing on gables, gaping duct-work in crawlspaces, improperly controlled condensation drains from air handlers and incorrectly installed HVAC equipment—that are the most common errors that lead to water intrusion. With all the concern about building wraps that block air intrusion and vapor barriers to control water movement, we've forgotten that the barrier's effectiveness is reduced when 260,000 nails affixing the siding each make a hole!

At the end of the day, the building envelope—the walls, windows, doors, roof and floors—designed by the architect is only as good as the work done by the $8.00 per hour plumber's apprentice, or a supervisor's hurried inspection of the plumbing connections before the walls are sealed up.

Have you noticed that developers are building homes just about everywhere? This lack of buildable land creates increased flaws in design and construction. Developers force-fit homes into land unsuited for building; they are asking for water problems.

There is no "one formula" for building homes that fits all terrains. Flaws in design, say building a home at the bottom of a hill and not the top, independent of cut-rate construction, certainly will ruin the work of even the finest craftsmanship. When a house foundation is cut into the side of a hill, the ground water pressure on the upside of the foundation will create the constant potential for water intrusion through porous materials such as brick, concrete and block. Ground water intrusion into basements, crawlspaces, or any structure below grade is going to happen. Water will flow downhill.

There's no doubt that correcting the flow of ground water takes a lot of land, time and thus add costs. In most cases, after the developer's made a substantial investment in time and materials, and then finds out he has to construct swales or French drains—special features that limit ground water from entering a home—he simply decides not to build them. It's not his home!

Basements—better named as "mold caverns"—are often used for living areas, with a recreation room and a wet bar right beside the partition for the washer and dryer. For the homeowner, it's like having a "free room." Add some seepage from the in-ground side of the walkout room ... and some more from the carpets on the porous concrete slab ... and a little more from the paneling over the porous concrete block wall, and is it any wonder that mold thrives in basements?

Porous becomes "Poor us."

It gets worse. Without any extra land per home, where does the HVAC equipment and ductwork usually go? Out of sight in the crawlspace, where no one can see the flexible ductwork sagging in the middle, with water condensing and puddling in the low spots. Just add a few spores to the ducts from the crawlspace, entering through a joint in the ductwork that isn't as tight as it should be, and the contamination is off to the races, using the HVAC as the universal distributor.

And how about the popular second floor rooms over garages? They might be luxurious master bedrooms or bonus rooms. Yet in the garage below, the concrete slab has water wicking up through the concrete, providing cover for mold growth. And what if the garage door was open during the sudden rainstorm last night? Where does the water go?

The extra room, complete with its Jacuzzi, walk-in closets and bay windows, gets an *un*healthy dose of spores every time a door in the room is opened, creating negative pressure, sucking air and spores from the garage below through every air leak in the room above.

If the extra room was roofed in before the concrete for the slab floor of the garage had a chance to dry out, look out. It takes *months* for concrete to dry. If ground water forces moisture into the slab, drying is slowed and the concrete becomes a moisture generator! Where does the moisture go? Straight into the walls. Buildings with metal siding can become a condensation factory capable of supporting many happy molds.

Could John Q. Public Be the Cause of His Own Soggy Building?

You bet. Failure of routine maintenance comprises a big source of mold buildings.

Anything made by man will fail eventually. Plumbing fixtures and pipes break. Caulk dries out and flashing corrodes. Sump pumps burn up. Unfortunately, Forces of Nature won't call "time out" when something man-made is installed. While we call deterioration inevitable, we don't often consider, what is the life expectancy of appliances, gutters and flat roofs? After 15 years in one place, why am I upset that there are new failures in mechanical systems designed to last 12 years?

Every time I see a teacher from another flat-roofed school with water intrusion from the leaky roof, the same question comes up: How much would effective roof-maintenance cost the School Board? What will we do differently to take care of our schools when we see the legal, health and learning problems that mold growth creates? How many of the limited public school dollars go to routine maintenance when there isn't enough money to buy new books or fund the art and music teachers?

The real question is: what is acceptable suffering? Is it really an acceptable school "cost" to have large numbers of teachers and students regularly ill?

Mold Loves It When People Cut Corners on Maintenance.

Just look at the costs of not taking care of water. A simple example: when we built our ponds along the wooded wetlands of the Pocomoke River, we had to install "water control structures." Wetlands construction is tricky, and artificially constructed ponds have a natural tendency to drain, returning the water to Nature.

The storm surge from the occasional torrential rains here can rip out dams. Water will follow any defect in the dam, whether caused by a tree root or a muskrat, eroding the soils of the dams. Instead of watching the ponds wash into the Pocomoke, a lot of routine maintenance is required.

The same idea applies to homes and buildings. There *never* should be a gutter dumping water into a basement or a condensation pan overflowing *anywhere*. Take 10 minutes to check. Are the water control structures in good shape? Is the caulking old? Are flashings around windows and doors intact? Does ground water flood into the basement? How about the integrity of your foundation or your slab-on-grade? This is especially important if you're anywhere near the water. If you don't have a water maintenance chore checklist, start one.

Mold Loves It When People Don't Take the Time to Prevent Water Intrusion

My suggestion, after treating several thousand mold victims: When there's a water emergency that's not the result of lack of preventive maintenance, fix it promptly. Mold growth can be a problem in 48 hours after water intrudes. When a water heater bursts, flooding the office building, hire a cleaning outfit right away. Don't wait for mold to grow only to react with horror two days later when it's too late. Don't wait for an insurance adjuster to give you the go-ahead; get the water under control before the mold begins to make you and your loved ones sick.

Mold Loves It When People Postpone Taking Care of Immediate Needs

One more thing about moldy homes and buildings: These days, insurance companies are quick to say, "Not responsible." Instead of arguing about liability, and paying all the remediation and repair fees, not to mention the legal costs, the insurance company usually will have written exclusion for water damage, pollution or mold. Why fight an insurance company's latest approach at cost reduction? Only you will put the needs of the "insured" first. Insurance companies have one goal: Minimizing losses. Think about it though: with a little prevention, would there be any horrific loss?

Now that we can diagnose and treat mold-induced health problems efficiently in most people, before the cascades of adverse biological effects become irreversible, we should be able to begin reducing the high medical costs associated with mold litigation.

And if you realize you have a biotoxin "hit," we can identify you as an affected patient. The chances are you can be made whole following our screening procedures. Forget all the expensive litigation and worthless diagnoses of "ghost illnesses" like fibromyalgia or depression. Get diagnosed and get treated.

A potential problem, given that most building owners' insurance companies would rather fight than settle is that mold-savvy physicians might end up expanding the potential pool of potential litigants. That would create a shortage of plaintiff's attorneys.

Somehow, I don't think that shortage will last very long.

CHAPTER 21

To Build the Safe House

Why Don't We Protect the Children?

Craig Horsman's medical complaint seemed ordinary. His left heel hurt when he walked. Forget trying to run after work, hence the office visit.

For an avid surfer, runner and all around athlete like Craig to skip part of his day off to visit the doctor, well, whatever was wrong probably wasn't routine. Over the years, Craig had always found time to feed his passion for riding the waves and running the roads. His law practice came second. Now his pain took away his passions.

His office visit was no different than the scene in countless doctor offices every day. In three minutes his diagnosis is clear: Achilles tendonitis. And in less than six minutes, a physician might diagnose overuse, order an X-Ray (we always order X-Rays on lawyers), and begin treatment with stretching and meds. See you later, next six minute case. If you aren't much better soon, we'll send you to an orthopedist (and let him worry about your law degree).

This time, though, the inflamed heel cord was the least of his problems. He *hadn't* been overusing his calf muscles, so where did the pain really come from? If only physicians asked the few extra questions to show that Craig's tendonitis could be due to inflammation, initiated by environmentally acquired biotoxins.

Yes, mold illness often causes inflammation in the tissue between muscle and tendon (called the "enthesium;" Craig's illness is called an enthesopathy).

He'd been working extra hard and spending long hours at the office, so his fatigue wasn't a surprise. Others might see he had too much work and immediately call his problem stress (caused by that big court case and all the time away from the beach). As can be the case, that knee-jerk assumption would be wrong.

Maybe it was just stress, he himself had thought, causing the headaches and difficulty with concentration. The shortness of breath and cough had to be because he hadn't been working out. How about his short-term memory? Poor, but just more stress. The weight gain? Surely it was due to his recent inactivity, but 12 pounds was a lot in just four weeks. And wasn't the abdominal pain just irritable bowel syndrome, like his secretary had? But what about the night sweats and sensitivity to bright light?

Maybe it was time to get a medical opinion. So many symptoms appearing all at once from so many assumed causes didn't make a lot of sense.

Part of the clinical benefit I have from working decades in solo practice primary care is that I am often like one of the family. The doc in a HMO, or the practitioner in a large group often will never know his patients as anything more than a name in a chart. For me, I know what my patients look like and how they think even before I walk into the exam room. Just like countless other family physicians, sometimes I know an illness by the change in the patient's appearance.

Craig just didn't look right. Oh, sure, he didn't show much difference physically, but he was not himself.

His problems began when he started working on the old kitchen in the back part of his house. Craig wanted to open up the kitchen area, featuring the 200 year-old walk-in fireplace, and renovate the two adjacent bedrooms before he put the house on the market. After he opened the first wall cavity and he found the heavy black and green garden of mold, it was time for a change in plan.

A microscopic identification of his sample found *Penicillium* and *Aspergillus*. He was in trouble.

How long had the lush growth of *Penicillium* mold been going on? And who knew when the *Aspergillus* started growing under the flooring and behind the walls?

He tore out the old drywall and ripped up the cupped floorboards. He used a painter's mask to guard against inhaling minute spores (not good enough, Craig) and even bleached a few of the contaminated boards. Instead of effective negative pressure, he just had a small exhaust fan blowing toxin-laden air outside. He sealed off the area with plastic to make sure no mold spores went elsewhere, but he was still heavily exposed.

That is why he got sick with those 15 health symptoms—the tendon problem was only one of them. By the way, the same inflammatory process caused by mold and other biotoxins often goes on at other joints. Biotoxin inflammation at the elbow is often misdiagnosed as tennis elbow, and the knee, misdiagnosed as jumper's knee, bursitis, laterally dislocating patella, arthritis or whatever.

By now, the observant reader already knows what Craig's evaluation will show—his VCS was positive, HLA was mold susceptible, MMP9 sky high and MSH dead low. He took CSM, with prompt reduction of his MMP9 and symptoms, together with improvement in VCS and MSH. Perhaps of greater significance before treatment for Craig was an elevated level of his PAI-1, an alarming problem for a guy who's had coronary bypass surgery. Even that level fell with CSM alone during his treatment.

Don't get me wrong, not everyone with unexplained tendon pain will actually have a systemic illness due to mold exposure. But, if the symptom says there *should* be overuse, but there *isn't* overuse, question the diagnosis. Asking a few more questions about Craig's symptoms took about 60 extra seconds of physician time.

Craig's presenting symptom began the diagnostic process. Ignoring the rest of his problems didn't end his underlying illness.

The impact of exposure to toxic mold and his health problems wasn't lost on Craig.

"I'm still going to sell this place," he said. "But I have to be concerned about who might move in. This is a great house, in a good location, with good schools and low taxes. If I do say so, that back section of the house will be a big plus when it's cleaned up. Someone will buy it fast and I might even have several buyers bidding for it. I have an obligation to make the house safe.

"Once I get this house remediated, I'm really looking forward to building my new home. I have a lot of suggestions from my architect friends about how to make the perfect house. For now, I've got the chlorine dioxide running to kill the mold and I've got the old portion of the house isolated from the rest. We've got the remediation pros on board and so far, the environmental results look good. So even though the old portion of the home was a problem, it's safe now. I just couldn't stand it if the new owner set up the back rooms for the kids and then they got sick like I did."

• • •

What Would You Do If Your Kids Became Ill?

Ophthalmologist Ricky Jones MD knew all about mold too, after his house became a "Castle in the Virginia Black Mold Forest." His dream home was beautiful, just the spot for his bride, Joan and their two children. Enclosed in synthetic stucco, their home featured a gorgeous floor plan and a stunning view.

But the stucco didn't create an adequate barrier to water intrusion. When the mold started growing inside and Joan got sick, her illness was blamed on Lyme disease "reappearing" after it had been successfully treated years before. Ricky thought a 100-hour workweek was a vacation, so when his fatigue appeared, it made sense that he'd be tired. Tall and slender, with wingspan three inches longer than height, Ricky blamed the rest of his symptoms on the usual made-up reasons: overwork, stress, allergy, fibromyalgia and the others used so often by mold patients to rationalize their illnesses.

The Jones's two children, five-year-old Elizabeth and three-year-old Benjamin, were sick, too. But don't little kids always have a runny nose and cough? Evaluating young children is different than adults. Because the neurologic function of contrast sensitivity usually develops between ages six and seven, we can't use one of our main diagnostic tools in younger patients. Because both children were exposed and had no other sources of symptoms, we could move directly to the secondary levels of the case definition of Sick Building Syndrome. And the second tier requires that a substantial amount of blood testing. Adults often have 15 vials of blood sent for analysis. How many vials are required for a minimum evaluation of a child?

Part of that answer comes from a study of 100 children with mold illness seen at my Pocomoke clinic, compared to 50 healthy children. Most of the labs we need can be drawn with just a few tubes of blood. HLA genotypes are just as important in kids as in their parents, so check the HLA. MSH deficiency is vitally important and we check that, too. MMP9 elevation is rarely present in kids, but when it is, it's the single most dominant finding in the lab array. Always check MMP9. Pituitary hormone abnormalities are nearly impossible to analyze in children when puberty is still years away, so we skip the androgens, ADH/osmolality and ACTH/cortisol.

Another critical diagnostic test for children is the presence of autoantibodies to gliadin and myelin basic protein. By themselves, these antibodies, rarely found in unexposed children, won't diagnose mold illness. But in combination with HLA, MSH and MMP9, the antibodies point solidly to mold illness. So do we have a new definition for mold illness, based on availability of blood for testing? Sure enough, if children have symptoms, exposure, no confounders, HLA susceptibility to mold and MSH deficiency, that's enough to diagnose with 90 percent confidence. Adding antigliadin or myelin basic protein antibody positivity makes the diagnosis positive greater than 99 percent of the time.

By the way, the anticardiolipin antibodies mentioned in Chapter 4 are only occasionally found in young mold illness patients and almost never found in healthy kids. When you find the cardiolipins, however, it's time to be worried because those antibodies aren't the friends you want for your children.

Back when Ricky and Joan first came for care, I didn't have all those facts about kids sick from mold, so I couldn't have prepared Ricky for the possibility that autoantibody formation was as serious and as common as it is. When the HLA, MSH and autoantibody test results came back on Elizabeth and Benjamin, I think Ricky and I were both stunned.

It was bad enough that he himself had so many autoantibodies —that wasn't any surprise, based on his genotype and body habitus. We knew he was quite ill and Joan too, but now they had to see their active, intelligent children facing a life with antibodies to cardiolipin, gliadin and myelin basic protein? What did these findings mean? Maybe treatment would correct the problem.

Treatment of the mold illness with CSM makes a big difference in adults, with levels of autoantibodies often falling, sometimes to undetectable levels. Joan didn't have a long wingspan. As expected, she didn't have the autoantibodies. Long-armed Ricky certainly saw a decline in his autoantibodies with treatment. And the kids? Both Elizabeth and Benjamin showed remarkable improvement in symptoms following treatment with cholestyramine. What would their autoantibodies do?

How long would the Pandora's Box of attacking autoantibodies remain open? And would the opening of Pandora's Box by mold exposure just once be enough to give the children a lifetime of active autoantibodies?

• • •

Ricky went through the five-step repetitive exposure protocol to prove that his home made him sick. He was adamant that he, and not his family, be the one re-exposed to the poisonous indoor air of their home without protective cholestyramine. The children would be protected at all costs. He got deathly ill from his re-exposure, but got out in time to start a new life.

I think finding the autoantibodies common in high wingspan adults with mold exposure shook Ricky deeply. Look at his lively daughter, Elizabeth, with her antigliadin antibodies, anticardiolipin antibodies and the antibodies to myelin basic protein: What would she be like five years from now? And Benjamin, with his antigliadins? Don't forget that any food, especially the gluten-containing foods, these kids eat that brings gliadins into the intestine will also bring them into the bloodstream. Gliadin, acting as a foreign protein, will bring antibodies and cytokines surging into blood, further attacking MSH pathways and predisposing the kids to lower VEGF and all the other complications of the Biotoxin Pathway.

Put the child on Marty Clarke's diet, right?

Think about it. What would the kids eat when their classmates at elementary school had pizza, spaghetti, hash browns, muffins, biscuits or bread? The truth is that we have no data on kids and autoantibodies following exposure to toxigenic molds. We know a lot about an illness called celiac disease that has proof of long-term complications from gliadin consumption. But mold patients don't have celiac disease—what will happen to them?

When in doubt, protect the kids from mold attacks before the autoantibody assaults begin.

How about Ricky himself? Would the myelin destroying antibodies affect his work in his practice? If he were going to stay in his home, how could he make it safe?

• • •

Knowing that people with mold-susceptible genes comprise about 25 percent of the total population, how many other kids were like these children? How many kids had homes with water leaks or schools with flat roofs and water intrusion? How many were walking autoimmune time bombs like Elizabeth and Benjamin?

And the parents of these children with mold-induced gliadin antibodies would have no way of knowing that the normal American diet could be poisonous to their children. Until now, there was little reason to test normal kids for unusual autoantibodies.

• • •

Here we have two different situations: Craig will make his home safe by repairing it, but will then leave it. Ricky won't move; he'll fix his home and protect it from further damage initiated by unwelcome water. He had no interest in seeing mold growth setting off a predictable string of metabolic and immune complications again. To make his home safe, Ricky would take the crash course in water ingress, condensation, air filtration and modern HVAC techniques we'll meet later on.

Location, Location, Location

Some concepts in mold prevention seem obvious, but they're often ignored in home construction and home repair. For example: water follows gravity. Mold follows water. Therefore, avoid having gravity drive water into your foundation. Avoid condensation in the external wall cavity such that the water runs down to the foundation or slab. If water contacts your foundation materials, it can wick upwards, defying gravity. Don't let water into your understructure! If your home-site sits at the top of a hill, with drainage away from you on four sides, great. For Craig's new home, site work should be done *before* building a foundation.

Like Craig, most of us don't have such a desirable location. If you're going to build into a slope the "walkout basement" idea will be high on your idea-list. Cutting out a chunk of soil from the hill, basically to slide in your home like a boat into a docking site, means that you're asking your foundation to face ground water pressure on three sides, with the uphill side under the greatest pressure. If you do nothing to protect your basement walls from water intrusion, you will have water in your basement.

If you are committed to this site and this design for the site, before you lay the first concrete block, ask your foundation crew to dig out an *extra six feet* on the uphill and the slope sides. That extra void can be used in a variety of ways to direct the downhill flow of ground water away from your basement walls. If you fill the void with pebbles surrounding perforated pipes, you can collect the downstream flow into conduits and let it flow away from your walls safely.

You could use sloping sides on each side of your void, making a berm, using gravity to push the water to each side of your house. If you fill the berm with pebbles, they can be covered with mesh as you get close to the top of the uphill grade, so that you can put a nice thick bed of flowers and shallow rooted shrubs over the void. (You don't want to plant trees or use the void area for vehicle traffic, though, because the weight of the vehicle could crush your protective berm and it wouldn't work right.) You can make your new foundation "perk" safely.

Be sure to block or divert the uphill surface water away from your foundation as well. You can make terraces or swales that will ease the pressure from the potential torrent of water into your basement walls. Sometimes building a pond (goldfish or water lilies anyone?) uphill, long and not too wide, can provide you with year-round beauty as well as protection from surface water intrusion.

Don't even think of building at the bottom of a hill. We call those sites, "birdbaths." Building where water must collect equals mold growth. Trying to keep water out from the bottomland just can't work, no matter how many sump pumps, French drains and elaborate attempts to protect basements from water intrusion you work into your plan. At the bottom of a hill, you'll have moisture, and any man-made protection from water intrusion into those basements is going to fail. And those pitiful attempts to move water out of a basement after it's intruded just never work well enough. And no, putting waterproof paint on an interior wall in a basement won't eliminate any potential for mold growth.

If your site is level, look at where the water will flow in heavy rainstorms or snowstorms. There has to be a downstream site where the water goes naturally. When I hear about mold growth from downspouts that funneled water into a basement, I'm especially saddened. They could've used the same gutter to protect their home site by attaching a diversion pipe to the downspout to carry the water 20 feet from the foundation!

Now your new home is safe from surface runoff in storm conditions and from ground water pressure. What about water from above; in other words, where do you channel the surface water from your roof? This issue sounds like a simple siding and gutter question, but problems in this area show up frequently in the homes of sick patients.

"The gutter was clogged," "The gutter directed water inside the siding or into the basement," the patients said. In each case, there was a simple maintenance failure. It's difficult to say that a gutter dumping water in your basement is anyone's fault but your own. If someone else is required to do the gutter maintenance, double-check that it gets done. It's so much easier to prevent mold illness from starting in the first place.

• • •

Craig is going to have a basement surrounded by rapidly draining soils, with perforated pipe installed and directed into his "Nature Pools." His neighbors will marvel at how his home site supplies free flowing water, even in the dead cold of winter, providing a unique pond habitat on either side of his home. His ponds are small, but they are a continuous source of pleasure.

Craig isn't stopping with simply diverting water away from his basement. What if something goes wrong and he has a sudden rush of water into his home? He will try to avoid water intrusion, but he has a new idea to use. He is going to adapt commercial drying technology for home use!

Talk to experts like Tom Desmond about drying out big buildings. He's done the job for Commercial Drying Technologies from his home in Toms River, NJ to Guam. The guts of his system are a "giant hair dryer," called Dri-Max. Basically, he sets up his huge machine to exhaust air out of a flooded area. The negative pressure pulls drier air in, moving 1000 cubic feet per minute. The wet air is forced over special desiccant containing grids, letting water attach to the silica gel inside. When the gel is saturated, Tom simply reactivates the silica (no different than from those little desiccant sacks you see in cameras and medicines) by heating it. When combined with dehumidifiers to take out water and HEPA filters to take out the spores of early blooming molds, Tom can stabilize a flooded area quickly, before mold can ramp up its growth rate.

Craig laughs, "If any water dares drip into my basement, I'll blow it out or grab it before any mold can use it as a thirst quencher."

If you're going to build without a basement, knowing that going without one eliminates a major mold source, all my comments about water under your home still apply. Any crawlspace created by a foundation must be ventilated. Some experts suggest the crawlspace be treated just like an indoor room. Simply leaving a few small vents open year round (it only takes a little bit of air pressure to ventilate a crawlspace) makes sense if there's a mechanism to force air through the entire crawlspace. There are a whole series of joist-mounted fans, new to the marketplace, which can accomplish this job quietly for a minimal cost for electricity.

Just about every new home with a crawlspace will have some plastic put down over the exposed soil. When the plastic is put in haphazardly, I cringe. The builder is guaranteeing the potential for mold growth. The purpose of the plastic is to be a vapor barrier. Just like in a concrete slab foundation, you have to *stabilize* the vapor barrier. Spread it out evenly and then cover it with about 4–6 inches of gravel. Secure a "lip" of the plastic at least 6 inches up on the perimeter walls and seal the edges. Don't forget, the dry-looking soil in a crawlspace remains a constant source of moisture intrusion.

Never put any HVAC equipment in a crawlspace. Don't even think that you can save some backyard space by putting the bulky air handler unit in a wet area under your house. I can't tell you the number of mold patients I've seen whose sole exposure to mold was from spores blown at them from their crawlspace.

Those flexible HVAC ducts? They are disasters. *Forget them.* Flexible means they can sag, creating pools of condensation in the duct itself. Make sure ductwork is securely fastened around intake and outflow from all air-handling equipment. Just a little air leak can cause trouble, with outdoor spores intruding into the hospitable environment indoors.

If thick beds of mulch surround your home site, make sure surface water drains away from your foundation. More than one person has been affected by poorly drained foundation plantings that became a source for indoor seeding of mold.

Let's assume you took care of water pressure from the soils around your home. What kind or roof, entryway, gable or porch construction will you use? When I drive through new housing developments, it looks like one neighbor is trying to outdo the next by having more gables, more elaborate entryways and more changes in rooflines. How about making your building a little simpler and safer? Every fancy addition increases the chance of building error and water intrusion.

Every time you break into a sidewall for a window or roof deck, there will be the potential for flashing failure. Water loves to wick up under siding like a siphon, travel along structural members and follow gravity down into your living spaces. Make sure the flashing is put in correctly, and all potential passageways for water are blocked.

If Craig can pick his construction site and control the ventilation before the home is built, what can Ricky do to protect Benjamin and Elizabeth? Don't forget that they're now primed for mold illness. He can't protect them from the mold in the school, the shopping center, the church or on airplanes, but he hopes to make their living environment safe. He'll start by filtering the air entering and recirculating inside his home and then exhaust air out of his home using positive pressure. Thus he will blow air out of his basement, pulling outdoor air into his home through filter devices.

Some Filtration Basics

So how can you separate the tiny indoor mold spores from the vast expanse of indoor air? Toxin removal from circulating air is a major problem in any mold-prevention project. Fortunately, modern filtration technology has made massive strides that come to the home owner/building manager's rescue. Maybe we should listen to some words of advice from Wells Shoemaker, a chemical engineer who first encountered filtration 60 years ago. He has spent his career in this industry, and recently received a Lifetime Achievement Award from the American Filtration and Separations Society.

I have known this expert for all of my 53 years. He's my father.

"Ritchie, you and others have pointed out the dangers of toxins made by fungi and other organisms in the circulating air stream. No surprise, HVAC equipment suppliers have great interest in this emerging need. But given the new interest in selling technology, the homeowner and building manager must be wary.

"While there's a mix of well-designed systems that work well, be sure to tell your readers that there are also many over-hyped simplistic devices, with no controlling authority or technical society to make sure they can back up all their claims for efficacy. I've seen mold being described as a vegetable or a bacteria in some ads! There are claims extolling wondrous devices that have minimal supporting evidence.

"Pity the manager of a building or the homeowner who realizes that there is a mold problem and something has to be done, but doesn't know where to start. Perhaps we can offer some hints to control mold and mold spores.

"The main objective of the manager/homeowner: Eliminate or control those spores! You stop mold by stopping spores. But, when mold is there, you can choose several approaches: kill the spores or remove them, or both.

"The manager must assume the spore is viable, and must be killed so that it will not spread the contamination. Techniques include UV radiation and ozone. Both are readily available in the industry. The UV radiation will cancel the viability of the spore. Ozone is a radical treatment, and not suitable for the average householder because of possible rare side effects of lung toxicity and nerve damage, but can be used when major mold and spore population is present.

"An alternative is to physically remove the spores. This approach eliminates the danger posed by non-viable spores that are still bearing toxins. Electrostatic precipitators have been used to remove spores. In this system, a positive electrical charge is applied to a plate or a grid of wires and the spores in airflow over the plate/grid. Zap! Like as a commercial bug killer, the spores are removed. Later the collected particles will be carefully removed for disposal"

Note: some experts, like Dr. Harriet Amman, toxicologist for the State of Washington worry that the electrostatic charge releases toxins from spores and that the precipitator simply collects leftover debris.

"Devices with negative ion generators and air ionizers have been advertised heavily, and must have some benefits, but I haven't seen the engineering results that support the product claims. Too bad there isn't a 'Consumers Union' for all these systems to which one could turn for unbiased comparisons. Certainly each has pros and cons. I'd like to see some data on maintenance needs, service life and efficiencies. One item missing from ads is any mention of performance certification or comparisons.

"No performance data are given for these devices. Can you imagine a drug company offering a new medicine with just qualitative puffery? The FDA would have a fit! This industry isn't yet regulated, so maybe the comparison isn't fair. If the devices are as good as we are led to believe, let's see the facts.

"I see the industry evaluating and recommending various combinations of these systems. You can find units, especially small in-room devices that put together filtration, UV and electrostatics even adding activated carbon for odor removal. As you know, when spores are created there are also volatile organic chemicals emitted that cause the undesirable odor from mold. That's where the carbon helps.

"There is much developing technology that's becoming available. Ingenuity is at work in this trade. With the increasing awareness of mold problems from the health and legal standpoint, as well as explanations such as provided by *Mold Warriors*, we should expect continued performance improvement and cost reduction. From the homeowner's standpoint, you may well see a requirement for any house being built or resold to obtain a certificate for no-mold, similar to termite and radon inspection, with the issuer to be liable if mold is discovered. That will be a hot potato!"

Filtration Courses for Upperclassmen

"Now let's get into the specifics of filtration techniques that are undoubtedly the most popular to control mold spores. The system has the following basic components:

1. Duct work for directing the air
2. Blower for moving the air
3. Suitable hardware for holding the filter in place
4. The filter medium itself, referred to as the 'filter.'

"There can be upgrades that complement the system, like a differential pressure gauge to indicate when the filter needs to be replaced; or a humidifier and a dehumidifier, mostly for the comfort of the occupants and control of extremes of humidity, not for spore removal.

"The ductwork and the blower are usually undersized, especially if the system was constructed years ago, when the knowledge was less developed. The result is insufficient air exchange, and less efficient removal of particles and spores. Mold problems were not recognized, although no doubt there were health problems that the local physicians didn't realize. Go back to the dank and drippy dungeons and stale air, putrid prisons of yore—no air circulation, but certainly mold!

"Dirty ducts are a source of many problems as mold spores accumulate. In the darkness, the humidity, and the dust, the spores create more mold, the mold creates still more spores, and the air moves them into the rooms. I have a friend who had a profitable company just devoted to duct cleaning, and he couldn't keep up with the orders from schools. There are many such duct-cleaning companies; even a trade association!

"The devices for holding the filter in the system are critical, as bypass and leakage have been found to be the rule, not the exception. Maintenance is critical, and a high quality of labor is required, ensuring that the filters are inserted snugly and carefully, and then inspected regularly.

"Now let's focus on the most important element of the system —the filter itself. The filter is made from a fibrous filter medium with wide variations in construction available. The first filters used cotton fibers, now replaced by glass and today's synthetic fibers, formed into sheets using paper-making and textile techniques. The fibers are individualized and formed into a random arrangement, then bonded together and supported in a suitable device.

"Molten glass can be extruded and air attenuated for household filters. Some have wondered if small glass break-offs might be a health hazard, but the producers insist there is no problem. One well-known technique applies a sticky adhesive to the fibers to increase trapping and removal.

"One valuable modification in the filters is the use of electret fibers—synthetics that have been treated during production to provide an in-built electrostatic charge. I use this type in my HVAC, perhaps you do too. Criticisms have been made that such electret fibers lose their advantages in high humidity or when too much debris has been picked up, thus insulating the charge. So perhaps in some settings the filters should be changed more frequently.

"There are many kinds and many suppliers of these filters, as the potential market continues to increase. The removal of cigarette smoke, pet dander, dust mites, bacteria, pollen and allergens has become an important health issue. Very retentive filters are needed in 'clean rooms,' popular in pharmaceutical and electronic applications. The total industry is huge; there are many trade associations and fortunately, many research programs in the suppliers and universities. Mold spore filtration is probably a minor area dollar-wise now, but watch it grow!

"The action between the particles and the fibers is the same during the removal process. Here's how these filters work. When it's new, it has high airflow and is reasonably retentive. As air with particles, including spores, is forced through the filter, four mechanisms may take place simultaneously.

1. Some particles are trapped on the incoming surface—'cake filtration'
2. Smaller particles work their way through the maze of fibers and get caught between them—'direct interception'
3. Some are bound to the fiber surfaces by 'hydrogen bonding,' e.g. Van der Waals forces
4. Some particles pass through all the fibers.

"The particle removal can be viewed as an incremental process through successive fiber layers. A percentage of particles are removed in the first layer, more in the next layer, then the next layer. And so on. Thus a 2-inch filter for your home unit works better than a 1-inch. It's difficult to find a 100%, or absolute, air filter. HEPA states 99.7%, so it's not absolute!

"As the process continues:

Particles accumulate on and in the filter
The resistance to flow increases
The pressure drop across the filter medium increases
The total flow of air decreases
The efficiency of particle removal is increased.

"Selecting the filter, especially for the large volume installations, brings an interesting compromise. First, if you want a high-efficiency removal filter, you have to sacrifice rate of flow. Next, if you want to come up with a lower cost of filtration, another compromise has to be made between filtration performance and total cost. There's no free lunch.

"As an example, you can use a filter of twice the thickness and weight, it lasts longer, maybe more than twice as long, depending on the nature of the particles. It will also save the expense of having the operator do the regular filter changing, which can be considerable when there are huge banks of filters.

"The filtration industry is able to set some particle removal (not spores) standards for air filters—the Minimum Efficiency Reporting Value (MERV) ratings. They run from 4, the lowest, to 16, the ULPA filters (You don't want them for spores, they're for super-clean rooms.) The HEPA's are 14, with the classical standard being the 99.7 percent removal of 0.3-micron particles.

"Consider that 0.3 microns size. This page is probably about 4 mils thick, or the equivalent of 300 particles of that size stacked on top of each other. Now, if the HEPA is that good on 0.3 micron particles, why isn't it perfect for catching spores, with their diameters ranging from 1.5 to 20 microns? Well, it should be, but several things happen. There's a phenomena called grow-through in which spores or bacteria captured on a filter surface continue to live, breed, and gradually work their way to the back side of the filter, and from there are whisked by the breeze into the room supposedly protected.

"HEPA filters are being widely and wisely used to improve indoor air quality. As you know, there's a lot more in the air besides spores that can cause health problems. Particulates such as bacteria, pollen, actinomycetes, viruses, other allergens, plus vapors from the mold, cooking odors, plasticizers, emanations from insulation, aerosols, all can be a problem.

"One thing people need to understand is that the safety of building occupants is really not a new idea. You recall a paper I gave in 1988 at a filtration conference that discussed the 'sick building syndrome' and was greeted with 'ho-hum.' It's more pertinent now I'm sure.

"Filtration in HVAC initially was motivated by the desire to keep the cooling coils free from dust to reduce electricity costs, not to protect the occupants!

"Coming back to our building manager, he is making sure that the filter is properly installed in the hardware. Studies have been done of large installations by collecting small test samples downstream of filters, and the bypass has been substantial. The building manager should be familiar with the filter system; he should check it out frequently, look at pressure gauges and at the filters themselves. The manager needs to check up on his operator's accuracy in inserting the filters and ask to have the factory man come in for an inspection. Get a mold expert to test for spores before and after the filter, and see what kind of job the filter is doing.

"Cost is extremely important for large systems, especially when it's suggested that you're pumping spores through your system. Go through the math of figuring out filtration cost. Get an estimate of electricity cost, filter replacement cost, labor to install and inspect filters, depreciation of the entire system. Then if you're asked to use a more retentive filter, with a higher MERV, get some data on life and purchase cost and redo your math. And, most importantly, be prepared to defend your system's operating design and costs, because there will be the inevitable bean-counters who wonder why your costs are so high?" You can give them a table of data comparing retentive quality versus operating cost. Be sure to have a footnote suggesting typical legal fees for defending mold damage suits."

• • •

"OK, Dad, I think I've got it: the rate of filtration is balanced by the efficiency of filtration. You can't have both increased rate and increased efficiency. And better filtration is more expensive. So, as a consumer, how do you have to decide what you want?"

"Ritch, just as you don't want your patients shopping for prescription drugs without a physician's OK, so it is with air filtration. It's not a job for the local hardware store. There really are many competent companies who will do your planning for you. ASHRAE is an outstanding national organization, and as you know there are many other fine trade associations with an interest in spore filtration. By the way, your prospective supplier ought to be a member. If not, ask him why. On your computer you can go to Google.com, type in 'air filtration mold spores' and you will have a huge library available, with names of major suppliers.

"First question that will arise is what mold spores do you have? The size of the spores varies considerably with the species. You need to have somebody check your mold and tell you which species you have, and you likely will have a mixture of the relatively few species of mold we see indoors. When your air filter supplier knows what he has to filter, he can tell you what retention level, or MERV, is needed. Remember that the HEPA filters are so retentive, MERV 14, that they might not be suitable. They could be too expensive or too slow flowing.

"One major supplier, Camfil Farr, shows on its promotional literature the following sizes of spore particles from various molds, and the MERV suggested as the initial estimate for the filter:

Phialophora	1.5 microns	13 MERV
Acremonium	2.5	9
Penicillium	3.3	7
Aspergillus	3.5	6
Stachybotrys	5.7	6
Cladosporium	9	6
Epicoccum	20	6

"You no doubt have noticed more and more ads on TV and the print media for systems for providing better indoor air in homes. But you probably haven't seen many ads for schools and other big buildings. The considerations should be the same whether you're the homeowner or the building manager—you want the best product for the money, and you want to buy a product that works from a technical standpoint the way it's advertised.

"So, how does Craig or Ricky avoid getting cheated, or sickened, from inferior products? Simple answer: do your homework; check several potential suppliers and their references. Be sure to ask how they will evaluate the success of the system they propose to install, and what happens if spores are not retained. Don't believe them if they guarantee 100% removal!

"Ritch, you asked if there's a need to develop an understanding of filtration science if you're desirous of reducing mold spores and thereby reducing toxin exposure. That's a subject that always invites discussion. Do you care whether spore control is accomplished by direct interception, by hydrogen bonding on fiber surfaces, by cake filtration, or by electrostatic charge? Do you want to consider the incremental dynamics of particle removal in successive mini layer of the filter fibers, and variations with time? Chances are you won't want to get into the subtle theories, but rather just want to insure reduction in spore count. When you are at the mercy of a filter salesman you may hear a wide variety of explanations of filter mechanisms. The major suppliers do have a firm handle on these theories, and their salesmen are usually well prepared to review them. If you go to a large retail store, however, you may get a wide variety of pitches, provided you can ever find a salesman. Sometimes knowing a little about filtration can be a dangerous thing, as the science is complex. But, Ritch, your Biotoxin Pathway is complex, too. The reason you want patients to know the science of toxin illness is no different from having them know enough about indoor air filtration: they gain strength through knowledge.

"Don't rule out techniques such as UV, alone or in combination. Perhaps ozone can meet its potential for home use. Look at all possibilities; I'm sure there will be many coming into the market shortly.

"This should be an intriguing industry to follow in the coming months. The recent court decisions seem to indicate that judges consider the health implications of mold to be dangerous. Your studies have shown that biotoxin disease from mold can devastate those with HLA problems.

"Now the industry will have to be preparing to offer improved spore removal systems. Schools, offices, arenas, restaurants, gyms, factories will be preparing to recognize the risk/reward factor of providing proper mold management compared to doing nothing and keeping fingers crossed. The medical profession will be preparing to diagnose biotoxin disease from mold and taking steps for treatment of the patient, while pointing a finger at the mold makers. The lawyers will have ammunition to support plaintiffs. Net result: the danger of mold will be reduced.

"Product development for spore control will come. You asked once about possibilities for treating fiber surfaces with agents for capturing and killing old spores. There has been activity in the past for applying bactericides to filters, perhaps sporicides will follow. That's an intriguing idea. It will be interesting to see how the technologies of engineering and medicine can be melded to improve the quality of life of students, occupants of buildings, and families. When this happens, perhaps the waiting rooms of all the family practice docs won't be so crowded with mold patients!"

• • •

Dr. Jones had considered most of Wells' thoughts previously. He installed a variable speed HVAC unit that would continuously monitor the pressure in ducts, and compensate for changes of pressure caused by an in-line electronic filter in the air returning to the HVAC unit. He installed a UV light unit around the coils of the HVAC to damage bioaerosols. Rather than put the UV unit in-line, say in a large duct, in which the air would have little "dwell time" around the UV, he protects his unit from being a toxin-production source.

He adds a media filter on the upstream side of air delivery, knowing that this unit would be rapidly clogged if it were on the downstream side. He added a humidity device that will either add water vapor to air or reduce it through condensation. If he wants 38% humidity, he simply pushes a button and bingo; the machine will do the job. This equipment requires quite a bit of maintenance, as it can become a source of water intrusion in humid months of the year.

His air intake into the house is through ducts attached to the basement windows. The basement itself is under positive pressure, continuously exhausting upstairs air out through the lower levels. The air intake through the window next to the exhaust creates a counter-current air exchange that helps save on heating/cooling costs. The intake air is passed through filters, dehumidifying units and several sequentially smaller pore size HEPA filters.

Sounds complicated and expensive. But he's a successful ophthalmologist, so he can afford to protect his kids that way, right?

They'll stay on toxin binding medication as a fail-safe plan in case some toxin escapes his elaborate filtration system.

Given what we know about mold illness in children, we don't have a choice when mold has harmed their brains or their immune responses. Who can afford not to take care of their kids?

• • •

To make an absurd, hypothetical point, let's take Ricky's example of protection from mold a little further. If we're serious about protecting young brains and bodies, shouldn't we have some mechanism to protect the indoor air in our schools? If the expense of having all these filtration units is too great, couldn't a school district set up special schools for those 25 percent of the total student population at risk for mold illness due to their genotypes? Those schools, call them "bubble schools," could be staffed by mold susceptible teachers, too. Our society permits the use of "magnet schools," why not have a few bubble schools, too? Cash-strapped school districts could avoid lawsuits for mold illness by segregating people based on their genes. The schools would make HLA testing mandatory before school assignments were made. And just imagine what a powerful research data set that would create when the schools had access to the individual genotypes of the students.

Sounds ridiculous, doesn't it? This is 2004, not *1984*. Our society won't tolerate discrimination against anyone for anything. But when we recognize the inherent unfairness in letting some children and teachers be protected from mold illness, we must also recognize the importance of protecting all students and staff from mold.

Hear it now: That day is coming soon. In the meantime, when it comes to our children, I don't think we can put a price on safety.

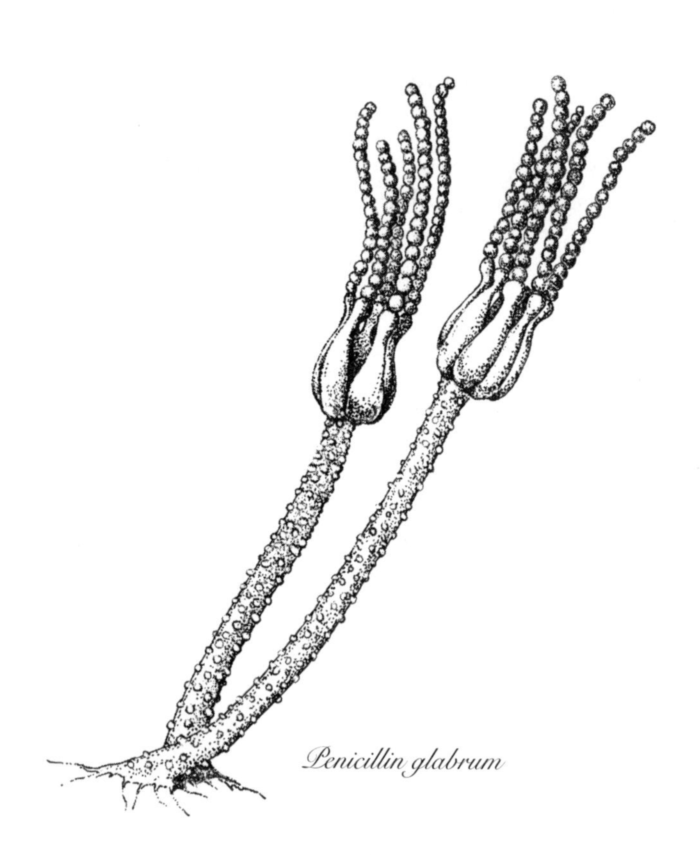

Penicillin glabrum

CHAPTER 22

Remediation
A Routine Failure

"Thank you, doctor. I'm so much better now. The insurance company says the mold clean up is finished except for some painting. We can move back in next week. So I'll let you know later what's happening in my life. I'm so thankful I met you and I'm so glad this ordeal is finally over. Good bye, and thanks again."

Sounds like the end of another mold case, right? Probably not. When a mold susceptible person has suffered one cytokine hit too many from their immune response to mold toxins, they usually *can't* go back into a remediated home safely without added protection. Once that HLA priming happens, they're going to get sick again, unless they take some toxin-binding medication. And if their MSH didn't rebound back to normal, I can almost guarantee you they won't be completely safe from mold, even with preventive medication.

Based on what I see every day in fulltime mold practice, it's a rare home that can be made safe. Why can I say that to you? I'm not trying to make you unhappy and I'm not trying to sell kits used by fire departments to burn down vacated, moldy homes. I'm telling you what I have observed since I first started treating the first of 2000 mold illness patients back in 1998.

The sad facts are that despite all of the advances in HEPA filtration for vacuum devices, topical treatment of mold populations with powerful chemicals like bleach, quats, phenols, peroxides and others, and all the other nifty things remediators can do, mold usually *doesn't move out* after it moves in. Once mold gets a foothold in *your* home, growing on *your* cellulose and using *your* moisture, it will be incredibly hard to eradicate. That means if you go back to visit your "remediated space" *after* you're primed for mold illness, chances are you will get sick again.

Just think for a minute about the effectiveness of biological eradication programs for kudzu in the Southeast, hydrilla in Florida, phragmites in wetlands, nutria in the Blackwater Preserve in Maryland and zebra mussels in Eastern waterways. There are countless additional examples of organisms growing where we don't want them. Too bad for us. We've tried to burn, poison, harvest and destroy them by any chemical means, with one predictable result: failure.

What an oxymoron: effective eradication! When we give an exotic species everything it needs to grow and reproduce, and with no predators in sight, it's going to be a very successful newcomer. What do you think is different about survival of toxin-forming, indoor resident fungi in the face of poisons and chemical controls? Even if you kill the parent fungi, the spores can live for years!

There's something different about the biology of "indoors" improves survival of organisms that aren't faced with competition. Maybe our failed cockroach eradication programs parallel what we do for indoor fungi.

To me, the only feature of successful remediation is effective health treatment provided to the affected individual before it is too late, and the rapid and effective drying of water damage once it occurs. That means really aggressive medical treatment and really aggressive building treatment. The earlier the better in both cases. As is true in medical therapy of mold illness, until you've fixed *all aspects* of the patient's physiologic problems caused by mold, from VCS deficits to high C3a and low VEGF, you haven't fixed the illness. And as you know, if your MSH doesn't go back to normal, then you are going to be in trouble.

Face it: You're not going to sterilize a building by using half-hearted, cost-cutting measures for mold removal. Don't even think that splashing some bleach on mold or painting over some contaminated drywall, 2 by 4 framing or plywood will do anything other than create tons of financial liability for failure to remediate properly. And don't let some bean-counting insurance adjuster get away with trying to send in a cut-rate crew who bring dollhouse-sized fans and no remediation equipment to dry out a flooded building.

Can't you just hear all the remediation companies yelling when they read these words? Effective remediation is a serious business. It is expensive and an expert crew is needed for the job. Too often, the inexperienced remediation company can actually become part of the problem (and a target of the lawsuit!), stirring up spores and contaminating more of the local environment. When I hear of the rare successful remediation, I cheer. When I hear about the disasters of failed remediation, I wonder what more we can do to help prevent mold illness in the first place.

What I said is especially true for those patients who have become sensitized to mold. We don't want them to ever be exposed again to amplified growth of molds living indoors. For primed patients, once they become reactive, even a few spores will trigger a relapse. We must toss the idea of an "acceptable level," into the nearest trashcan. The dose/response idea applies to medications like digitalis, but it doesn't apply to people and mold toxins. Everyone is different!

And if you decide to listen to me, you'll learn that you can't let the water intrusion into your home or workplace go on *too* long, since guaranteed growth of mold will follow. How long is too long? If you learn from the lessons of the Mold Warriors and the Biotoxin Pathway, and if you took care of the water problem and the mold problem *before* your genetic susceptibility kicked in forever and *before* you're too low in MSH, then you will avoid mold illness. Most people don't realize that if you wait to clean up the water and your MSH falls, then the mold was growing too long.

Those are a lot of ifs and toos.

It's my belief that many indoor fungal species have developed resistance to commonly used indoor fungicides. Unfortunately, there are no commercially available tests to confirm or refute my contention. There's no proof that the overuse of industrial strength fungicides has created a race of super-fungi, spewing toxins in every room they meet.

There's no proof *against* that hypothesis, either. And it's biologically *implausible* that resistant strains of fungi wouldn't emerge from the cauldron of "effective control agents." And when we really want to get worried about the safety of our indoor environments, just consider the reservoir of potentially toxin-forming fungi living outdoors that are bathed in sprays, baths and drenches of fungicides routinely. Given the existence of at least 14 different species of *Stachybotrys* listed on the roster of fungal plant pathogens, and not just the famous indoor black mold, *Stachybotrys (atra) chartarum*, when are the outdoor poisoners going to travel on someone's shoes or through some open window to become the new, improved son of Stachy?

So, let me try to follow a simple argument. Here comes the moisture indoors in a building. For the sake of argument, let's call the building the Ritz Residences in Washington, DC. The process by which mold growth follows moisture intrusion is the same there as in the Social Services Building in Accomack County, Va., or anywhere else. Anywhere we find sick patients, we're going to find sick buildings. Mold gets introduced to an environment that fulfills its every need to grow, reproduce and flourish in the absence of predators. The mold makes spores, loading them with toxins to defend these progeny against unknown future predators.

Mold colony counts of 250,000 per square inch aren't unusual in heavy growths of indoor molds. For sake of this argument, let's cut down the density of mold growth to 250,000 colonies per square foot. Each of the 250,000 can make 1000 spores and each spore can have 1000 molecules of toxin on it. Let's see, 250,000 times 1000 times 1000 per square foot is a lot of toxins. Now let's make the remediation process unbelievably efficient, eradicating 99.99% of toxins. That means that the remediated square foot could house 25,000,000 toxin molecules *after* the work is done. Now let's recognize that the remediation process might be supernaturally good, with the ability to eradicate 99.9999% of all toxins. That brings us down to a mere 250,000 toxin molecules per square foot possibly being left behind.

Now let's bring a sobering bit of reality into the square foot example. Look at a typical second-floor plumbing leak problem in a bathroom vanity in a standard two-story home. In just a short time, the water will spread across the drywall of the downstairs ceiling, limited in its distribution only by time. As the water inexorably wicks across joists, it can easily contaminate 10 sheets of drywall in a day or two. Multiply the spore counts by 320 square feet. Now turn on the HVAC. By next week, you will have multiplied the spore counts by 2500. Put the spores into Grandma's purple comforter that you take to bed with you in the RV parked out back and the multiplication process starts anew.

Now, discover the water leak, tear out the vanity, sterilize the subfloor around the vanity with phenol, throw away the contaminated drywall, with proper isolation of the worksite, using proper respirators and Hazmat suits, of course. All follow-up mold cultures from the water-damaged site don't show any bad-guy molds. Case closed, right? Time to let the insurance contractor put in a second rate, knock-off vanity, after the floor is repaired with a cheap imitation of what was there before, right? It happens every day.

Wrong. Excuse me, I think you forgot the 60 billion spores strewn around the rest of the house. And don't think the RV will be a sanctuary!

• • •

How many can fit on the head of a pin?

A good example—and possibly even answers—again comes from my *Pfiesteria* experiences. Back in 1999, Ken Hudnell and I went to the laboratories directed by Dr. John Ramsdell at the NOAA Marine Biotoxin Research facility in Fort Johnson, SC, near Charleston. This lab had an extraordinary array of talented researchers working full time, trying to solve the incredible scientific riddles posed by *Pfiesteria* toxins. One particular researcher, Dr. Peter Moeller, was talking about the difficulty in isolating a dose of toxin adequate to cause disease, the so-called "minimal dose." He was finding toxicity in animals (but back then, no one knew to look at HLA) at levels of 10 to the minus *12th power* grams. Picograms, he called them. Imagine how small that number is. Peter desperately wanted to measure zeptogram quantities of toxin to answer the minimal dose question. Zeptograms are 10 to the minus *20th power.*

And there's Peter *measuring* a gram of toxin divided into picograms: 1,000,000,000,000 pieces to a gram. I honestly can't imagine how small that is. I do know how sick people get without much exposure to much mold for short periods of time. But that amount of toxin is way more than enough to take away the energy, brain and breathing ability of a mold victim. Just because some insurance company-hired "expert" testifies that I can't measure the amount of toxin in a sickened person doesn't mean the toxin isn't there! Every time I hear that obvious smoke screen argument, I get annoyed.

At the end of the day, I asked Dr. Ramsdell's group how many molecules of *Pfiesteria* toxin total would be enough to cause human illness. No one could give a proven answer, of course, and these are top-flight scientists who won't speculate without data. But after a while, the number 6000 kept coming up. *6000* molecules fit nicely on the head of a pin.

So if the size of the mold toxins are about the same as *Pfiesteria* toxins and the toxins seem to do the same kinds of things, could a remainder of 250,000 molecules of toxin per square foot, left over after super-efficient remediation, possibly make someone ill? And what if the area were larger than one square foot? Maybe 12 square feet? That would mean 3 million toxins. But maybe not all toxins are so exuberantly produced; say with only 100 molecules per spore and 100 spores are made (divide by 10 and divide by 10) per day? The number is still really big, 30,000. And I only gave you toxins per 12 square foot at 10 percent production in *one day*. 6000 molecules looks like a small number now!

And did I ask what the numbers might be if the toxin production was going on longer than three years in an area bigger than 12 square feet and the toxins are distributed evenly and routinely by the HVAC system? If the house is 2000 square feet, and all but 40 are contaminated, how many toxin molecules will remain after super remediation? If the HVAC serves 161 upscale condominiums, say only 75 are vacated due to mold, how long will it take for re-contamination of a super-remediated condo once the whole-building HVAC is turned on?

Enough to make Carol Anderson sick within five minutes of re-exposure? And how about the plush comfy side chair left uncovered during remediation? Is that chair likely to harbor undetected spores and toxins? Of course. Throw it away.

• • •

Let's build this hypothetical moldy Ritz into a real example. An alert building manager screens all the exposed occupants with symptoms, HLA and VCS and determines that 25 percent of them are ill from mold exposure. The water intrusion is stopped and the building is dried out. All the contaminated materials are taken out of the apartments or and are burned. Every surface is scrubbed with 10 percent bleach or quaternary ammonium compounds (quats). Repeat testing shows no fungi growing anywhere in the condominium. Now the *"experts"* say the building is safe. But these experts don't treat patients. And it's clear to me from years of experience *treating* patients that if *they* don't know what the illness is, then they can't state *with any academic basis* that the building is now safe.

And if their recommendation says, "OK, go back into the Mold Castle, it is safe now," that decision would be *worse* than worthless. Someone might actually believe them. Carol Anderson did. Now, that's really dangerous.

The occupants of the building are re-introduced to their newly sterilized indoor environment. A reduction of 99.99 percent is the best remediation money can buy. Even Ivory soap isn't that pure! Never mind the dings in the furniture or the dust that escaped some cleaner's eye. The mold is gone; we know that because the building report said so. Oh, did anyone look in the ventilation shafts? What, you forgot that?

How can anyone say the building safe if the vents are Mold Conduits?

Safe? Not for Carol and not for me. The remaining toxins might just as well be carrying a banner for Carol: Welcome to Brain Damage Central. Glad to see you again.

• • •

Remediation needs can sometimes run into blinded loyalty to a job or institution. One of the Occupational physicians for the Smithsonian Museum challenged me about my report on one of the employees there. Using the 5-step repetitive exposure protocol, we proved the illness was caused by occupational exposure to the museum. Talk about a Mold Castle, this building is *called* "The Castle." He agreed the building was old, had leaked for years through the roof and the porous walls. And he agreed that the building probably had some mold contamination, not to mention amplified growth of bacteria and actinomycetes. But a little mold didn't make anyone ill; his organization said so in 2002 in a *consensus* statement. I have already talked about how useless consensus statements are. Here's more proof.

By now, you know that assertion is pure nonsense, but said by reputable and powerful physicians. It is still nonsense.

In all fairness, this guy at least knew what an ABAB study design like the repetitive exposure protocol meant. He was willing to learn from the data, even though he didn't know how to interpret the labs I use.

As we talked about a building-wide health survey, the next logical step after finding sick people in a building with water damage, I told him he could stonewall like most occupational docs and say there was no illness, the option employers usually preferred. Or he could find out whether other workers were ill. If so, they could be treated and removed from the building. When I told him to forget trying to make the Castle mold free, I could feel his resistance.

Smithsonian Doctor: "Fine, let's say we get the patients treated after screening identifies them. But we can't just move them to another location. We have to do something proper about the building so that they have a safe place to work."

RS: "Make that old water-logged building safe? OK, do you have a match?"

Maybe that comment was a little too flippant for an initial conversation with a doubting physician who had a long history with the Smithsonian. But it was true. Buildings like the Castle should be staffed by non-mold-susceptible workers only. You'll never fix the toxin load in that place enough to make it safe for people sensitive to mold. Think about it. If only the 25% of workers get sick from mold, well, let the other 75% work in the Mold Castle. Get the susceptible people OUT! And then continue to monitor health status of those still exposed.

So much for my advice. They got rid of the worker and didn't burn down the building. Oh, by the way, to my knowledge, they never did the health survey either. And to my knowledge, they did nothing to protect the rest of the 25% of the at-risk population exposed to water damaged work places.

When an employer *knows* there are health hazards in the workplace and does nothing to protect the workers, is that a morally responsible posture? Does the Smithsonian know that their workplace could make people sick?

You and I both know a few buildings have been remediated successfully. What should you do when you find yourself working or living in a mini-version of the Smithsonian Mold Castle?

Besides burning down the building, that is.

Control, Contain, Remove, Replace

I can't tell you how many times I've seen remediators named in mold lawsuits because they've done really stupid things. Can you imagine trying to fix a mold problem without stopping the water intrusion? Or trying to correct a localized area of mold growth by going in ripping and a'snortin', with the dust a'flyin', but without protecting the rest of the indoor area from the particles in the air?

"Yes, the patient wasn't sick until the floorboards were torn out, without isolation. The pulse of Stachy toxins in the air was enough to stop watches, suck your brains out and make you so tired you didn't care," the old-timer said on the witness stand.

Or in a local school, where they tried to correct the mold

problem in air ducts by sweeping out the dust from the ducts with a hand broom.

If you think that's a stupid thing to do, I agree. But it happens so often that I wonder if the idea is characteristic of the personality of school maintenance supervisors. Then, after they moved all the contaminated vent dust into the air and onto the floor, they finally realized that the dust might be dangerous if they swept it into the air again. The maintenance supervisor was consulted.

Solution? Sweep the moldy dust into a rug to disperse it. Spray an oil-based cleanser onto the rug, so the dust wouldn't leave and couldn't be seen. No one would see there was a problem if it was hidden in the rug. Oh, that rug was the one used by the pre-k kids for a play area and learning center. "They are just kids, no one will know."

Almost as ridiculous is the practice of taking contaminated porous furnishings and fabrics out of the moldy environment and moving them into an adjacent area *in the same building*. We call that practice cross-contamination. And then that bit of idiocy is topped by taking the moldy building materials out of one room and placing them under the return for the HVAC until the dump truck arrives to haul away the debris. Finally, you should see what happens when a properly rehabbed room is re-contaminated when a moldy set of curtains is put up again or the contaminated carpet is re-laid or previously soggy insulation or drywall is recycled and re-installed.

As Carol will tell you, heat pumps add an additional way to trap mold spores in patients living areas. At her Ritz location, the noisy heat pumps were quieted by insulating them, covering the insulation with cardboard and duct tape. Where was the moisture coming out of the heat pump supposed to go? Carol's heat pump was directly adjacent to her bed.

I'm not making these examples up! How do those guys think they're going to avoid being nailed in a lawsuit?

It seems obvious that you would turn off the water if you were close to drowning in a closed room with a spigot. Let the water run out; you can stand up and breathe. And yet, one of the most common failures in rehab is to try to fix an indoor area when the outdoors is an open spigot, adding life-giving water for the mold. Fix the leaks, change the downspouts, correct the plumbing, and tear off the fake stucco cladding that is causing the condensation, install a vapor barrier—do whatever you have to do to stop the bleeding. I think the military uses the term, "secure the perimeter." In medicine, we say, "Secure the airway, establish circulation and stop the bleeding!" Isn't that the first order of business?

Next, you must try to vent any air from a contaminated area outside, at the same time you prevent contaminated air from infiltrating the rest of the structure. On one of my last site visits to a remediation project, the contractor decided that the best way to ventilate a contaminated structure was to turn on the HVAC. That sure sucked the bad air out, all right and spread the toxins everywhere. Where did he think the toxins were going?

Around Pocomoke, using big fans in chicken farms to exhaust air to the outside is a fairly common practice. This one fellow was very careful to tape down 6-mil thickness plastic all around his work site. Meticulous taping; not an air molecule in the world could get through that vapor barrier. But then, neither could he.

And then when he finally cut a line for a doorway to get inside the enclosure, when he re-taped the "door" and turned on his oversized exhaust fan, the whole structure, pulled towards the window, collapsed on him. I wish I had that bit of slapstick on video.

After the work site is isolated, now the contents must be removed. Or is it the other way around? I have seen it done both ways. For my rehab nickel, I want the particulates in the air stopped before I start stirring up dust by moving furniture around. And no, don't keep the contents in the same place when you begin to remediate. And when you want to put contaminated contents back in, make sure you test them for contamination!

• • •

One of the problems with remediation is the lack of standards for remediators. Who are these guys any way? A professional outfit with years of experience technically hasn't any way to prove they do a better job than the guy who took a correspondence course in molds, graduating three weeks ago.

When they say a UV filter or an ozone machine or something that hums and whistles and emits a blue flame is going to protect you, RUN away, holding your checkbook tightly.

Another problem is that there are so few studies on what has been accomplished by remediation. Dr. David Straus at Texas Tech has been working in the Sick Building field for years. His group recently reported what happens when mold/toxin-exposed paper, cloth, wood and carpet were treated with gamma X-rays, detergent/bleach wash and steam cleaning. With each remediation method, known amounts of toxin-formers were tested on each surface.

The X-Rays killed spores but did nothing to toxins. Washing helped, unless you had *Chaetomium* on paper—no help there. Steam reduced spore counts some, but did nothing to toxin burden. So what do we know about the Guidelines for Remediation? They are useful for toilet paper, but if there is no discussion of human illness, demonstrated by the Biotoxin Pathway, forget it.

What should be done?

• • •

Pocomoke Answers

First, know that mold spores and toxins are everywhere, including the papers in your file cabinets and that amplified growth sources will hurt you if you have a mold-susceptible genotype.

So what's your HLA DR? What's your baseline VCS, MSH MMP9, C3a, VEGF and nasal culture? What do you mean the doctor hasn't ordered those tests? They define the illness!

If you have a symptom roster (see Appendix 1), check it weekly. If symptoms begin to steadily increase, take the time to do a VCS test. If you live in a house with the *potential* for mold growth, like a wet basement or an area with roof or plumbing problems, look for mold. While mold might hide from direct view, biomarkers for mold illness won't hide. Check some biotoxin labs just like you would a cholesterol test. If you are exposed and susceptible, don't wait to be treated: the HLA DR priming from illness will appear soon enough. And then it's too late.

CHAPTER 23

Mold at Ground Zero for CFS

History Doesn't Remember the Names of the Critics

The history of medical practice is filled with ideas that failed the test of time. Medicine evolves as new information arises; some ideas don't last.

As a simple example of theories of illness that were accepted practice but found later to be wrong—and there are many—do you remember the bland diets that were prescribed to treat ulcers? That therapy dominated medical practice until stress was blamed for excess stomach acid production. That was the era of the diagnoses of stress-induced stomach problems, as wrong-headed then as the idea now that mold illness is stress-related. When we found out that many stomach conditions were actually an infectious disease caused by Helicobacter pylori, goodbye stress theory. Fortunately, the discovery of an infectious source of ulcers, as opposed to stress and other psychological contributors, has meant new treatment for ulcer sufferers.

The history of medicine is also full of individuals who single-handedly blocked the development of new ideas. The victories of knowledge achieved by medical greats like Louis Pasteur and Joseph Lister, came only after intense battles and vicious personal attacks by so many of their "peers" were rebuffed by truth and science. The Listers and Pasteurs were right. Their many detractors were wrong. History doesn't remember the names of the critics.

The history of Chronic Fatigue Syndrome (CFS) begins in Incline Village, Nevada in 1985. In the medical history of CFS, each of the concepts applies—failed theories and failed criticism. One victim, Erik Johnson, told everyone who would listen that mold was a cause of CFS. He came up with his theory at the wrong time in the politics of medical opinion, as a unknown viral cause was blamed instead. Johnson tried repeatedly to get the attention of leading CFS researchers then and now to look at what he knew about mold sensitivity. None of the heralded CFS physicians would listen.

Johnson tried to tell all that mold avoidance helped him return to normal life after his repeated bouts with mold illness. No one in an authoritative medical role listened.

We now know the mold connection to CFS is incredibly strong. The resistance to that idea is also incredibly strong, as the long-standing ideas about CFS said often enough and loud enough and supported by the CDC, are believed by many. And if there's a tragedy of the CFS story, it's that Erik was right from the start. No one listened.

Mold hurts us. Until now, few have listened.

• • •

Two practicing physicians, Drs. Dan Peterson and Paul Cheney, saw something completely new in their practice in Nevada in 1985. Suddenly, they were seeing numerous people with multiple health symptoms, especially a strange "bone-crushing, devastating fatigue," but there was no obvious cause. They felt that an infectious disease, a virus, was responsible for the outbreak. Just look at all the people made ill in one area: It had to be infection. (I wonder if they discounted the possibility of exposure to a toxin, or just didn't even think of the possibility?)

At first, an epidemic of Epstein-Barr virus infection was blamed, leading to the "chronic mono," and "Yuppie Flu" concepts. As Peterson and Cheney pursued an academic basis for CFS, they discovered significant clues in esoteric biochemical research that suggested an unknown virus was responsible for the chronic symptoms experienced by sufferers of CFS.

If only those caring physicians, had listened to one sufferer, vocal proponent of mold as the source of the CFS epidemic, how would medical history been changed? As you'll see, biotoxin exposure, especially to mold, was a gigantic player in the development of CFS and continues to be so to this day. But with the viral source of CFS eventually accepted as plausible by much of organized medicine, who'd listen to alternative theories? No one.

In the true story to follow, you'll read how one person found a way to reduce his multiple health symptoms by avoiding mold exposure. He represents a group of people who likely don't have *biotoxic illness* from mold, but their illness is still caused by mold. For the 25 percent of people with a genetic basis for mold toxin illness, and therefore, absence of any significant self-healing once ill, the approach of the "Stachysterian," Erik Johnson, to restoring health will never work. But for many of the remaining 75 percent of the population, some of whom have illness from mold and initial sensitivity to mold toxins, Erik's story makes him a Mold Warrior.

Erik remains bitter that his correct insights about mold and CFS were ignored for 20 years. But like so many people in medical history who were right but ignored, he must be content that his ideas are now validated. We will have many more years of clinical CFS research to follow 2005, beginning by educating the "CFS establishment." That will involve swimming upstream against the CDC once again but *Mold Warriors* will offer a new set of ideas and data for the Chronic Fatigue Syndrome experts. That is, if they read it.

The Stachysterian
By Erik Johnson

"What's wrong with you? What's your problem, Johnson? This is the Army! You're expected to follow orders," the Captain yelled.

I knew I was in big trouble. The Captain respected my work as a nuclear missile launcher specialist, but he couldn't tolerate my lack of enthusiasm for following orders. In a simulated missile attack or on maneuvers I could do things my way, because our unit worked well together and had a high performance rating. But now, back in the garrison, I was in hot water.

I felt sick inside with that sinking feeling that I was about to hear some bad news. Standing stiffly at attention, I waited for the onslaught. The uniformed man stood up and walked around the desk to discuss the matter in my face. He continued his diatribe, loudly. While the captain respected my technical ability, that wasn't going to get me out of a disciplinary action for my failure to follow orders.

If you want to know why I was in the Captain's office in the first place, it was for trying (not really) to bend the fender on a five-ton truck. I had just unrolled the canvas over the bows of the truck bed and was getting off the truck when a sergeant yelled, "Hey, don't climb down the fender of that truck. You'll bend it."

My response was, "We've been doing this for years and it hasn't bent yet. I don't think I could bend it if I tried," and did a quick test of this by jumping up and down on it.

"Disobeying an order. Attempted destruction of government property. What were you *thinking*?" he yelled. "And what the hell is your *excuse*?"

In an absurd situation like this, I could generally guess my punishment by the Captain's actions. It was like he had some kind of high standard for disciplinary actions to show the Major on his rounds. If the Captain was a strict adherent to doing things "by the book," it was the Army way: he was a good person underneath the layers of authority.

If he just sat there and screamed, it was going to be extra duty or a loss of privileges. If he got up out of his chair and put his hands on the desk, it would be Sgt. Major's detail and restriction to base. But if he walked around the desk and yelled in your face, it was certain to be an Article 15, which meant loss of rank and pay and probably a lot of nasty extra duty. This was one of those times he'd come around to yell in my face in the best tradition of Drill Sergeants.

Damn it, I hate polishing those toilets.

This last blast of anger clearly called for a response, so I locked my eyes squarely forward, looking into a point in the distance, and in my best attempt at military tradition yelled, "NO EXCUSE, SIR!"

"NO EXCUSE?" he yelled back, his face inches from mine.

I really didn't have one, so what else was I supposed to say?

My response only seemed to make him madder; I saw the pupils of his eyes get as big as the latrine I expected to soon be cleaning.

"NO EXCUSE?" He demanded again, almost apoplectic, but not screaming like before. I thought his voice sounded funny, but I couldn't look to see why, or else he'd yell at me for looking at him.

Suddenly, his speech sounded like unintelligible gagging noises, and I had to look. His face got bright red and the veins bulged out on his neck. I began to sweat; he'd never gotten this bad before. His eyes seemed ready to pop out of their sockets. It looked like he was trying desperately to come up with some appropriate response or say something really important. Oh, God, I'd never seen him this angry. What should I do?

Before I could decide what to do, he fell over backward, bounced off the wall and crumbled to the floor in a heap.

Good God, I thought, I'd made the Captain so angry that I gave him a heart attack. I stood there, wondering if they could give me the firing squad for this.

Luckily, while I was contemplating my impending ritualistic death, the sergeants who had brought me in and were standing outside the door ran in and helped the wheezing Captain into a sitting position on the floor. The Captain was alive, struggling less to breathe with each minute away from me.

Sergeant Sampson turned to me, demanding, "Johnson, what did you have for lunch?"

What kind of a stupid question is that? I get in trouble for the stupid truck and now I get in trouble for eating lunch at the mess hall, I opined to myself. Tersely, I answered, "A sandwich."

"What kind of sandwich?" he roared back, angry at my lack of specificity.

"Peanut butter," I yelled. You idiot, I was thinking, what the hell does that matter?

"That's it," he said, relieved.

He turned back to check on the Captain, still gasping for air. He wasn't as desperate now; it sounded like he might survive.

I was swept out of the office, and was taken back to quarters, with orders not to move. I sat there wondering why the Captain had such a fit and why the Sergeant seemed to think that my lunch had something to do with it. Soon the Sergeant returned, much mollified, even a bit subdued.

"Johnson, promise you won't tell anybody about this, will you?" he asked.

"About what? What the hell is going on, anyway?" I was clearly missing something.

"The Captain is allergic to peanuts," blurted out Sergeant Sampson, who seemed surprised that I hadn't already figured that out. "Your breath could have killed him."

He instantly looked like he regretted telling me, but it was too late. And his look told me what had happened. They'd thought that I knew what dropped the captain in his tracks.

And then the bloated enormity of the situation hit me.

When the Captain crumpled, I'd instantly been transformed from a lowly, subservient enlisted soldier to the possessor of knowledge that could terrorize my military superior. A word from me and everybody in the unit would be chomping on peanut butter sandwiches, with peanut butter crackers for snack and peanuts—forget the popcorn—during the movies. Peanut butter and bacon on toast for breakfast, too. Booker T. Washington, the Peanut King, would be proud of us. We'd even play Skippy to My Lou on the boom box.

Peanuts as frag. Incredible! Military times had changed since the war in Viet Nam, but they weren't *that* much different. Imagine the charge against me the next time he stuck his face into mine. Assault with a deadly weapon: my breath.

Discipline and maybe morale would disappear every time the Captain had a peanut-induced collapse. Our Captain's life and career would be destroyed. Meanwhile, I'd be a hero among the ranks. His future rested in my hands.

• • •

I can only imagine how he felt in this predicament. In the space of a moment and for a bit of bad luck on his part, I'd found his Achilles Heel.

I've always had a strong sense of fairness, so I told Sergeant Sampson that since I really had disobeyed orders, I would accept my punishment, whatever that turned out to be. And I wouldn't tell anyone about the Captain's life-threatening weakness.

The Captain was evenhanded and just in his actions towards me. I couldn't destroy someone who was doing his duty protecting the truck and trying to be fair in the process. My sentence was far lighter than I deserved—just a few days of restriction and extra duty on the guard post. I don't know if my punishment was reduced out of fear that something harsher would cause me to spill my guts about the Captain's weakness. Maybe my promise not to tell made some points with the Captain.

I never told anyone. I'm glad. I couldn't have stood for knowing I had maliciously harmed another human being, especially my commanding officer. Underneath the military blustering and bravado, he'd helped me tremendously. I owed him. More importantly, the Captain had given me a clue that later saved my life.

• • •

It was unbelievable to me that such a small amount of anything as harmless as a peanut could have such a profound effect on somebody. My thought was, "How many exposures could the Captain possibly take before he died?"

Normal standards of what a person could stand clearly didn't apply to him. If he had a response like that to my peanut butter breath, just making sure there were no peanuts around wouldn't be enough to keep him from having a problem. If other items contained any amount of peanuts, or were cooked in peanut oil, or if peanut residue contaminated anything he got near, there would be a problem. Anything that had the slightest amount of residue on it could cause a response. Peanut oil and soybean oil are everywhere and in everything. What kind of life would he have?

And even though he'd responded to an unbelievably small exposure while yelling at me, he hadn't responded for a long time although we'd shared the air in his office for a time. It seemed impossible that he wouldn't have been exposed to at least a few peanut molecules within a few seconds of my entering the room, yet he didn't respond until I'd already been there for a few minutes.

So the amount of peanuts that was involved wasn't important. It was the nature of his response to a specific exposure that did the damage.

He could only be helped if he avoided peanut products completely. I'm sorry to say this, but anyone less reactive wouldn't have any idea what was wrong with the Captain. Some physicians and some people wouldn't believe the story. His survival would depend on his ability to perceive exposure. If he knew he was exposed, then he had to act quickly to get away from the exposure before his response stole his breath first and his life next.

This "reactivity" concept became useful for me years later when I identified my own reactivity to "unbelievably" low levels of mold.

• • •

Erik Argues with His Doctor and His Doctor Doesn't Listen. Chronic Fatigue Research is Misguided for Years

"When am I going to get an answer?"

This time it was my turn to be the angry person yelling in front of the desk. I leaned over and shouted, "You've been stringing me along for a year. Are you going to help me or not? I need an answer and I need it now. Just tell me *YES* OR *NO*."

I slammed my fist on the receptionist's desk for emphasis as I added, "I hate to put *you* in the middle of this, but I have to yell at someone because I'm just not getting any help and I'm not waiting any longer."

She glared at me.

I'd waited months for this appointment. When I arrived, they told me, "Oh, I'm so very sorry. Dr. Peterson is in an important conference and won't see you today. This is a very important Chronic Fatigue Syndrome (CFS) meeting and he had to cancel at the last minute. He's discussing the possible treatment of the viral cause of CFS. We didn't have time to call you to reschedule."

That's when I lost my temper and started yelling. It wasn't just that canceling my appointment without notice that made me so angry. I'd asked Dr. Peterson to *help* me find out why mold had such a devastating effect on me, but he showed no interest in my insistence that there was a *specificity* to the problems that CFS patients had with certain molds.

I'd become painfully aware that mold was a problem I couldn't ignore. My CFS happened just after I inhaled a blast of mold spores from some rotten wood. I'd felt the same strange sensations at my old High School in Truckee and at various places in Incline Village. All along, I've maintained that mold triggered my illness, which went from being known as "Yuppie Flu" to Chronic Fatigue Syndrome.

Dr. Cheney and Dr. Peterson had diagnosed me as "the perfect case of CFS" during the Incline Village epidemic. In fact, they used my blood to help prove that the new mystery illness wasn't "Chronic Epstein Barr Virus Syndrome," since I was one of the few patients that was EBV negative. I reminded Dr. Peterson of my need to find out why mold is such a driving force in my symptoms and that I needed to find out what I could do about it.

It was the same question I'd asked during the outbreak in 1985 but no doctors back then were interested in pursuing the mold connection to CFS. Years had gone by while I waited for any physician to show he had the slightest awareness that mold was having an effect on people.

My illness worsened every winter and abated during the summer months, no matter where I lived or how well I tried to take care of myself.

By 1997, I was *totally* desperate. My mold exposure was worse than ever. My illness was the worst ever, too. Even worse, now my sensitivity to chemicals was disabling to the point where some places, like the fertilizer store or the carpet store, would put me on my knees. I had to leave if someone wearing perfume or cologne walked into my room. I couldn't drive close behind a diesel truck because if I did, I could only drive for about a half an hour more before being a threat to all God's creatures on the highway. My feet would go to sleep and I couldn't feel the gas pedal. The chemicals in grocery stores would make me so sick that it took me hours of resting in my truck before I could drive home.

When I'd asked Dr. Peterson for a honest assessment of my condition, he'd told me "You're at a point where most people with CFS commit suicide." Yet, despite the onslaught of bizarre chemical sensitivities, there was one irritant that stood out above all others and I was determined to find out why.

Luckily, my angry outburst got results. Dr. Peterson came out to see who was making all the noise. I asked him again if he would help me research mold, and this time I got a definite answer: "No."

Dr. Peterson had been my last, best hope to find an open-minded doctor who would help me with this particular problem; so many others had refused.

• • •

Living with CFS hasn't been easy, but some times and some places were far worse than others.

I've been in plenty of moldy places; most didn't bother me. Yet sometimes there was something in the air in those buildings that knocked me flat. I started calling these exposures "mold hits from mold plumes." My hang-gliding experience showed me that these mold spores rose high into the atmosphere. It was incredible. I could pick up mold plumes in the thermals over specific areas. And I could tell you just where you could go in Incline Village to get whapped by mold nearly all the time.

Still, no doctors believed me. They didn't even check out what I was saying. Despite their assurances (on what basis?) that mold simply couldn't do this to any person, and that even if it were possible that I had a problem, "then it must be from some chemical," I stayed ill. I never felt comfortable bearing the brunt of the innuendo that I was crazy or worse. But that's what I got from doctors for years.

I tried to understand why this reaction happened in one supermarket but not in another of identical construction that contained the same products. Or the way I felt it in certain parts of some schools and public buildings. It wasn't chemical use that was different in only one part of the building.

When I became so reactive and sick that a trip to one store with a leaky roof could put me down for hours, I also was able to point at my torturer: mold.

I hired an environmental specialist to identify various molds for me based on what happened to me. We went around to different locations with visible fungal growth; he identified them while I assessed my response. *Aspergillus, Cladosporium, Alternaria* and a number of outdoor molds *had no effect*. Then we disturbed a swath of mold that was growing on some wallboard. Almost immediately, I almost collapsed, shouting, "That's the one!" My symptoms were the ones that people had so often told me were "the product of your own imagination." [Note from RCS: if Erik had measured levels of C3a before and then two hours after his exposure, he could have proven that C3a was a biomarker for why he felt so bad. And maybe, just maybe, if he had blocked the C3a skyrocket, he wouldn't have had the mold hit at all.]

Before that day, I'd never heard of the fungus he found: *Stachybotrys.*

The air sampling showed no airborne spores. The tape samples lifted from horizontal surfaces revealed nothing and the amount of mold wasn't "known" to make anybody sick. I thought of my Captain when I was told by experts that there was a "level" of mold that caused illness. So, was anyone supposed to believe that *Stachybotrys* up to that level was safe?

Yet, just walking into water-damaged buildings hurt me. I had a similar reaction from my clothing afterwards as well. As I learned my response, I could tell what buildings had mold even without seeing water stains or smelling the mustiness that I knew too well.

You know what makes me want to puke, even to this day: all those people, especially the doctors, who told me that mold couldn't make anybody sick. Then later the same people said, well, maybe mold could make people sick, but it would take an unprecedented amount from the moldiest building in the world to possibly make anyone ill. Now I see the CDC/corporate funded group, the Institute of Medicine, admits that yes, mold causes respiratory problems, but no other symptoms, not a chance, of course. Puke. No, double puke.

And how about all the idiots like me who listen to this "expert" garbage and we're expected to believe it? Triple puke.

Yet here I was, pointing directly at a mold colony and feeling the symptoms that had been messing up my life for years. There had to be a way to make mold exposure fit the obvious facts.

• • •

If I were in a highly contaminated building, I think it makes sense that someone like me could be overwhelmed by mold. Yet how could a *small amount* of contamination give the same massive discomfort?

Unless I was reacting to mold in the same way that my commanding officer had reacted to peanuts that day long ago. It sure didn't take a bushel of peanuts to drop him in his tracks.

And how did someone get so reactive to peanuts, anyway? The theory people were comfortable with was, "Too much exposure and you become very reactive."

So had my commanding officer been a peanut butter addict and at some point he just hit the wall?

I started reading about children who were reactive to unbelievably small amounts of peanuts. Many of them weren't even old enough to have had an overwhelming peanut exposure—so much for that argument.

My type of reactivity didn't fit the too-much model of reactivity, either. As far as I knew, I'd only been exposed to mold in the same amounts as other people in the same places, yet I responded to mold in the same inexplicable way that the Captain had reacted to peanuts.

As I looked at my response to mold, it all made sense and the symptoms that "seemed to come out of nowhere" fit perfectly into a now-solved puzzle. If just a small peanut hit nearly killed the Captain, he wouldn't have any choice but to avoid all peanut hits. Which small peanut exposure would be his last?

So it was for me and mold. I needed to avoid all exposures! Perhaps if I learned the precursor symptoms, then I could quickly recognize the lesser "mold hits." By conducting a more extreme level of avoidance, I might give my immune system the healthful reprieve throughout the year as I felt during the summer months.

I believed that there was a specific source of my reactivity that I needed to identify. If I could avoid symptoms by reducing mold exposure, perhaps I could improve my overall level of health.

And here's where I ran into problems, because Dr. Peterson told me that CFS had turned me into a "Universal Reactor," (what is that?) and he discounted my claim that there was something special about mold that caused my illness. I stood my ground.

He insisted that there was nothing special about mold and that my reactivities were probably due to "toxic metals." Metals? Where did that idea come from? He agreed that avoiding anything that triggered a response was a good idea and cautioned me to, "stay away from those toxins. It'll make a big difference in your life." Talk about mixed messages. I felt like I was back listening to the CDC again.

Stupid double talk like, "Mold isn't proven to hurt anyone, yet stay out of moldy buildings." I paraphrased that ridiculous statement (it's still quoted today by esteemed medical groups) by saying that fire isn't proven to hurt people, but we think it's a good idea to stay out of burning buildings. And those jokers are the authorities? Based on what? If they had data, then bring it!

On the other hand, Dr. Peterson reaffirmed his concept that CFS is a viral illness. His research with an enzyme, RNase L, clearly showed a link from viruses to CFS. He rejected my insistence that there was something peculiar about the way mold affected me that was *worthy* of investigation. If only I had known then that mold also causes the *same effects* on RNase L.

He referred me to a dermatologist who told me to get rid of my *cat*. This doctor said that he'd seen hundreds of cases just like mine and that there wasn't even any need for testing since everything screamed "CAT." He told me, "I've even seen one person turn blue when a cat was brought into his presence." This didn't exactly sound like me. I thought about asking for a CAT scan, but he was mad enough at me, so I let the joke go.

I'd heard "the cat concept" before, so I was prepared to defend my cat.

"I've been in plenty of houses with cats that don't bother me at all," I said. "I'm not even allergic to my own cat unless we're in a moldy house. So tell me why I am only allergic to cats when I'm in a moldy house?"

I even offered to bring in my cat and demonstrate that I had no reaction. Looking back on that day, I wished I had brought some Stachy with me for him to see.

But the doctor angrily said, "Mr. Johnson, trust me on this. I've seen your illness hundreds of times and the cat is the cause every time. Get a dog if you need a pet, but lose the cat. That's my advice and that's all I have to say."

I wasn't even tempted to believe him. Trust him? Forget it. The contradiction of having no reaction to cats except when I was in certain houses was something no one could ignore. His arrogance—telling someone to get rid of their pet without even testing for a reaction—was the clincher. I wonder how many other cats received the blame for people's mold problems. Or even worse, how many people are told their mold illness is an allergy to anything? You see why the allergy idea makes me nauseated to this day.

The visit with the dermatologist did have some benefit—it helped to completely destroy my confidence in doctors. And thanks to the dermatologist, I can accept that Dr. Peterson would never listen to me.

And he was wrong the whole time. Man, what a tragic loss for all those like me who have searched endlessly for answers, understanding and maybe a little compassion. Yeah, I would have liked some help too, but I went in a different direction from Dr. Peterson. I guess that's the way things worked out.

Looking back over the whole experience, I see that I could have responded to the clues for mold toxin reactivity many years before I did. I had been refusing to look at them because I still believed it when the doctors told me that the things I complained about were irrelevant or impossible. Doctors are adept at projecting an aura of confidence that helps "reassure the patient."

I made the mistake of believing these doctors, even to the extent of denying sensations I could clearly feel, denying my own opinions and feelings. It wasn't until I heard them forcefully reject ideas that my own perceptions told me were true that I could flush their opinions down the toilet.

Abandoning me was probably the best thing they did, because now I knew that they had no understanding of what I was dealing with and no intention of trying to help. I knew I had to work this out on my own.

I planned to devise an avoidance protocol based on the concept that I was reactive to mold in the same way that the Captain was reactive to peanuts.

Perhaps if I could get the kind of relief from a mold-avoidance strategy that I had usually felt during the summer months, then I might just stay alive until some cheaper treatment became available. I was fading fast and I had nothing left to lose.

I moved out of my house and put everything in storage. I got a sample of mold from that Stachy colony and by experimentally controlling exposure, I learned to recognize subtle symptoms of reactivity. If I were in a place where those subtle symptoms increased, by acting in accordance with my perceptions and leaving before a mold hit turned into a mold "slam," I gradually started to improve.

I often returned to the mountains for peace of mind. There I really noticed the benefits from avoidance. Fresh air, cool winds and spectacular views of mountain peaks 100 miles away all gave me strength. Everyone said it was the low stress that made me feel so good. It wasn't until much later that I found out that it was physiologic effects begun by the low oxygen and not the restfulness of the isolation of the mountains and the mold-free air there that made my energy jump up ten steps.

Blaming stress for any illness makes about as much sense as saying mold won't hurt you.

It was a problem that the more I practiced mold avoidance and recovered my health, the more sensitive I became to the presence of mold. I learned to make the distinction between "sensitivity" and "reactivity." They aren't the same. I gradually grew able to tolerate longer times in places that used to nearly kill me, but my sensitivity increased to such an incredible degree that many times I'd have to step away from CFS patients I encountered because mold spores on their clothes slammed me.

In some CFS group meetings, I felt like a bloodhound because I could smell and sense the poisons on their clothes, but they just looked at me. Like, *if they* couldn't smell it, then it didn't exist.

The Recovery From Consistently Practicing Mold Avoidance was even better than I dared hope for. Within months, I was exercising an unprecedented amount of control over my symptoms. I kept up my mountain climbing because I felt even better when I returned. I started telling my story at CFS support groups. Many of the people there were complaining of the same symptoms that had led me to mold avoidance, and yet when I tried to suggest that they might get some relief by mold avoidance, my proposal was instantly dismissed by most people. CFS was a virus disease; don't worry about mold.

The few who even considered that mold might make them sick asked their doctors for an opinion. They were told that mold reactivity of the type I described was impossible. Meanwhile, I was continuing to improve and made repeated efforts to interest Dr. Peterson in researching the reasons for my "unexpected" recovery. I dropped off information about mold at his office, but the only person who was willing to read was one of the receptionists who saw my improved appearance and heard about my control of symptoms.

She finally stopped me on the street one day and told me, "It's terrible that Dr. Peterson isn't looking into your information." When doctors won't listen to what everyone else hears, it isn't just the CFS community or the mold community that's in trouble. It's *all* of us.

Nobody else was taking any interest in my story, either. I'd returned to Dr. Peterson's office to take part in his CFS study funded by the NIH and told him and his research assistant of my amazing recovery and how I could now exert control over my symptoms with a deliberate strategy of extreme mold avoidance. Her response was, "How nice for you." I felt that because my case didn't fit their model of a viral source of CFS, they ignored my story!

After many years of looking for help, I was amazed that I had finally found something that directly affected my symptoms and my ability to function. This clue was dismissed without the slightest consideration. If researchers and doctors are truly looking for ways to help their patients, and not simply to justify their own peer group's beliefs about illness, it seems to me that refusing to listen to someone who demonstrates the benefits of an alternate treatment strategy isn't a very good approach.

Forget them. In the end, what they said was of no importance. I had taken control of my life. I felt that I had experienced a miracle as big as if I had climbed Mt. Whitney again, which I try to do every year to celebrate my recovery. For so many years, I hadn't been able to predict when I would be overwhelmed by fatigue and illness, and now by simply acting in accordance with what my body had been telling me all along, I was back in control. This information was worth knowing. If I could achieve a remission of symptoms through a careful strategy of avoidance, someone had to understand this was an important clue.

Every time I went up into the mountains, I always felt better. Every time I spent time at higher elevations, even if I didn't climb a mountain, I was refreshed mentally and my fatigue fell like rocks tumbling into a gorge. The energy jolt lasted longer the more time I spent at altitude. I thought it was the freedom that the mountains brought. Freedom from toxins, freedom from mold, and freedom from having to explain what I felt. Jonathan Wright, another mold victim, listened to me, but hardly anyone one else.

But how could I get a doctor to think or read about what I knew to be true? People at support groups who formerly commiserated with me about our symptoms now told me, "If you got better from mold avoidance, then all you had was mold allergies." The dermatologist wasn't even preaching that idiocy!

If I found more people with CFS complaining of mold reactivity, I could demonstrate that I wasn't the only person to be affected. Surely, some physician would be willing to listen.

But even though I found many CFSers complaining of exactly the same problems, none of them considered avoiding mold when their doctors told them there was no reason why they should. My experience was dismissed so completely that people with CFS would tell me, "Since everyone knows that nothing helps with CFS, if you found something that helps, it just proves that you never had CFS."

And when I would say, "But I was one of the people that Dr. Paul Cheney, who worked with Dr. Peterson to define CFS, used to demonstrate the syndrome," they would respond, "Then Dr. Cheney must have made a mistake."

It was like reliving the Incline Village epidemic when nobody would believe me. Only this time even the other sufferers stopped listening. So I just practiced my avoidance strategy and kept telling my story and watching the phenomenon of Sick Building Syndrome emerge. People accused me of trying to spread "*Stachybotrys* Hysteria," so in fun, I started calling myself "The Stachysterian" and tell them that I'm promoting the cult of Mold Avoidance, or "Stachysterianism."

After years of successfully escaping my worst symptoms by my mold-avoidance strategy and having everyone discount the importance of mycotoxins and mold, I told my story to another person diagnosed with CFS. She asked if I'd ever heard of Dr. Ritchie Shoemaker. I hadn't, so she sent me a copy of his book, *Desperation Medicine*.

• • •

The Sound of One Hand Clapping on Another

As a "CFS" patient, you can appreciate it's routine for people with this illness to experience their quality of life deficits being ignored by doctors. You face arrogance and dismissal from the medical profession. You're forced to fight for recognition that your illness even exists. I had contacted some of the most prominent CFS and MCS researchers. Many doctors, allergists, toxicologists and mycologists weren't interested in my strange mold experience, yet here was a doctor who knew exactly what I was talking about. He even described the Incline Village CFS epidemic as "a biotoxin-associated illness."

How did he know that CFS was a mold illness? He wasn't there. But I was. And I know that mold was on the list of causes of the Incline Village illness. Maybe a virus was also involved; lots of smart researchers think so. But Dr. Shoemaker said mold is number one on the cause list. And those researchers who say "virus," without even knowing what mold does to people need some "not so gentle" education.

From the sound of Dr. Shoemaker's experiences, it seemed like he'd been "out there" alone, just like I'd been, saying things that were true, but were ignored by "authoritative physicians."

I'm not one of Dr. Shoemaker's patients, but I'm glad that as a researcher he's looking at the phenomenon that has made such a difference in my quality of life. I think it's sad that so many doctors refuse to consider that toxins are the cause of some of their patients' problems.

Jonathan Wright Goes to Washington

September 22, 2004 will be another of the remembered days in the Mold Wars. Representative John Conyers of Michigan held a press conference and then a Congressional briefing as part of the kick-off of his new mold bill, HR 1268. Joining him was Bianca Jagger, an internationally recognized human rights activist, as well as a mold victim, and a number of mold activist groups. Dr. Simone Sommer and many others had worked tirelessly to organize a week-long series of events to personally bring the truth about mold illnesses to Congress. As a family practice physician and "mold expert," I was asked to discuss human health issues and Dan Bryson was asked to talk about legal issues as part of the briefing.

The press conference included a series of statements from many mold victims, including an eight-year-old boy sickened by Carol Anderson's condominium complex. Another victim, Jonathan

Wright, shared his story of illness beginning with mold exposure in 2000 in Salida, Colorado. Many in Colorado don't think that mold illness could exist in that dry climate: Jonathan has news for you. Water leaking indoors there gives mold a home in Colorado just like they do in humid climates.

Jonathan's mold illness story is no different from the thousands I've heard. Multiple symptoms involving multiple body systems. Ignorance and arrogance of his doctors. Loss of ability to function as a wildlife photographer. Loss of his livelihood, health and confidence in the future.

But he had spoken with Erik Johnson about extreme mold avoidance and began to regain his life doing just that. Erik's approach worked for Jonathan.

What Jonathan was also saying was how much better he felt when he was at altitude. Given that Colorado averages more than 5000 feet above sea level, it wasn't just a little bit of high altitude exposure that made a difference to Jonathan. He was working above the timberline frequently (higher than 10,000 feet) as part of his new life after mold. Clearly, he felt better at great heights.

His testimony really impressed me. He'd listened to Erik, gotten rid of all potentially contaminated possessions and had lived away from the indoor contaminants that we all take for granted. And by working at great heights for a brief time, coming back to the "lowlands" intermittently, he looked healthy.

Was he telling me something? What was he really doing to his mold illness by spending time at reduced oxygen concentration?

Simple. He was pushing his erythropoietin level like crazy by staying in areas where there wasn't much oxygen. Jonathan and Erik both were using low oxygen to give themselves added production of erythropoietin. When I gave patients shots of erythropoietin, was I just reproducing what Erik and Jonathan were doing to themselves without the needles?

Just for fun, I measured my VEGF levels and my erythropoietin levels before and after several days in Gunnison (8000 ft) and Montrose (6000 ft), Colorado in October 2004. After just 48 hours, my VEGF began to rise! Oh, great, another study of one. Think about it: maybe mold avoidance and auto-dosing of erythropoietin makes sense for a whole lot of people and not just Erik and Jonathan.

When a mold victim with low VEGF and low MSH asks me where they should move to get away from the mold environments of Florida, Maryland, Oregon and Washington, and anywhere else, the answer now is easy. Be sure to go regularly to the "erythropoietin" heights!

• • •

Going Round and Round

Twenty years passed before Erik's mold opinions were vindicated. I wasn't in Incline Village; I have no idea what really happened there. I don't see any proof anyone recorded a biotoxin history. Was mold the problem? One part of the problem? Did the virus of Incline Village *unveil* mold susceptibility? No one can say now.

Erik remains upset that his ideas weren't respected. I haven't heard the Incline mold story from the side of the docs involved. From what I've seen, physicians like Dr. Peterson and Dr. Cheney remained true to a standard of high-quality, compassionate, ethical care.

Erik raises some important points, however. There's never any time that we can afford to ignore insights from those who have a solid experiential basis for their opinion.

• • •

Most of the CFS docs who wouldn't listen to Erik way back then still won't listen to him, perhaps because they're so committed to the viral cause of CFS. Indeed, a recent drug trial involving Ampligen, a compound that stabilizes the enzyme RNase L, showed benefit in treating CFS patients. Both Drs. Peterson and Cheney are prominently associated with the Ampligen research program.

The Ampligen study data were announced at the recent scientific meetings of the American Association for Chronic Fatigue Syndrome (AACFS), held in Madison, Wisconsin in October 2004. The drug had increased VO2 max by 19 percent after 40 weeks of intravenous use in patients with an impairment of VO2 max that wasn't severe. That's all you got? Nineteen percent in a relatively not-very-sick population? We had far better results from creatine (40 percent increase) and anaerobic conditioning (50 percent increase) and our erythropoietin data showed an increase of over 75 percent! I was told the cost of the intravenous dose of Ampligen was not small.

As I stood in line at the conference to challenge the trivial improvement seen in Ampligen patients, a CDC-funded physician, one ostensibly controlling access to the microphone which allowed questioning of the speakers, hissed at me to make my question short. What? This was the same guy who said in a lecture that Lyme disease isn't a big problem in Chronic Fatigue Syndrome. As if my anger at his insult to me personally wasn't enough, this guy somehow *forgot* what Chronic Lyme illness does to so many patients. While waiting, I asked him how many Chronic Lyme patients he'd treated: He said he didn't have to treat the illness, as it didn't exist.

Was I in some kind of Kafka novel here? This guy, who knows nothing about diagnosis and treatment of his topic for an international audience and knows nothing about mold and knows nothing about the Biotoxin Pathway, is called an expert by the CDC and the AACFS? And those organizations control the media spin about Chronic Fatigue Syndrome!

OK, short and sweet: "How can you justify use of an expensive medicine that gives marginal benefits in the physiology of Chronic Fatigue Syndrome, and how can you fail to control for the role of mold exposure turning on the RNase L enzyme activity in your clinical assessment?" I asked.

My CDC friend actually told me to, "Sit down, now," as if telling me to sit would stop the demand for an intelligent response? What was he *really* worried about—fifteen seconds of mike time? Or was the presented data too weak to withstand queries? Why didn't the CDC want toxic mold issues raised at an international CFS meeting?

You know, if CFS patients, many of whom *do* have biotoxin-associated Post-Lyme syndrome, actually knew what their public servants said and did, they'd be more than disappointed. And if their Congressman knew what a waste of money it can be to give those same people misdirected grant money, maybe we would see some decent scientific work in the biotoxin area funded by the government.

Answers from Abroad: Vindication for Erik

One of the researchers who was well represented at the CFS meetings was Dr. Kenny De Meirleir from Belgium. He has published extensively on the different mechanisms that contribute to Chronic Fatigue Syndrome. Before I could talk to him, however, Dr. Robert Suhaldonik presented his data on the enzyme complexity involving RNase L. He had been working with Drs. Peterson and Cheney for some time.

I asked him about mold and CFS. Making me fall to the floor, he said, "Oh, yes, we know a lot about mold exposure in the original cohort in Incline Village. The source of activation of the endopeptidase that cleaves RNase L is increased response of a cytokine, alpha interferon."

"Now wait," I said, "our data is very clear that alpha interferon levels are increased like crazy in mold patients compared to controls. Can we say that mold exposure doesn't change RNase L like you have reported in putative viral CFS patients?"

Dr. De Meirleir chimed in, "We know that cytokine increases are important activators of the subsequent increased activity of these enzymes. Given your data, we need to look again at our data in which we clearly see changes in innate immune responses in CFS. Mold could be the common denominator."

The issue for me was that while De Meirleir knew about MSH, innate immune response activation, including complement, coagulase negative Staph and changes in exercise tolerance and reduced VO2 max, he didn't know about VEGF.

But he knew about mycotoxins binding to Toll receptors (no one else did); in fact, he said that a Toll 3 receptor, a mycotoxin receptor, was critical to the abnormalities in the innate immune response abnormalities in CFS. When I said that the interferon and IL-1B increases induced by mycotoxins, binding to the Toll 3 receptor, activated excessive cytokine responses that then altered genes expressing autoimmunity, VEGF and erythropoietin and lowered MSH with all its downstream physiologic changes, Kenny just smiled.

"Yes, when we can put the changes in the other findings (increased elastase is just one example) that I know to be true, alongside yours, then we might begin to understand Chronic Fatigue Syndrome," he said.

• • •

Let's look again at the chronic fatigue issue from a mold perspective, looking at patients like Erik. Was there the potential for mold exposure at Incline Village? Sure, but no one looked properly. Was there a distinctive grouping of symptoms? Sure. Was there a disciplined "ruling out," a differential diagnosis of all variables that led to the viral diagnosis? No. Now that we know the biomarkers for the viral cause of CFS are *not specific* for viruses and indeed they're also affected by mold, can we rule out mold as the source of the illness? Of course not.

In the End

Erik Johnson is a Mold Warrior. He'll never give up demanding that physicians recognize an obvious truth: Mold hurts people. He follows no one else's ideas because his ideas are uniquely confirmed by his time and his experience. A few fellow sufferers, like Jonathan Wright, know first-hand what happens when you practice Erik's brand of mold avoidance. Whether or not we can prove that high altitude helps correct low VEGF and raises erythropoietin isn't the point of this chapter. That idea is a viable hypothesis.

For 25 percent of our population with mold-susceptible genotypes and exposure to indoor resident toxigenic fungi, no mold avoidance protocols will reduce symptoms. For those like Erik, extreme mold avoidance is a successful strategy for a better life.

The point of this chapter is simple. We've come a long way since Erik's Incline Village experience. We can show susceptibility to chronic mold illness that won't improve with mold avoidance with a genetic blood test. Mold illness, like Erik's, adds to the public health burden beyond the genetic basis of chronic cytokine effects from biotoxins.

Regarding Erik's insights, I remain in awe. Despite the fact that his physicians weren't interested in his helpful insights and the incredible personal abuse he took for his beliefs over the years, in the end he was right.

Mold makes us sick. Let us define what mold illness is, as we did in Chapter 4, and use that information to protect mold victims from a lifetime of mold illness. And let us also insist that our CFS researchers grow their insights beyond what they're saying now; that they listen to treating physicians, to one another and to patients, all of whom have valuable contributions to make.

In the big tent that is CFS, all researchers must make room for those people who have biotoxin-associated illnesses, even if that means giving up the stranglehold they have held on the academic thrust of CFS research and the singular direction of funding.

CFS patients deserve as much.

CHAPTER 24

21st Century Medicine

It's the Inflammation, Stupid

Quick, write down three risk factors for heart attack. And add two risk factors for obesity to your list. Did you choose high cholesterol, cigarettes, high blood pressure and diabetes as the most common heart attack factors? And how many wrote down over-eating and under-exercising for being too fat? According to nearly everything we hear in the media and scientific literature, those choices would be in lock-step with accepted expert medical opinion today. These "traditional" risk factors, ones that still dominate contemporary medical thought, are stale leftovers from 20th Century thinking. They perfectly describe about *half* of those with heart attack and weight problems.

Everything we hold to be true about attacks to hearts, brains and bellies are only correct fifty percent of the time? Who isn't telling us the whole truth? Let me tell you why all the "experts" are wrong. Your list and the list of the experts didn't include the *real* source of the illnesses of the 21st Century: Inflammation.

Inflammation is a general term. It includes what you see around an ingrown nail in your big toe, a swollen knee joint and the inflammation we see in the Biotoxin Pathway. 21st Century Medicine must include the treatment of innate immune responses. Those primordial defenses that descended from the first living things on Earth can be activated to hurt us in ways the traditionalists have never discussed.

Just look at what we know now about inflammation in the 21st Century, thanks to the lessons from the Biotoxin Pathway. Right in front of us are answers to two modern enigmas—heart attack and strokes with no "traditional" risk factors; and obesity that never goes away despite exercise and calorie restriction. Look at the new insights we have for new sources of cancer and for new treatments too. And after getting this far in *Mold Warriors*, you already know about multiple sclerosis and brain scars caused by toxins from the environment.

Now that we know what toxins do and how they are linked to genetics, cytokines, complement, autoimmunity, VEGF, MSH and so much more, we see the illnesses of the *21st Century* in a new way. Maybe now other practicing physicians will understand what we see everyday, bringing an end to some of the outmoded traditions and ushering in a new age of medicine.

• • •

David Selby Saves YOU from a Heart Attack

David Selby didn't need to have a heart attack. But he had one last month.

David Selby will be treated with the standard meds most cardiac patients in America wear like expensive choke chains: cholesterol lowering drugs in increasingly higher doses, angiotensin converting enzyme inhibitors, beta blockers and platelet antagonists. Can calcium channel blockers be far away? Actually, David needs *none* of these medicines prescribed to protect him from cholesterol and another heart attack. But he'll take them because the standard of care in 20th century medicine says those drugs will help him. For the fifty percent of people who *do* have high cholesterol and go on to have heart attacks, the traditional risk factors still apply.

But David *doesn't* have a cholesterol problem or a blockage in any of his coronary arteries. Why is he being told that his treatment must include cholesterol reduction, as if he did have coronary disease?

And he'll follow a diet that takes away cholesterol and fat, as if that diet had *ever* worked for anyone to prevent heart attack number two. It doesn't! Prescribing low fat, low cholesterol diets is just more wrongheaded 20th century thinking. With a landmark study in the Journal of the American Medical Association, we reached a milestone of common sense by seeing that reducing elevated cholesterol in women without heart disease makes no difference in any health measure. Hopefully, the "cholesterol-reducing diet," with its complete lack of supporting logic, will soon disappear from America.

Any person who had *normal* cholesterol levels when their heart attack happened needs to learn from David Selby. Your treatment for the next heart attack must include correcting in you what David had: mold exposure at his workplace, multiple symptoms, VCS deficits, sky high MMP9, incredibly high PAI-1 and C3a off the charts. Could it be you *have a heart problem* and you *don't know* what your MMP9, C3a or PAI-1 is? Not testing for these compounds is like riding a bicycle blindfolded on an interstate highway; something bad is probably going to happen.

When retired President Bill Clinton had cardiac bypass surgery the airways were filled with expert commentary condemning junk foods and carrying a few extra pounds. Fortunately, in Clinton's case, we didn't have to hear the exercise zealots preach their spin on the benefits of regular exercise as a nostrum for prevention of cardiovascular disease since Clinton followed a "standard" exercise program. And we were all grateful for that silence.

We know David's prescription drugs aren't the right ones for David's true heart health risk. His problem isn't cholesterol: it is *inflammation*. And his health risk isn't going to come from eating junk food, not having time to exercise or living a high stress lifestyle: diet, exercise and lifestyle have *nothing* to do with inflammation. His high levels of MMP9, PAI-1 and C3a came from inflammatory innate immune responses to fungal toxins.

That's right. David's health risk comes from inflammation *caused* by mold exposure. David had a heart attack caused by the inflammatory sequence initiated by the mold in his workplace.

Before David went back to the hospital cardiologist for his follow-up, I asked him to have a battery of 21st Century Medicine blood tests done. Sure enough, the cardiologist looked at David's C3a of over 1100, his MMP9 of 1450 and his PAI-1 of 34.

"Mr. Selby, I have heard of several of these tests, but I have no idea what they mean. They aren't traditional markers for cardiovascular disease. I have no way to use them in your case. If you want, I can send you to a specialist at the NIH. You are teaching me something here."

"Well, then, doctor, if I am teaching you, does that mean you are paying *me* the $220 for this office visit?" asked David playfully, knowing he was asking the cardiologist for a lot of help.

Fortunately, cholestyramine fixed David's toxin problem in three weeks. Actos fixed the mild residual PAI-1 problem and finally, a burst of high dose statin drug treatment dropped his C3a to normal. And then his medications were *stopped.* No lifetime prescriptions for cholesterol reducers, angiotensin converting enzyme inhibitors, beta blockers, platelet drugs or anything. But he does need to COMPLETELY clean up the mold (good luck).

• • •

Real Fat, Real Sad and Now, Real Mad

Remember the terrible chronic trio of biotoxin illnesses. Patients are too tired, in too much pain and are often too fat. We have talked about leptin as the lead compound that turns on production of MSH. Now you need to know how leptin makes us fat. I have told you more than once that if you have high leptin from a cytokine response to biotoxins, you will never lose a significant amount of weight until the biotoxin illness is fixed. Meet my friend and patient, Real Pudgy.

Real Pudgy (pseudonym for a real person) didn't need to weigh 280 pounds. But he didn't stop gaining weight until he finally came to see me for his shortness of breath last year.

Real's past physician treated with him every weight loss medication known. He was diagnosed with sleep apnea, low thyroid, laziness, gluttony, edema, gallbladder problems, depression, food addiction, arthritis and more. His doctor even wrote in the chart that she had nicknamed Real as a bird, the "*American Waddler*!"

He was on a downward spiral. Every time his doctor said, "Eat less," Real did. He didn't lose weight. And when the doctor said, "Go to the gym and exercise," he did that too. He didn't lose weight. Only he felt so bad after he exercised a short period of time, he didn't want to go back. He hired a personal trainer who told him, "You can do it Real. No pain, no gain." Real had a lot of pain and no loss. The trainer clearly had no insight that Real's low anaerobic threshold prevented adequate training benefit from exercise, just like Phil Ness in an earlier chapter.

Real didn't lose any weight because what was wrong with him was *inflammation*. And his demeaning doctor didn't even think about Real as a person. She didn't even ask where he worked or anything about the building. If Real's doctor had to explain her nickname for Real to a jury in a failure-to-diagnose malpractice case, it wouldn't be a happy time for her. It is bad enough to be arrogant as a physician; documenting that attitude in a legal document like a medical record is reckless and uncaring. She made Real really angry and hurt when he read it.

Every person who has a weight problem that doesn't respond to 20th century instructions like "eat less and exercise more" and every person who pays big bucks for a CPAP machine to correct their sleep apnea problem and every person who is saddled with a doctor like Real used to have, needs to learn what was *actually* wrong with Real Pudgy.

And every physician who insults their overweight patients by *not understanding* the effect of the physiology of inflammation on leptin and weight needs to understand the lessons from Real.

Let's go back a bit in time to understand inflammation as part of 21st Century Medicine, the medicine that is based on the lessons from the Biotoxin Pathway.

• • •

It's Day One of Pathology class in 1973. For me, it's Day One of medical school at Duke University, Durham NC.

First lesson: Inflammation is the key to medicine. Redness, heat, swelling, pain and loss of use are the cardinal signs that inflammation is active.

Back then, we knew inflammation was present when the patient had gout. There was inflammation in those joints! Pain, swelling, redness, heat and loss of use; just look, listen and feel, Mr. Student. And the patient on Long Ward (Duke's teaching area for Internal Medicine) who had a vasculitis, an inflammation of blood vessels, could also be diagnosed by *thinking* about inflammation. We knew there was an inflammatory process causing that illness because we could feel it and see it. And if we listened well enough, the patient would tell us.

There were other diseases to learn about besides inflammatory illnesses: Chronic lung disease, diabetes, heart surgery, hormone problems, degenerative arthritis, anemia and dementia, too. These illnesses didn't have the same cardinal signs of inflammation, but they were still important to Mr. Student. And cancer was anyone's guess. Every day of the second year rotation in Internal Medicine was a 20 hour work-treat on Long Ward. Dr. James Wyngaarden who went on to run NIH was our attending and with

Rick Klausner who went on to run the National Cancer Institute (he now works for Bill Gates) and his wife Cecile as fellow students, there was no end to the fascination I felt trying to contribute to solving the problems that brought patients from "all over the world" to Mr. Duke's Hospital.

Truth be known, while we received terrific instruction from excellent teachers, there was a lot of grunt work that must be done in a teaching hospital. Somebody had to do it. Med students rarely see the celebrities from all over the world; you better believe that if there were a patient who didn't present as great a diagnostic challenge as others, the med students surely were called in first ("This is the 45th Duke Hospital admission for Mr. Ernie B. for bleeding into a joint caused by uncontrolled hemophilia, Factor Eight deficiency").

And the people that showed up in the emergency room with twenty different problems, they were the med students' primary responsibility. In a medical history, we record what is called a "review of systems," basically a series of questions about symptoms that help us understand the patient's overall health. When the patient says yes to most of the review questions, a physician will often write in medical shorthand, "ROS diffusely positive." That is doctor language to warn other physicians, "Look out, this guy is a crock, a gomer." Gomer is short for, "Get out of my emergency room." Heaven forbid we would get one of those gomers admitted to the hospital. If the gomer did get in though, we could almost hear the attending thinking, "Get the med student on this case *now* and get me out of here."

By now, you should suspect that many of the gomers had biotoxin illnesses. Of course the systems review is diffusely positive; that's what is wrong with them! Probably a lot of them were mold victims. They all had a million complaints and nothing ever showed up as abnormal on the lab tests. If a Duke physician had discovered the importance of MSH, VEGF, C3a, leptin, MMP9 and all the rest of the Biotoxin Pathway, *then* I bet the gomers would have been exciting teaching cases, because Duke would have aced Stanford, Hopkins and Harvard in proving gomers had real illnesses. But no one at Duke did. And more important than competition between schools, we lost a great opportunity to help heal and rebuild damaged lives.

We just weren't trained to listen to patients with a litany of complaints like we were trained to take a good heart history. Sure, we were taught that 90 per cent of diagnosis was from history, obtained by listening and talking with a patient, but no one could show us what a good biotoxin history was because we didn't know about biotoxins. And we didn't know to order the defining lab tests that would attract the attention of most academic physicians on the planet, because we didn't know that the tests would show us the way to an easy diagnosis and a helpful treatment.

And here's the arrogance of 20th Century Medicine: because the *physician didn't know* to order the right tests to show the (now) obvious biotoxin diagnosis, the *patient* had to have a psychiatric problem or was less than a medically acceptable human being. Am I too harsh in my judgment? Ask a gomer or a crock.

Besides, back then at Duke when there was *disease* to stomp out, and diagnoses to prove, proof that no physician in the community had made before, what were those chronic complainers doing taking our time? Later, when we had a socially acceptable grab-bag diagnosis of "Chronic Fatigue" or the wonderfully contrived diagnosis, "fibromyalgia," we finally could provide those incurable people with a label. "Learn to live with it, here's the psychiatry clinic referral," wasn't uncommon back then.

Sadly, physician attitudes are no different today. Ask any of the thousands of patients who come to Pocomoke for an additional opinion after having been given a series of wrongheaded diagnoses.

We didn't know those patients were telling us the truth. They were *pleading* for us to be smart enough to understand what was wrong with them and we didn't even listen. Truth be known, it wasn't just the med students at Duke who didn't understand what the gomers were saying. The problem was then and remains now a nationwide disgrace.

Boy, were we *blind*. And *wrong*. From my point of view, the saddest part of the last 32 years of advances in learning about the role of inflammation in disease (we'll talk about immunity shortly) is that supposedly well-trained doctors are *still* making the same close-minded mistakes that we made back in the 1970's. Shame on us.

And now, we are only just starting to see that inflammation really is the core problem in chronic lung disease, diabetes, hormone problems, recovery from heart surgery and all the rest.

• • •

You have seen that biotoxin illnesses are incredibly complicated. From the first toxin molecule emitted by a moldy building turning on inflammatory cytokine production to the last bit of complement silently eating away at our health like an immortal colony of termites chewing on the foundation sill plates of the finest mansions, biotoxin illnesses are all about inflammation.

You've also seen how inflammatory responses interact with immune functions. HLA susceptibility to toxin illness exerts its effects by shutting down proper antibody formation. And then there is the controlling Prometheus of the whole system, MSH, the body's great regulator, which should be supervising inflammation, hormones and immune function. Instead it is being attacked and consumed daily by cytokines and biofilm forming Staphs colonizing our noses. Biotoxic patients cannot survive these onslaughts and they are relegated to become the gomers of 1973.

We now know the physiology of the mold illnesses. We now know what is wrong and how to treat what is wrong. With all our knowledge, what is the basis for arguments about mold illnesses? Are the tests like MSH, known and used since the early 1970's, still too *new*?

You've learned how Mold Warriors have fought for their rights to justice in medicine, law and career. Underlying their ongoing struggles have been the inflammatory responses of their illnesses that weren't revealed until a new language of inflammation became known in medicine. Like the trip through the door in the rabbit hole in the Introduction to this book, the language of biotoxin illnesses has helped us to understand how little we have actually learned about inflammation in daily medical practice. Those diseases we learned about that didn't have redness, heat, pain, swelling and loss of motion, diseases such as diabetes, obesity, heart disease and dementia, they are *all* linked to inflammatory processes that are shared by biotoxin illnesses. Knowing one illness will help us learn the others.

If we have learned so much about biotoxin illnesses, what do we now *think* we know about the diseases of the 21st Century? Look at the illnesses of America today: heart disease, diabetes, obesity, Alzheimer's and cancer. Each one of these illnesses is tightly linked to inflammatory problems created by mold and other less common biotoxin illnesses. Are we able to say that mold illness *causes* heart disease, multiple sclerosis, diabetes, obesity and plaques in the brain? Hear me out before you *assume* not.

The medical education experts say that medical knowledge doubles every five years. Judging by what we know about mold illnesses now compared to two years ago, I'd say the medical knowledge doubling cycle is closer to 18 months. A reasonable question for all practicing physicians is, "How can you keep up with the new information?" As an example, I have talked a lot about pro-inflammatory cytokines, including TNF. TNF, the first important cytokine to be discovered, was only reported to be important in the mid-1980's, but by the mid 1990's, there were over 10,000 references to TNF in the Index Medicus. And now, we have whole textbooks on newly named cytokines, with endless identifying numbers and endless compound prefix/suffix words, like chemokine, lymphokine and more. The list is almost endless. Every time I pick up a new copy of the journal, *Nature Immunology*, there will usually be articles introducing *more* new acronyms for new compounds that are either inflammatory or immune related or both.

We are now in a new era of exciting discoveries in inflammation, just like in the days of discovery of bacteria and viruses back on Long Ward. Back then, we were just learning the now-old words and now-old concepts about medicine. As a medical student, every day brought a new world of new findings in medicine. And one thing has not changed. Inflammation is still the key to medicine; we are just learning more about how complex inflammatory pathways are. Back in 1975, we just didn't know what we know now. But we could have learned a lot by listening to the gomers and we didn't.

• • •

NO, IT'S THE IMMUNITY, STUPID
Innate, Acquired—Both!

Pick up a copy of any current journal of immunology and chances are good you'll see the new buzz words: acquired immunity and innate immunity. Neither of these arms of the immune response is new, but *attention* to these different systems has grown incredibly. Stated simply, acquired immunity refers to antibody production and innate immunity refers to the manufacture of cytokines; the cascades of responses that cytokine production brings; and complement. Acquired immunity involves the antigen presenting cells, called dendrite cells (see Chapter 4!) and HLA DR to activate the string of lymphocyte reactions to make antibodies. Innate means that foreign molecules in the blood, including those from foreign invaders and biotoxins, bind to a receptor on a particular cell (fat cells; cells lining blood vessel walls; and special nerve cells), to send a message into those cells. The message is translated into another signal that rapidly turns on certain genes (the NFkB pathway; see Chapter 4). Those genes respond within minutes to external attackers.

Innate also means certain circulating compounds, especially complement, which recognize and attack foreign molecules immediately following their entry into our bodies.

You have to be impressed that the "higher animals," those with backbones, have such wonderfully developed acquired immune systems. Backbones, immunoglobulins, T cells and HLA all seemed to evolve with each other. That would make sense: just look at all the microbes there were trying to attack them. They needed acquired immunity. Pity the poor jawless fishes and all the spineless bacteria, fungi, dinoflagellates, cyanobacteria and all the rest. No HLA DR in them. And no antibodies either. All they had was innate immune mechanisms.

Just think about those organisms. They are all toxin formers! Of course the innate immune response involves a method to detect and respond to the primordial toxins, each amazingly built around the shape and size of water, toxins which these ancient organisms all made!

Question: How did *the primitive organisms* survive? Answer: they had *innate* immune responses, beginning with receptors on cell walls turning on engulfment (use the word phagocytosis to talk to an immunologist) of competitors since the dawn to time. We have already speculated on the role of phagocytosis on the evolution of mitochondria in animals and chloroplasts in plants, with the amoeba phase of *Pfiesteria* the best example. And if you look at how our white blood cells work to destroy invaders, once the foreign antigen is recognized, coated with all the trappings that the complement system brings, the invader is just engulfed and destroyed inside special engulfing cells.

These tiny organisms are teaching us about the illnesses of the 21st Century! They survived in part because they could trick their competitors into turning the innate immune response of the attacked organism back onto itself. The toxins were simply like the unseen guy who threw a rock into a crowd of 25 angry people. Person 1: "Why did you hit me with a rock?" Person 2: "I didn't, I don't know where the rock came from. Maybe Person 3 did it." It isn't too long before everyone is fighting each other, while the real rock thrower walks away. Toxins do the same thing: they turn on the innate responses that are supposed to protect the victim, but instead harm the victim.

If it is true that the innate immune response has been conserved over time, it would make sense that the few receptors involved with the innate response would also be conserved. Sure enough; the so-called "Toll receptors," remain few in number but are able to respond rapidly to compounds held in common by many types of offenders. The Toll receptors are found throughout the kingdoms of organisms that have no backbones and in those with backbones as well.

And if the innate immune response is so important, then we should see the importance of NFkB. Sure enough, that versatile nuclear factor is activated in our fat cells after biotoxins turn on the Toll receptors. NFkB then rapidly makes the nucleus go wild producing cytokines. All our evolution, complete with all the sophisticated, specialized antibodies from the acquired immune response, left us with the innate immune response as our defense against the primordial toxins conserved over eons of time.

Don't forget that those toxins are made by *spineless* creatures. The defenses turned on against those toxins are leftovers from the warm seas and swamps of Archaea!

Look what happens when the acquired immune response can't protect us from biotoxins! And look what lessons we can learn by understanding what the innate immune response does. In biotoxin illnesses, MSH levels are dead low: without MSH there is no control exerted over the innate immune response. When the toxin throws the first rock, there is no calming force to stop the riot. If we take a little time to look at biotoxin patients as teachers, we can learn how to help David and learn how to help Real. In many ways, their illnesses are simple examples of innate immune responses turned against us. And with no protection from the acquired immune response, the inflammation caused by the out-of-control innate immune response becomes the driving force of 21st Century Medicine.

21st Century Medicine begins by remembering lessons from the ancient history of immune responses!

• • •

Well, this is an interesting idea. Do you have any facts to back up your speculation, Dr. Shoemaker?

The molecules critical for innate immune function should be found in many "old" organisms too. We would predict that compounds like superoxide dismutase, tyrosine kinase, C-reactive protein, lectins (in complement) would be highly conserved in the primitive organisms. They are. Just when we thought we were so advanced.

So, if we have two main kinds of immune responses active in the development of inflammation, with the more specialized acquired immune responses coming later on in evolution, wouldn't it make sense to look for the *interaction* of these systems in maintaining health? And if there is a reasonable interaction in health, wouldn't it make sense to look for disturbances in that helpful interaction of the innate and acquired systems when disease strikes?

In essence, that is what 21st Century Medicine does. Moreover, 21st Century Medicine shows us what goes wrong with health when innate immune responses are over-reactive at the same time when control of acquired immune responses is lost. Here is the real importance of the Biotoxin Pathway for David Selby and

Real Pudgy and all the rest of the millions of patients in the US with biotoxin illnesses.

• • •

The critical bridges between the two immune systems are the "dendrite cells." These are the cells that engulf (innate immunity) foreign antigen invaders, label them with a piece of HLA-DR, and sequentially present the processed antigen to special lymphocytes that cause antibodies to the antigen to be made (acquired immunity). We should see primitive NFkB mechanisms in the dendrite cells and we do. We should see these cells located in areas with close contact with the outside world and we do. Skin, gut and lung are richly invested with dendrite cells. That's where we see the Big Daddy Controller, MSH, as well. The bastion of defense initiated by dendrite cells primarily is in the lymph nodes in vertebrate organisms. Here is the main source of immune processing (acquired) in us. Defects from cytokine production (innate) can easily disrupt normal dendrite cell function.

You may have heard a lot about particular lymphocytes, "natural killer cells (NK cells)," being highly responsible for controlling virus infections and cancer cells. Sounds like they would be part of the *acquired* immune response, but they are innate. They have close ties to dendrite cells, with each playing off cytokine responses to change the other's function. If there is a problem with the innate immune response and with NK cells, will we see more cancers in biotoxin patients? Good question.

It is the interaction of the innate immune response, complete with all of its receptors, genes and cytokines, with the acquired immune response, complete with all of its HLA, lymphocytes and antibodies, that should form one of the essential platforms of 21st Century Medicine.

• • •

Anyone who has ever had hives or angioedema or anaphylactic shock from a bee sting might know about the important role of other white blood cells besides lymphocytes and macrophages in the immune response called mast cells and basophils. These cells are mini-factories that produce huge quantities of cytokines and chemicals that, when released by contact with invading substances, change blood flow, shift body fluids and help fight off the attack by outsiders. Most mast cells are found in skin, gut and lung, with more in the blood stream. Curious locations! Remember those locations from the discussion of the innate immune system? Sure enough, these cells are innate system "multipliers." We will meet them again when we discuss complement C3a. Mast cells can hurt us if their weaponry is turned against us.

• • •

Complimenting Complement

Complement is the biggest chemical player in the innate immune system's effects on health. Based on what I know now, I can't read an article on infectious disease or cholesterol or autoimmune illnesses for example, without saying, "Did they measure C3a or even consider the alternative pathway of complement?" And of course not. So, I simply toss the study. Without controlling for complement, the study is incomplete, maybe worthless. No logical conclusions can be drawn, but you'll still see the study published somewhere if the funding source is strong enough.

That attitude sure saves on reading time. Studies on heart attack and obesity that do not measure complement are fundamentally flawed, yet often those same studies are being touted as "state of the art." They tell us that David's illness comes from eating too much cholesterol and being stressed. And Real's illness is the result of being a lazy slob.

Thankfully, there is room for hope in the heart disease literature. Like a slow moving turtle, the word "inflammation" has crept into the atherosclerosis discussions. Use of C-reactive protein (CRP), a wonderful primordial innate immune response element, is now part of a cardiac risk factor history. And we know that the statin drugs provide additional benefit by reducing the inflammation, as well as lowering cholesterol. A reasonable question would be, "Do the statins reduce abnormalities in the innate immune response?" Sure enough, high dose statin treatment knocks the dickens out of elevations of C3a, without a re-rise of C3a after the statin drugs are stopped, *but only if* the biotoxin illness is gone.

We know that complement activation, with increased C3a, occurs almost immediately following biotoxin penetration of our surface barriers of skin, gut and respiratory surfaces. We can measure levels of C3a in normal people and show they are normal. We know that acute exposure to biotoxins, mold especially, causes a rapid rise in C3a, often measurable within four hours. Our data on acute mold exposure and subsequent rapid rise in C3a will be published soon.

We did a simple study this summer looking at C3a in acute Lyme disease. People without tick bites were our "control" group used for comparison. They had normal C3a. Another group had a tick bite without rash or illness. They didn't have high C3a either. If the patient had acute Lyme disease, however, with a typical bull's-eye rash, we showed their C3a rose to levels too high to measure within *two days* of appearance of the rash.

What happened to the high C3a? In the Lyme patients who didn't have the Lyme susceptible HLA DR genotype, use of antibiotics successfully dropped the C3a. In patients with HLA DR genotype conferring susceptibility to Lyme toxins, however, the C3a didn't fall with antibiotic treatment. Amazingly, the standard antibiotic treatment for Lyme, endorsed by infectious disease experts all over the country failed to correct C3a. Fortunately, application of the principles of the Biotoxin Pathway leads us to use Actos, followed by CSM. This protocol successfully corrected their abnormally high C3a.

If the patient has one of the dreaded genotypes, the 4-3-53 or 11/12-3-52B, the C3a won't revert to normal even though the patient feels well following CSM. It is this group of patients who are at highest risk for irreversible Chronic Fatigue Syndrome (CFS). As an example, I presented a group of 89 patients with CFS at the October 2004 meetings of the American Association of Chronic Fatigue Syndrome. Of the 89 desperately ill patients, 79 had one or two of the dreaded genotypes. Based on the occurrence of these genotypes in "normal" people, we would have expected to find no more than four. And yes, the "dreaded genotype" patients had sky high C3a as well.

The reason I am so insistent about demanding that complement studies be performed is not just because complement is the link to almost every part of the immune response; it also is because our data on complement are so strong. Complement, as I told you, is the patroller of the perimeter. It activates immune responses following foreign antigen invasion. It lets loose the immune Dobermans to hunt down and tear into any foreign antigen invader. The problem is that complement also controls the destruction of "self" antigen. That means that if some autoantibody is being formed, say to cardiolipin, complement *should* shut down that antibody. Over-reaction of complement, especially C3a, shows that shutdown doesn't occur.

There is an incredibly complex interaction of complement with the clotting or coagulation system as well. Bill Green, a hydroponics tomato farmer, had a mold susceptible genotype; he was exposed to mold growing in his greenhouses. He developed a rare gangrene of much of his small intestine following a bout with pneumonia. No cause was ever found. Then several years later, he started having pain in his legs after walking a certain distance. It wasn't the result of a blockage; his arteries to his legs were clear. But blood wasn't getting into his right foot. Was he another Sue Donahue (Chapter 6)?

No one knew back then that Bill had developed antibodies to

cardiolipin after the cytokine response to his pneumonia unveiled his HLA. Now, his C3a was too high for the lab to measure. Could it have been the combo of C3a and the cardiolipin that stopped blood flow to his intestine like that combo did to Sue's foot? And were those responses to toxins now trying to shut down circulation to his right leg? Sure enough, after treatment of his mold toxin problem his symptoms stabilized. But removing toxin alone didn't completely clear his Pandora's Box of anticardiolipins. Happily, treatment with heparin, a blood thinner, in low doses, corrected the high C3a and removed his typical claudication symptoms. Just like Sue Donahue.

What about David? If innate immune responses to toxins are hurting his heart, did he have anticardiolipins too? No, in David, C3a and its fellow anaphylatoxins cause spasm of coronary blood vessels! Does C3a cause the blood vessel cells to make extra adhesion molecules that cause the temporary appearance of platelet plugs in coronary arteries? Yes, they do! Does the C3a cause local release of those packets of cytokines and chemicals from mast cells and basophils? Yes, they do!

If David's C3a went up from mold exposure and susceptibility to toxins set off all the derangements in the innate immune response, then did his C3a fall when he was treated first with CSM and then all the rest of the Biotoxin Pathway drugs? It sure did, just as fast as his 18 biotoxin symptoms vanished and his VCS deficits returned to normal.

David's case is more complex than just the whacked alternative pathway from complement. His MMP9 was incredibly high as well. What does that do to David's blood vessels? Remember MMP9 in Marty Bernstein (Chapter 11)? And the role of MMP9 with antibodies to myelin basic protein associated with scar formation in the brain (Chapter 14)?

Don't forget that MMP9 is the delivery boy. It delivers inflammatory chemicals out of the bloodstream into the tissues just beneath the blood vessel walls. Illnesses with high MMP9 will result in inflammation in brain, nerve, muscle, lung and joint. So if you see a red hot joint in Lyme disease, think of MMP9. And the same is true in Lyme when it causes a brain infection: look for MMP9 to be high in the spinal fluid. For the patient with terrible lung symptoms after mold exposure—high MMP9. Numbness, tingling, weird muscle pains all common with high MMP9 too.

I get a lot of raised eyebrows from rheumatologists and pulmonologists when I use Actos to lower MMP9 in hot joints and cold lungs respectively. Patients don't complain, as they feel better when MMP9 falls. Trying to convince a pulmonologist that mold causes hypersensitivity pneumonitis can be difficult; getting him to admit that MMP9 is one of the reasons for the breathing problem is a real challenge. You would think that seeing patients improve, even ones with common biotoxin symptoms might make them increase their insight and turn on their academic interest.

When MMP9 is high and plasminogen activator inhibitor-1 (PAI-1) is high, like in David, we have a prescription for enhanced delivery of oxidized LDL cholesterol out of blood vessels, across several membranes, through the gristle of the connective tissue that supports the blood vessels and then into the swamp of proteins that separate the blood vessels from the underlying muscle. Could the delivery combo of MMP9 and PAI-1 also send C3a to set off spasm of the smooth muscle in David's coronary arteries? Sure could.

I have long argued that most of what we say about cholesterol and health has been biased by the flawed conclusions made by the NIH cholesterol panels in 1986. Those esteemed experts, convened into a consensus panel ("Look out, there's another consensus panel trying to deceive us!") looked at the five big studies on cholesterol and heart disease which clearly showed a break point of increased disease with cholesterol levels over 240. The participants in the studies were white males between the ages of 40–65. The expert panel concluded that *all people*, without respect to age or race, therefore be held to a standard level of cholesterol no greater than *200*. If a diet that eliminated fat and cholesterol didn't accomplish the desired reduction (they had to know that diet was of no benefit; the diet discussion was pure lip service), then drugs, especially the new wonder drugs, the statins, should be prescribed.

Many people assumed that the manufacturers of the statins influenced the illogical recommendations, but we'll never know. Fortunately, some of the cholesterol hysteria is being corrected by big studies, like the one in the Journal of the American Medical Association that showed no benefit from lowering cholesterol in women. Now we just need the same kinds of studies in kids, persons of all races and people over 65.

And we have plenty of data that shows the diet was illogical and wrongheaded to begin with. But, I guess if enough authoritative organizations say the Emperor's clothes look wonderful, we will be forced by the weight of the experts to agree. Just put this idea in the Daubert perspective and you'll see the political and legal strength that cholesterol traditionalists have.

And they are just as wrong now as they were in 1986.

Here the story gets much more interesting. As much as I object to people taking statin drugs just because they are alive (do you think I am kidding about the attitudes of some cardiologists?), when I heard about the impressive improvement in one-year survival in patients after their first heart attack *if* they took *high dose* statins, I asked another question. Could statins help us in another way other than reducing cholesterol? Are high doses anti-inflammatory?

It didn't make sense that if cholesterol were delivered under artery walls for, say, *45* years, that the statins could then reverse the blockage in *one* year. Could the undisputed benefit from statins be due to simple reduction of C3a? *Low dose* statin therapy does *little* to C3a. It is certainly possible that high dose statins will help heart patients, though I can't agree that cholesterol is the reason when the cholesterol proponents haven't looked at C3a, PAI-1 and MMP9.

David certainly will. And he only charges $220 to teach you what he knows.

• • •

The Leptin Factor
Obesity is a Symptom of Cytokine Damage To Insulin and Leptin Receptors!

Real begins each day with a cup of coffee while he reads the morning paper. These days the odds are high that he'll find himself reading an alarming health story about "the growing obesity epidemic in America." The stories in the paper all sound the same: we need to exercise more and eat less. Maybe the authors think that if they say the mantra enough, people will buy it. Just look at the successful blinding of America in the cholesterol story!

Unfortunately, most of those breathless, page-one articles have plenty of fuel for their opinions. Take a look around; the proof of fat everywhere is everywhere. You don't need a Ph.D. in human physiology to understand that Americans (and especially American *children*) are getting fatter with each passing year.

How bad is the U.S. "obesity problem," and why is it becoming worse?

This bad: According to the latest data from the number-crunchers at the U.S. Centers for Disease Control and Prevention (CDC) in Atlanta, fully one-third of America's 280 million citizens are currently classified as overweight.

Even more ominous, say health researchers, is the fact that "clinical obesity"—a chronic disease with potentially life-threatening consequences—is now affecting more than 56 million Americans (about 20 percent of us), day in and day out. Just 10 years ago, the numbers were significantly lower.

America's "fat epidemic" is rapidly spiraling out of control, and the negative impact on grownups and kids alike can be observed from the rocky coastline of Maine to the towering redwoods of California. More than 300,000 Americans are now dying each year of diabetes, hypertension, colon cancer and breast cancer —diseases directly linked to obesity. If heart disease is added to the toll, obesity is by far our Nation's biggest health threat.

Disturbing? You bet. Amazingly, however, our National Fat Explosion has been taking place during an era dedicated to "physical fitness" during a time in which we are bombarded daily with powerful messages about the vital importance of "eating right and getting plenty of exercise," if we wish to enjoy healthy, vigorous lives.

Wait a minute. Did we get the message wrong, or is the current accepted thinking about obesity out to lunch?

Why do most people who get fat tend to stay fat, even if they manage to shrug off a few pounds now and then, before promptly gaining them back again? And why are some people able to eat all they want, meal after meal, while remaining wonderfully, infuriatingly slender? Even more challenging, "Why do many people gain weight that they never lose, while eating less than their slender counterparts?"

Imagine my surprise years ago, when my treatment of obese patients—along with my continuing research on the causes of the disease—forced me to begin challenging the "conventional wisdom" on obesity that has ruled American medical opinion for the past several decades. Imagine the exhilaration I felt, as I began to analyze the latest data on human biochemistry and was able to be sure: *We've been wrong about fat from the very beginning.*

What if the advice from popular diet books, Sugar Busters, The Zone and even from Dr. Robert Atkins—the legendary "guru" of weight-loss in America since the 1970s—had been wrong, failing to understand the complex biochemical, hormonal and genetic factors of *inflammation* that actually trigger such a great percentage of cases of obesity?

If you really want to understand effective weight loss, read *Lose the Weight You Hate* which I published in 2002. The no-amylose diet is well-described, combined with success stories, insights from 25 years of helping people lose weight and dynamite recipes. 70% of my weight loss patients keep their weight off.

Although I didn't know it at the time, as I studied the research and examined my patients during the 1980s and 1990s, I was gradually discovering that the assumptions of medical science about how we gain weight and store fat were just plain *wrong*. The wrong ideas, repeated so often, had become accepted as Gospel, as the newspaper articles demonstrate. Just like with cholesterol; if enough high profile physicians and researchers said being fat was due solely to excessive eating, the media would tag along.

Only no one talks about the real problem in fat, just like in cholesterol: *excessive storing*!

But let's back up for a second. In order to understand why the conventional wisdom about obesity is flat-out incorrect, we need to take a quick look at how the process of getting fat actually *works*, even as we ask ourselves the key question that will solve the fat-mystery once and for all:

Q. What would happen if we looked at obesity as a *symptom*, rather than as a diagnosis? In other words, "What is going on with fat manufacture and fat storage in obese patients that isn't happening in those without extra fat?"

To answer that question, we need to spend a few moments reviewing our notes from ***The History of 20th-Century Fat, 101: Or, Why the Low-Carbohydrate Diets Alone Failed To Keep America Thin.***

The "Atkins Diet Revolution": A Primer

When Dr. Robert C. Atkins appeared on the national scene in 1972 with the publication of his first "Diet Revolution" guide to weight-loss, he was quickly hailed as a bold pioneer with a radically different approach to dieting by the public. The Medical Establishment was horrified, however, by the idea that eating fat and rich food actually was a good idea. Remember, back then, and continuing today, the cholesterol argument influenced the weight argument: low fat and low cholesterol foods were good and that's all there is to it. Until the arrival of the famous diet doc, whose books on "low-carbohydrate" eating would eventually be purchased by millions, most scientific thinking about weight-loss focused on the importance of measuring (and limiting) the intake of dietary fat, reducing total calories, along with the need for lots of vigorous exercise.

But Atkins had a different idea. He knew obesity was not as simple as a failure to eat less and exercise more.

According to the dieting guru, the *real* culprit in weight-gain wasn't fat or lack of exercise but carbohydrates—complex carbon molecules, found mostly in sugar and starches, which the human body rapidly stores in fat cells for future use.

Imagine the shock-waves that must have rippled through the U.S. medical community, when Dr. Atkins began insisting that the best way to lose weight was to avoid carbohydrates and that eating moderate amounts of fat would *not* make most people gain weight!

The Atkins approach turned out to be reasonably effective for millions of people, at least in the short run. By cutting back on "carbs," the adherents of the Diet Revolution collectively lost millions of pounds, even if the weight they lost was invariably re-gained by most dieters within a matter of a few months or years. Regardless of these problems with "weight-maintenance," however, the Atkins strategy was generally declared to be a success and the weight-loss author became a household American name as a result.

But now let's flash-forward 30 years and take a look around.

Let's ask ourselves: What ultimately became of the "eat-no-carbohydrates" approach to dieting that had made Dr. Atkins so famous?

Answer: *Over the long haul, it simply didn't work.*

Why did it fail? Although the biochemical explanation is complex, the fundamental reason for the failure is easy to grasp. That reason is based on the fact that although Dr. Atkins did a truly great job of focusing national attention on issues of weight-loss, he and most of his fellow-researchers were using the wrong "model" with which to understand how human beings get fat.

They did not have access to the new research on *leptin* and *inflammation*!

Dr. Atkins and his colleagues can't really be blamed for failing to understand that being overweight or obese is primarily a result of the patient's own unique biochemistry (that is, of the multiple interactions of his or her fat-regulating hormones) and is *not* a result of overeating or refusal to exercise. They simply didn't have the data on hormone interactions that in those days were "part of the future" in treatment of obesity. The future is now here, creating a model for understanding new treatments that are based on molecular physiology, and not counting calories or fat grams. In short, the powerful new model tells us that managing our weight successfully is primarily a matter of understanding and then manipulating hormones—based on their efficiency at transporting and transforming nutrients during the process of digestion and fat storage.

Don't forget that the brain centers that regulate hunger and satiety, the MSH producing areas of the hypothalamus, are tightly tied to weight. We can start to recognize that the only carbohydrates anyone ever needs to fear are amylose and excess glucose, and that leptin is the most important hormone in weight regulation.

All you really need to know in order to take advantage of the molecular research that has radically changed our understanding of obesity in recent years is one simple fact: Weight-loss is actually about *hormones* (such as insulin, leptin, resistin and several others) and specifically about defeating the "resistance" to the effect those hormones normally produce. Think about a hormone attaching to its receptor like a lock and key. If you coat the receptor, say with some "glue," like gumming up a lock, the hormone won't work. Neither will the key. We call that receptor problem, "resistance."

It is hormone mechanisms gone awry, including resistance, that cause us to become fat, prevent us from losing much fat when we try and make us gain weight on the same low calorie diets eaten by those without hormone abnormalities. Those same hormone defects cause our failure to maintain our fat loss.

The bottom line: If you're one of the millions of Americans whose "genetic inheritance" prevents your fat controlling hormones (especially leptin) from working efficiently because of hormonal resistance, the good news is that you can fix the problem with a new arsenal of medications (such as the "leptin-modifying" TZD drugs) designed to overcome the resistance and help you eliminate fat safely and eat normally without fear of re-gain.

For Real and failed dieters everywhere, the future of weight-loss and weight-maintenance has finally arrived. In the wake of the latest obesity research, it's now possible to reliably correct the problems of excess leptin due to leptin resistance, and the problems of excess insulin due to insulin resistance. And these simple but exciting facts are going to revolutionize the way we treat obesity in the future.

Now, here is where the recent research on hunger and hormones gets *really* interesting. Remember those photos of genetically fat mice that didn't make leptin? When they were given leptin supplements, they all lost weight. At first glance, you might expect that Real, who's suffering from leptin resistance (and thus from a *shortage* of leptin *effect* in the hypothalamus, even though there is plenty of leptin around that normally would do its job), could be coaxed into losing his excess poundage, simply by giving him periodic doses of leptin, right?

Wrong. Believe it or not, extra leptin just makes the Real's of the world fatter. Although research shows that this approach works effectively on mice, our "leptin replacement strategy" doesn't translate to Real. Why not? Once again, the answer lies in biochemistry.

If the patient has leptin resistance, leptin doesn't connect properly with its receptor in the MSH production pathway and the hormone never initiates all the later effects that result in turning off hunger and burning fat directly.

Proving Biotoxin Illnesses Lead to Obesity

Surprisingly, the major breakthrough in our understanding of how leptin resistance helps to make people fat came from work recently completed by our research group in Maryland. For several years now we have been investigating and publishing scientific articles on a threatening new family of chronic biotoxin-associated illnesses.

Interestingly enough, our research group recently discovered a key fact that affects leptin resistance. What we learned was that the *biotoxins* released by microbial organisms such as those responsible for chronic ailments like fibromyalgia, Sick Building Syndrome, Chronic Fatigue Syndrome and Post-Lyme Syndrome *cause a harmful inflammatory effect on the brain's leptin receptors!*

These biotoxin-linked diseases cause their multiple persistent symptoms (fatigue, headaches, muscle aches, blurred vision, short-term memory loss, and many more) because of the way they instantly set off alarm bells within the body's disease-fighting immune system. When the "alarm" sounds, the system quickly begins to churn out some powerhouse chemicals (known as "pro-inflammatory cytokines") designed to help neutralize and eliminate the toxins.

But even as the cytokines set about doing their important work, our "biochemistry plot" takes another astonishing twist. The twist occurs when the inflammatory agents begin damaging a key pathway to the hypothalamus, preventing the leptin receptors there from doing their proper job of allowing leptin to turn on the satiety center! The center normally makes you feel full when it is working right.

You can imagine the reaction among the members of our research group, when we realized that we were confronting a revolutionary concept in weight-loss and weight-maintenance—the notion that much of the resistance to leptin (and other hormones such as insulin) was actually due to cytokine-damage resulting from exposure to environmental toxins!

At first we were in shock. But when we stepped back and reflected on our find, we came to the startling realization that a high percentage of the patients who were overweight because of insulin/leptin resistance had actually *acquired* their resistance as a result of biotoxins from the environment. As matter of fact, there is now convincing evidence to show that about one-third of all leptin resistance is "environmentally acquired" in this fashion; the remaining two-thirds of leptin resistance is the result of "endogenous" internal genetic factors, inherited from birth.

Exciting? You better believe it. Based on our studies of thousands of affected patients, three obesity-related discoveries were now crystal-clear.

First: Being overweight or obese is not about food or "overeating"; it's about resistance to fat-related hormones, which shuts down the effect of leptin on the brain's satiety center, among other effects, so that it fails to tell the patient: "Stop eating!" For 98 percent of weight-loss patients, this scenario is a major reason why they can't lose weight and keep it off.

Second: For about two-thirds of these resistance-linked weight-loss patients (and that's about 60 million people!) internal genetic flaws of insulin resistance and leptin resistance account for the body's inability burn fat properly and to store it instead.

Third: For the remaining one-third, the resistance—and the failure to activate the satiety center—is "exogenous," meaning that it's actually caused by biotoxins left behind by chronic, environmentally acquired illnesses such as Sick Building Syndrome and the Post-Lyme Syndrome.

The implications of our research seem far-reaching. For one thing, we're now exploring an entirely new approach to the inflammatory cytokine syndrome that underlies not only obesity but also diabetes, to say nothing of cholesterol-linked atherosclerosis caused by similar resistance-triggered cell membrane failures.

For weight-loss patients, of course, these implications are nothing less than revolutionary. Among other things, they're now making one crucially important fact obvious to researchers everywhere: If we can defeat leptin resistance, we can defeat virtually all attacks of obesity!

To understand fully, leptin is made from fat cells, and it's released into the bloodstream after we eat fat causing the levels of fatty acids in the blood to rise. Normally, the rising leptin will tell the satiety center that we have had enough to eat. In the case of the leptin resistant patient, however, the effect of eating fat isn't to turn off hunger; the leptin is merely ignored. But the key thing to emphasize here is that the extra leptin affects weight-gain in *two* important ways.

First, if there is an extra amount of insulin in the blood because the patient ate amylose, the combination of insulin blocking and high leptin works to prevent uptake (storage) of fatty acids in fat cells and muscle cells, keeping them suspended in the blood.

Second, the leptin prevents fat cells from burning fatty acids.

One significant result of keeping these fatty acids in the bloodstream is that they dramatically worsen a different problem, insulin resistance, by shutting down intake of sugar by muscle. As the muscle cell starves for sugar, it begins to burn itself for fuel

(Chapter 17). At the same time the liver makes *more* fatty acids from the excess sugar in the bloodstream as a defense against diabetes. The extra fatty acids, meanwhile, simultaneously drive up release of more leptin!

Fat storage, without fat burn, is activated both by the rise of blood sugar and by the rise in fatty acid levels. Without this normal "feedback control" on appetite, Real, our leptin-resistant friend, just keeps on eating. At the same time, the resistance fouls up the storage system for fatty acids in the bloodstream. And the result? Poor Real gets hit *twice*, as a result of his leptin-resistance. He *eats more and stores* more of what he ate.

We now understand the "double whammy" faced by those of us who struggle with leptin resistance. And we remember that the development of leptin resistance is essentially no different than insulin resistance. Both disorders are deeply affected by genetic inheritance and the cytokines that result from the body's response to environmentally acquired toxins, with both impacting the brain's satiety center *and* the body's physiological fat-storage mechanism.

Are you beginning to see now why "eating fewer carbohydrates and getting more exercise" simply won't get the job done for 98 percent of those who are significantly overweight? Welcome to the brand-new Leptin Resistance Era in American dieting!

But if Real has no chance to control his weight-gain through sheer "will power or exercise" how do we help him?

Enter Panacea, the Greek inventor of medicines, who carries a hopeful message for all of us: the recognition that, once we properly identify the *real* culprit in obesity, which is inflammation causing hormone resistance, we can use appropriate drugs to rapidly offset its impact and restore a healthy biochemistry.

The really good news for all of us here is that these drugs will *work*—unlike the Atkins approach, which allowed people to eat fat each day, activating their leptin production and thus turning on the satiety center. (This was the *real* reason why Atkins sold all those books, by the way; his diet kept people from feeling hungry and miserable most of the time!) But Dr. Atkins' technique failed for so many in the end, because it didn't factor inflammation driven *leptin-resistance* into the dietary equation for fat. Nor did he properly understand the crucial role played by amylose, the key carbohydrate that triggers a rapid rise in blood sugar after ingestion and therefore triggers both insulin and leptin resistance.

In spite of his contributions in some areas, Atkins never perceived the crucial fact that the "amylose subset" of carbohydrates is all-important in provoking resistance, and that dieters should avoid the stuff at all costs! The rest of the carbohydrates won't hurt your weight!

Based on the latest discoveries at the molecular level, science now understands that the truly effective way to keep weight off is to *reduce leptin and insulin resistance*. For many people—those whose own genes cause the resistance—the solution to being overweight will likely mean taking a medication that blocks the overactive leptin response and overactive insulin response every day, in order to overcome their flawed internal chemistry.

For those of us whose leptin resistance is environmentally acquired, however, such as the office worker who's struggling with biotoxins from the "Sick Building," the solution to the weight problem begins with removal of the offending toxins from the bloodstream.

And what about the two percent of overweight patients who *don't* have either endogenous or acquired leptin resistance? For this tiny minority, the old rules still apply; to lose weight, they will have to eat less and exercise more!

Okay, time for the next question: Do I have solid, convincing research data on which to base my claim that about one-third of all leptin-resistance comes from the chronic biotoxin illnesses that are now spreading rapidly across America?

The answer, of course, is a resounding "yes." Quite recently, our research group, Dr. Ken Hudnell, Dennis House and I published an academic paper in the January 2005 issue of Neurotoxicology and Teratology on Sick Building Syndrome in which we analyzed 21 SBS patients who were working in five different buildings with water damage and abundant growth of toxin-forming molds. After we documented the numerous symptoms of chronic, biotoxin associated illness, especially chronic fatigue, leptin levels and deficits in visual contrast sensitivity (VCS), we prescribed the toxin-binding and toxin-eliminating medication (cholestyramine), which helps them eliminate the biotoxins, restore their health and correct their VCS deficits.

As we expected, the patients improved dramatically within a couple of weeks and all experienced an immediate reduction of their leptin-resistance, with resultant weight loss! In other words, as our published and scientifically verified findings made clear, these unhappy folks were actually being made *sick and tired and fat* by their toxic environments!

In order to test our hypothesis prospectively to show causation, we stopped the medication briefly to observe the effect of avoidance of the implicated building. Patient symptoms, VCS and leptin didn't change. We then watched as the patients returned to their toxic environments. Within three days, when we looked at them again, their biotoxin-illness symptoms had returned, VCS plummeted and their leptin-resistance was soaring again.

The fact that we confirmed the leptin-changes within three days didn't surprise us, since we knew that the adverse effects on leptin receptors from cytokine responses to biotoxins occur quickly. What was fascinating was the immediate onset of weight gain with re-acquisition of illness. We later discovered that when these patients left their sick buildings behind and then flushed the biotoxins out of their bodies once and for all with cholestyramine, they all *lost* impressive amounts of weight.

We're almost to the finish line now, in our effort to understand how controlling the body's hormone resistance caused by inflammation will help us to control our weight. But one obvious question remains: What medications can help us most here, and what's the basis for their effectiveness in shutting down resistance?

And while dozens of medications are being developed for the huge obesity market, the current winner is the thiazolidinedione family of medications, which we can simply call TZD. These medications are commonly used to control diabetes, consisting now of pioglitazone (Actos) and rosiglitazone (Avandia), and they work by turning on a string of valiant, fat cell-based genes (known as "PPAR gamma") that produce many powerful organic compounds. These compounds decrease *both* leptin and insulin resistance. By using TZD in combination with a no-amylose diet, we can now achieve solid results for overweight people dreaming for a solution. We can control body-weight without starving our patients half to death in the process!

This breaking news about the wonders of TZD will be especially welcome among millions of Americans whose overweight status is due to chronic, biotoxin-associated illnesses, such as Sick Building Syndrome and Chronic Fatigue Syndrome. For these individuals, a few weeks of cholestyramine therapy will start the complex process that underlies effective treatment. Once the toxins that are causing their leptin resistance are gone (hopefully, our therapy won't begin too late) and the hypothalamic pathways that leptin activates are working again, what we see is removal of the resistance that then allows them to shed their excess poundage.

The enormously hopeful news for overweight patients everywhere is that your hour of liberation is at hand. Instead of blaming yourselves (or being blamed by the talking heads on national television) for your fat, and instead of blaming Burger Doodle for serving up too many triple-cheeseburgers, we're going to put the blame where it belongs: on human biochemistry, and on the insulin/leptin resistance which is the legacy of our biological evolution.

We are creatures whose forbears lived for countless millennia as hunter-gatherers under conditions of "feast or famine."

The Obesity Epidemic isn't anyone's "fault"; it is simply a product of our history. Until the arrival of mechanized agriculture, only about a century ago, the human body had never been exposed to a continuous flood of sugars and fats—a flood that never stops. For at least 200 centuries, and probably much longer, the human response to such sugars and fats had been: "Convert this stuff to fat, so that it will carry us through the next famine!" Famine released a life sustaining rise of fatty acids in the blood and then control of blood sugar rises as well, from stored fat.

Is it any wonder, given this evolutionary reality that many contemporary humans with hormonal resistance, when exposed to an endless flood of excess nutrients, are turning into chunky, double-chinned folks who closely resemble Real?

• • •

Cancer from Biotoxin Illnesses

Carol Anderson tells me that at least 5 of her condominium co-owners in the Ritz Residences have cancer. In my 2000 mold patients, there are so many people with unusual cancers that any of us could be tempted to wonder, "Did the mold or the cytokine response to the mold cause cancer?"

I guess if we knew what causes cancer, we might be able to answer the question. All of our studies on cancer causation from environmental exposure are case-control studies. We have this many people exposed to mold and a matched group not exposed. Which group has more cancer? Those studies can tell us about *associations* of exposure with cancer, but they don't tell us anything about *risk* of getting cancer.

We don't have the studies yet to show what the Biotoxin Pathway does to cells that might make them turn cancerous. Maybe the first step in looking at the cancer issue is finally acknowledging that mold makes us sick. It makes sense that acceptance of mold as a pathogen will start us on the cancer causation path.

And maybe the answer won't be long in coming given the interest in high levels of VEGF in some cancer patients. VEGF is a hot research item now, especially since the first VEGF-blocking drug, bevacizumab, received FDA approval for use in treatment of cancer of the colon. It looks like the drug (sold as Avastin) might be useful in pancreatic cancer and kidney cancer as well. When given intravenously, it blocks VEGF receptors, preventing the enhanced production of new blood vessels and limits the increased delivery of oxygen to cancer tissues by action of VEGF.

We have talked about elevated VEGF due to receptor problems and elevated VEGF without receptor problems in earlier chapters. The vast preponderance of biotoxin illness patients have *low* VEGF. All I can say about mold and cancer is that mold toxins hurt us in ways we are just beginning to understand. Perhaps this book will become the starting point for a better look at the incidence of cancer in mold patients.

David and Real are now both healed from their inflammatory cytokine illnesses. They won't forget how they were mistreated by 20th Century Medicine or how they were cured by 21st Century Medicine. You need to know what they know.

It might be that our work with erythropoietin (epo) replacement will lead the way next year in management of chronic fatiguing illnesses. Epo has a fabulous upside to lower fatigue, increase muscle performance and restore respiratory function. We have already seen fantastic benefits of epo in patients being treated for cancer and HIV, as well as those on dialysis. Our data with chronic fatigue patients looks too good to be true. We have to be cautious! Time will tell us more about adverse effects and unexpected benefits.

For now, the Biotoxin Pathway is a road to healing. Let us learn from it and value what we know now about innate immune responses and MSH. We can't forget where we started: patient care.

The Biotoxin Revolution started with one woman telling me in 1997 her memory improved, as did her cough and headache, when I gave her cholestyramine to stop her diarrhea. Thankfully, that day I heard what she was telling me. The years since have brought me a clearer understanding of what being a physician means. Our work is just beginning in the 21st Century.

Listen to the patient, be kind to the patient and never forget that in the end, we are all patients too.

CHAPTER 25

High Noon in Hampton Bays

Hampton Bays, Long Island: April 23, 2004

It was an unlikely group of fighters poised for battle. And here it was, High Noon.

At one end of town were Pat Romanosky, Carolyn Haines and Carolyn's son Fisher. As they walked into the inevitable onslaught of Truth versus Defensiveness, Eva Williams and her two children, Matthew and Stephanie, courageously joined the seasoned veterans of the Hampton Bays Mold Wars. James Havens marched into the line of Mold Warriors too. Another teacher joined but prefers to remain anonymous.

Aligned against them were the power of the administration, school board and all the consultants they'd hired. Theirs would be a flimsy defense resting only on bravado and on suppressing evidence of the adverse health effects caused by the school's mold problem.

There was no turning back now.

Instead of the street dust and climactic shootout of a Western movie, this intense battle would finally decide whether there was mold in the elementary school making people sick. Around the country, there were many interested observers: The lawyers were watching; the newspapers were watching; other parents were watching. In many legal venues relating to mold, this case had center-stage attention. Everyone knew that school officials would be forced to make a final statement today.

School officials announced repeatedly that they had cleaned up the fungal contamination.

"Everything is perfectly safe, and we have the air sampling reports to prove it," the school officials claimed.

"Your building reports are hopelessly inadequate and worthless," countered Pat Romanosky, "No testing has been done to look at the ongoing problems with the old gym. You've presented information that's not related to human health at all. Sick people are the marker for sick buildings, not meaningless building inspection reports you're using in an attempt to obscure the truth. And where are the health surveys that Carolyn Haines has been asking for since Dr. Shoemaker's June 2003 report, repeated in October 2003 and in the spring of 2004? Why haven't you announced those findings? And why haven't you allowed us to wear the personal sampling devices Dr. Joseph Spurgeon suggested we could use?

"Those building reports are an insult to every parent who has a child in this school."

"Everything is fine," responded the administration, sidestepping any direct answers to the issues and questions raised. "Come back to school. Let's be one family again. Pat, you can finish your career teaching at the high school. Fisher, you can come back to school—you won't need a tutor. Our physicians, Dr. Cheung and our Hampton Bays pediatrician, have given their word that everything is safe."

Dr. Cheung was a familiar antagonist. He had put his medical theories up against my practical experience in many legal cases in Maryland when he had been State Medical Examiner, losing repeatedly. He now belongs to a group of physicians who serve as expert witnesses, rarely, if ever, treating patients. They make their money from consultant and legal work, usually supporting the defense of a diverse group of insurance companies and corporations. As far as I knew, he still hadn't treated his first mold patient, though he had plenty of opportunity to talk to and learn from those I had treated. He invariably said these patients didn't have an illness caused by mold, despite the overwhelming scientific and medical data that *shouted* that mold was indeed the cause.

He and I disagree. Fundamentally.

On the surface, what the Hampton Bays Elementary School Board said sounded good. They had retreated on some of their previous inane statements, and now were willing to admit the school had a problem. Earlier the administration had been forced to give up the absurd idea that the building could be made safe even while the toxin contaminated carpets covering asbestos floor tiles remained in place. Somehow, a State Senator came up with $300,000 from the State to fix that problem. And the school agreed to make some corrections because of the mold contamination—wall-board was ripped out in a few places, closets repaired and windows caulked. "Everything was fine; come back to school," they cajoled.

The truth of course, was that the administration had done no more than window dressing. They had no reasonable basis to say the school was safe, in fact no basis at all other than the opinion of their witness, Dr. Cheung, and the inexperienced opinion of a prominent local pediatrician. Since he was the leader of a powerful local group of pediatricians, few parents would willingly antagonize him in public by saying he was dead wrong when he lent his name to bolster the comments of Dr. Cheung.

Pat had been the first to take the plunge into the moldy air of the old school in March 2004. After treatment and some extended time away from school, she was finally healthy again, though her low MSH was an ominous portent of increased health risk from re-exposure. As the school ramped up the financial pressure on her to return to school, she feared returning to an environment that made her sick. But she was going broke. She had no choice but to return to work—the board had denied her access to the "borrow bank," saved vacation or sick days donated by other teachers for use by those who become ill and who need some additional paid days off. Pat didn't have any time off left and had been out of work on unpaid leave for four months. Her only income was her husband's meager mayoral salary. So, with just a few months to go to qualify for a 30-year retirement pension, Pat didn't have any choice—she had to go back into Mold Hall or lose the retirement benefit she had earned.

She came to Pocomoke with her husband Joe for a final checkup. We went over her dilemma: she could stay home and give up the fight, suffer the financial loss and admit she was beaten—even though she knew she was right. Or she could go back into the school with a full dose of CSM to protect her, hoping her previous health breakdown wouldn't happen again.

"Pat, either way you're taking a big risk," I warned. "You're primed for mold illness by your prior illness experience as well as by your specific mold susceptible genotype. If you go back in, trying to buy enough time to get to the financial safety of retirement, you run the risk of *permanent* health effects."

I tried to persuade her that the money was not worth becoming irreversibly ill.

"Pat, forget the money they owe you, forget finishing your career," I begged. "The school is a trap. You can collect your last few checks and sacrifice your health, or you can get out and get on with a *new* life. Why can't they let you finish your time by doing home instruction or something like that?"

She laughed. "Are you kidding? They wouldn't do anything to help me now," she said. "I've got to try to finish my 30 years."

"If you go back in again, you've got to promise to get out as soon as we prove you have reacquired the illness," I said.

Pat left my office with emotions in complete turmoil. She hadn't planned to be a troublemaker or to jeopardize her retirement;

she was fighting for the rights of her students, her co-workers and herself. She only wanted a safe workplace.

Within three days after she was back in the mold school, she was sick again. First her eyes became red and inflamed. Then her throat became sore, she began to cough, and then she started suffering from cognitive changes and fatigue—the familiar symptoms piled up on her like a truckload of bricks. She couldn't go back any more.

"I don't know what will happen to me now," she said tearfully.

Carolyn Haines took the High Noon concept one step further when she demanded that the school, Dr. Cheung and the pediatrician take *personal* responsibility for guaranteeing the safety of the children if they returned to school.

The lawyers were *really* watching now. Personal guarantees?

Dr. Cheung sent a letter back to the school, co-signed by the pediatrician. After a review of my statements, they wrote, "The school is safe."

• • •

It was High Noon. Everyone involved was on edge. Waiting for this fight was unnerving—just like what the hero feels in a high-tension movie thriller.

If the school officials were right that the massive fungal contamination was cleaned up, then no one would get sick again. If the school were still contaminated, then the volunteers would be sickened quickly. As it turns out, the school officials were wrong, nearly *dead* wrong, in Pat's case. And they were wrong to assume she would be alone in this shoot-out.

• • •

Fisher Haines had done well away from school. He'd had no allergy problems, no cough and so no medications were needed.

He'd received one hour each day of home instruction, but really missed seeing his friends at school. Carolyn tried hard to be both a Mom and a Teacher, but she also had two other young children who needed her. Fisher should have been in school, safe from mold attacks.

He came to Pocomoke for a final evaluation. Carolyn was scared.

"The board says Fisher can come back to school and everything will be fine," she said. "But I saw what happened to him before, and now again to Pat. It's all a bunch of lies!"

Fisher took the VCS test. It was normal, just like when he was treated with CSM and out of the school before. His physical exam was normal. Thankfully, his MMP9, a measure of inflammation caused by exposure to toxins, was 194, a marked improvement from the profoundly sick level of 450 he'd had when he was first ill. Carolyn didn't comment about the fact that his MSH had fallen to less than eight, a value so low it could not even be measured. Fisher had no protective reserve when the toxins attacked his innate immune response, as they surely would sometime.

"What will happen to Fisher if he goes back into the building?" she asked me.

"Carolyn, you're better off moving," I said. "Put him in a school that you can prove is safe. It would be easier than trying to fight the board. We just don't know what will happen to kids this young who take immune hits this profound."

"They've given me their guarantee that everything will be fine ... I have to take them at their word," she said firmly. "Besides, I can't just get up and move. I *live* in Hampton Bays—a half-acre lot with 3-bedrooms costs more than a million dollars! We can't just leave our friends and family. No, they'll have to fix the school."

"And what about all the other sick kids," she asked pointedly, "being pumped full of allergy medications and antibiotics all the time by their baby doctors. Believe me, I'll never take him back to those pediatricians. Someone has to make a stand."

We agreed on a plan. We had already done the repetitive exposure protocol twice. This time we wanted answers to a different series of questions. Could he return to school safely, with CSM? Without CSM?

Fisher would have his labs and VCS done now. I would record his symptoms: none. He would go back, without taking CSM, into his first grade classroom at Hampton Bays Elementary, located in an area away from the worst fungal growth. If he became ill, his new physician would record symptoms, perform a VCS test and measure his labs, especially MMP9 levels. Then he would stay in school, take a full dose of CSM, repeat the VCS, note symptoms and repeat the MMP9.

In other words, we would look at him away from school, off medication and compare him to being in school, off medication. If he responded badly and became ill, we would see what CSM did for him. If he recovered on CSM, with a return to normal VCS and MMP9, he would need to be medicated with CSM indefinitely to protect him *from the school*. Carolyn knew about some of my patients who had successfully used CSM for preventing illness reacquisition despite re-exposure.

If Fisher did worse, however, with solid evidence of illness breaking through, despite CSM then Carolyn could ask them politely to find a safe alternative to the school for Fisher. Who cared if a board or a doctor had "guaranteed safety"? No one can do that. If the school didn't volunteer help, including financial help, Carolyn's attorney had some suggestions for alternative schooling.

It didn't take long for Fisher to become sick. He had 10 health symptoms within three days. His MMP9 rocketed to 550 and his VCS score crashed.

He stayed in school, took CSM, with improvement in symptoms and VCS, but his MMP9 barely budged. The mold's toxins were ferociously attacking him, hampering his immune delivery system, even through the shield of CSM. Was the school cleaned up? Not enough for Fisher. He was holding his own, but at a huge cost of ongoing immune activation. Would he end up with antibodies to myelin basic protein and brain lesions from high MMP9 caused by mold exposure? Who could wait long enough to find out?

Carolyn had no choice. Fisher had to leave.

• • •

Eva Williams was a new Mold Warrior. But she was a fast learner, and she knew her kids were exposed and in trouble.

Her two children, 8-year-old Stephanie, and Matthew age 5, were already sick. Stephanie had more symptoms than Matthew, but Matthew was beginning the downward spiral Eva had seen with Stephanie. The kids had unusual headaches, abdominal pain, constant respiratory symptoms and moodiness and *new onset* learning problems—they just weren't themselves. Eva noticed that whenever the kids had to go into Pat Romanosky's old gymnasium, they got much worse, but the school's new wing also made them ill.

She brought her family to Pocomoke for a standard evaluation. She was horrified to find that her children had the "dreaded" HLA DR genotypes, the ones that convey an extraordinarily increased risk of *chronic, irreversible health changes* following exposure to biotoxins.

Even worse, they both had extremely high levels of antibodies to gliadin, the *real* toxic agent for those who can't tolerate gluten. Gliadin is in so many foods since it's often added to whey. Whey is a protein in cow's milk (remember Little Miss Muffet eating her curds and whey?) and it's added to an incredible number of commercial products. If you're looking for gluten and whey in the grocery store to avoid them, good luck. Start with the obvious sources—bread, flour, milk products—but then add the unpredictable ones: tomato sauce, fried onions, frozen hamburgers and cheese twists.

If you have antibodies to gliadin, each time you eat foods that contain gliadin, you'll absorb the protein (that's not the normal response), setting off a cytokine reaction (also not normal), making you sick. Gliadin antibodies will ruin your life!

Did that mean Eva's kids could never have a sandwich or a piece of pizza again? Imagine being a child raised in 2004 without wheat (and add amylose starch, gluten and whey to the list).

As if the news for Eva wasn't bad enough, MSH levels for both Matthew and Stephanie were incredibly low. But they were typical cases and they responded to CSM.

So with hope that any symptoms would be treatable, and after informing the school administration, she followed the 5-step repetitive exposure protocol. Her plan was to prove (or, hopefully for her kids, *disprove*) what she knew already: exposure to the school was destroying her children's health.

• • •

Quietly, the next member of the Magnificent Seven joined the Hampton Bays fighters. The teacher—we'll use the pronoun "she" for this discussion—was fine working in the remediated area of the school as long as she took CSM. But when she saw what the others were doing, she joined the rest of the Mold Warriors by simply stopping CSM, and monitoring symptoms, VCS and labs. To no one's surprise, she was another "case."

• • •

James Havens' Dad brought him to Pocomoke for the standard work-up that he couldn't get in Hampton Bays. His mother couldn't come, but joined the visit by phone. The tragedy here is that the evaluation is so simple. As his father said, "At first I didn't know what was making him sick, but now the cause of James' illness is so obvious. The tests are so easy to do. What is going on that I have to drive seven hours one way to get some results?"

In a way, it was completely absurd that these people had to risk illness to go to school. Here was a large group of sick people from the building, including Pat (who even had a *positive* blood test for *Stachybotrys* toxin), each exposed to ongoing water intrusion, lack of aggressive cleanup and no remediation of the *Stachybotrys, Chaetomium, Aspergillus* and *Penicillium*, yet the school said everything was fine. When I ask experts in remediation what should have been done to clean up the molds at Hampton Bays, they all say a whole lot more than what was done.

And then the board bet their whole case, relying on air quality consultants' and Dr. Cheung's opinions. Despite multiple chances to change their minds, the administrators wouldn't back down. Why be afraid of four young children, two Moms, two teachers and a country doc from Pocomoke?

In the meantime, Fisher Haines' story was covered in several articles in the Hampton Bays newspaper, *The Southampton Press*. Everyone in town now knew he, too, was quite ill after he'd attended school. His repetitive exposures trial data were incredibly powerful. How could any logical person—eliminating those "experts" whose opinions weren't based on science—not recognize his illness?

The school is silently destroying Fisher, despite protective use of CSM. Carolyn was so angry, she told me she wished the doctors for the Board would "just sit down and shut up."

The end came in a simple letter to the Superintendent of Schools. I told the woman in a letter written at 12 o'clock, April 23, 2004:

> "As per our agreement, I have monitored a cohort of individuals with illness familiar to you and your consultant physicians. You have obtained a court order demanding testing for toxigenic fungi in Pat Romanosky's home, even though we have shown her to be well in the home, off cholestyramine. You have had Ms. Romanosky and Fisher Haines examined by Dr. Cheung. These patients, and Matthew and

Stephanie Williams, as well as one of your staff, followed a diagnostic protocol to confirm that the school still contains illness-causing elements in the indoor air.

"Each of these patients became quite ill when re-exposure to the school was the only new variable in their illness. Prospective documentation of illness with recrudescence of symptoms, VCS deficits and changes in laboratory biomarkers for mold illness confirm the diagnosis.

"It is time for you to stop denying the truth about mold illness in your schools. Please initiate a proper clean up, together with a building-wide health survey to determine the extent of the illness in students and staff under your administrative responsibility. While I am glad that you have ordered a number of clean-up items, though each was accomplished only after unnecessary argument and delay, the limited number of items your school has undertaken never corrects the fundamental problem of mold contamination, in my experience.

"I will be happy to discuss the proper procedures for group health screening with you, as it is unlikely that either of your consultants will know what constitutes a sound approach to diagnosis, given that they couldn't recognize obvious mold illness in two cases previously."

• • •

I don't know who will win the legal shootout that will happen in Hampton Bays schools, but I do know that the students will be the real losers. Lawyers will soon begin the lawsuit process that could threaten the tax base of Hampton Bays.

Because the damage to the school has gone on for so long, with mold allowed to reproduce unchecked for such a long time, it may be now beyond repair. Destruction of the school might be the only cost-effective way to recycle the grounds and salvage some financial benefit for the school district. If that's the case, then the school needs to be destroyed, before it ruins the learning ability of another student or teacher, before another person becomes ill.

The dust is settling, even before the lawyers and the vultures descend to tear the carcass of what was a good school into pieces. Pieces that even the hyenas wouldn't eat.

Pat has her grandchildren and her health. Joe still works for a pittance. Eva is home schooling her kids and the Haines family has thought about moving to Fisher's Island, an island nearby, named for his ancestors. James Havens still doesn't know where he will go to school next year. Carolyn would not even consider the nightmare of exposing her two younger children to any mold in Hampton Bays schools.

These children, teachers and mothers are Mold Warriors. They sacrificed their careers, homes and reputations to fight for a truth they knew: Mold illness is real and it hurts people, damaging their health, sometimes permanently. Denial of mold illness in the face of overwhelming medical evidence by anyone for money or for undefined gain must be stopped.

There will be more mold fights in the months to come. Our academic papers will quickly convince enough medical skeptics willing to read to stop arguing about mold illness and start researching the issue. Experience in day-to-day clinical work will convince *practicing* physicians even faster that the Biotoxin Pathway is easy to use, and is medically and scientifically sound. Using it works, if we catch the illness in time.

For those with mold illness we consider irreversible now, like Carol Anderson, research is in its infancy. Perhaps MSH replacement will be the answer; perhaps a new approach to cytokine cascades is in order. We don't know yet if erythropoietin or some other therapy as yet unknown will help. Hope will never burn out for her; she'll keep on hoping for answers.

And as for me, I'll never give up looking for them.

EPILOGUE

Putting it all Together

"You're interested in the legal implications of health effects caused by exposure to the air inside water-damaged buildings," I told an audience of attorneys and mold experts at a recent conference in Las Vegas. "I'm interested in *treating* those health effects. Our interests aren't necessarily the same. I believe that litigation shouldn't obscure the paramount need to improve the health of mold-illness sufferers."

In Chapter Seven, we met Bishop May, a minister in the United Methodist Church. He demanded that his staff be treated for mold illness caused by exposure to a building his organization leased, even if it meant that he and his employer might be named as defendants in a subsequent lawsuit.

Now we'll meet Matthew Hudson, chairman of Maine-based Scotia Prince Cruises. Like Bishop May, Hudson adamantly safeguarded his employees' health when they found themselves in a dangerous situation that involved mold in their workplace, a building his business had leased for 35 years. Luckily, Hudson realized his employees and his customers deserved to work and do their business in a safe place. So he took the necessary steps to ensure their safety, even if those actions lead him to litigation.

The Scotia Prince Cruises mold story includes all of the things necessary for a good drama: compelling characters and cover-ups, mysteries and heroes, stonewalling and enough intrigue to make a major motion picture. Matthew Hudson's efforts to heal his employees and protect his customers—the public—shouldn't have been so complicated or so difficult: it should be an easy, everyday practice. But it wasn't easy, and it probably won't be for some time to come. After reading this book, you understand why.

While both Bishop May and Chairman Hudson offer exemplary examples for the rest of us to follow, there aren't enough of them. In more than seven years of treating mold illness, I've only met a few people like these two men. Regrettably, many other mold-injured patients don't work for enlightened, generous individuals; instead, they're forced every day to fight for their right to a safe workplace, their benefits and their jobs, all the while fighting to regain their health. It's an overwhelming task.

• • •

If you've ever been to the rocky coast of Maine in spring, summer, or fall, you've probably seen the 485-foot ferry that travels back and forth between Portland and Yarmouth, Nova Scotia every day. The *M/S Scotia Prince* is the pride of Scotia Prince Cruises, which offers daily ferry service and cruise-drive vacation packages too. Some people ride the ship over at night and come back the next day just to enjoy the food, the entertainment, the hospitality and the sea.

The ship's comings and goings are a familiar sight to those near the port cities where it docks. Each evening about 7 p.m. during tourist season for the past 35 years, the ship has glided past Bug Light and eased into its berth at Portland's International Marine Terminal (IMT). An hour later, the ship begins the eleven-hour trip across the Gulf of Maine, filled with passengers, vehicles and supplies. Since 1970, the ship has hosted 5.4 million passengers and has securely transported 930,000 vehicles. That's more people and cars than most cities can claim!

During the crossing, passengers usually enjoy an excellent dinner followed by entertainment and a comfortable night's sleep in one of the *M/S Scotia Prince's* 322 renovated cabins or suites. In the morning, a sunrise breakfast is a fitting start to the day. Afterward, passengers enjoy the immense beauty of Nova Scotia, Prince Edward Island, Cape Breton or other destinations in the Maritime Provinces. The return voyage features sightseeing over the picturesque North Atlantic, with whale sightings, bird watching and sunshine-soaking popular activities.

The *M/S Scotia Prince* employees do their jobs well, reaping praise from their many repeat clients, and earning top-notch scores from the government agencies that periodically evaluate them. That's because safety, courtesy and efficiency form the foundation of Scotia Prince Cruises' ferry operation. Safety demands that every detail be checked twice, despite the need to keep to a tight schedule. They take just sixty minutes to unload and load the *M/S Scotia Prince* twice each day, every day for six months, without slowing for occasional wind, rain, high seas or hurricanes. That means they load and unload onto the ferry in only one hour everything necessary to keep a floating luxury hotel and all its typical amenities, 200 crewmembers, 1,000 passengers, and a 200-vehicle parking garage operating at full steam.

As owner and chairman, Matthew Hudson believes he is ultimately responsible for the quality of service aboard the ship and for the safety of his clients and employees, whether they work onboard the *M/S Scotia Prince* or in one of Scotia Prince Cruises' land-based operations.

Bishop May and Professor Hudson showed us the honorable way to take care of mold issues as employers and as people. They refused to shy away from the medical, legal, professional or human responsibilities involved, regardless of the costs. Their actions make them champions for mold patients everywhere.

You've already read about Bishop May's dilemma and how he handled it. Now put yourself in Matthew Hudson's place and ask yourself: What would you have done?

• • •

It was a new millennium, and at age 58, Professor Matthew Hudson had recently completed a distinguished career as a corporate savior. Trained initially as a barrister and solicitor, he'd succeeded as a venture capitalist and "turn-around" specialist focusing on aerospace, electronics and transportation. While living in Scotland, he was asked by Margaret Thatcher's former Minister of Defence Lord Younger, then Chairman of the Royal Bank of Scotland, to save Prestwick Airport in Glasgow, Scotland, from closing. Under Hudson's leadership, the airport introduced low-cost airfares to Scotland, made a remarkable recovery and became a model for airport privatization worldwide. His moniker, "Professor," came from his subsequent tenure as visiting professor at the University of Glasgow's Business School.

In 2000, Scotia Prince Cruises was called Prince of Fundy Cruises. Its owner had died some years earlier, senior management was ready to retire and the business was winding down. No one with the necessary money had stepped up to buy the firm, and the ship needed too much in additional capital expenditure to meet the regulatory requirements that allow large ocean-going ferries to sail safely on international waters.

Happenstance brought Matthew Hudson into the picture, but love of a challenge kept him there. Hudson thought an upgraded ship and high-value cruise/drive vacation packages between New England and Nova Scotia would prove popular. The ferry business there had always been profitable. After all, there are only three ways to get from New England to Nova Scotia: drive a full day; fly from Boston to Halifax and rent a car; or take a ferry. That's why Scotia Prince Cruises is a key player in Maine's tourism-based economy. The economic benefit to Maine associated with the ship and its passengers easily reaches $55 to $75 million each year.

Portland's International Marine Terminal building has had maintenance problems since soon after it was built in 1970. Owned by the City of Portland, it's been the leased headquarters of Prince of Fundy Cruises or Scotia Prince Cruises ever since. It was also used over the years by staff in Nova Scotia's tourism department, the U.S. Coast Guard, U.S. Department of Agriculture and a few federal border protection agencies.

Portland officials were anxious for Hudson to buy the firm, but he insisted that as a prior condition to his family's investment —more than $30 million to buy the company, plus the needed additional millions to upgrade ship, software, computers and skills—city officials had to agree to enter a new long-term lease agreement, and to move the business and the ship to a new ferry terminal located in the city's center by the end of 2002. The present terminal is located in a run-down part of town and isn't conducive to growth. When city administrative officials told him the long-planned Ocean Gateway project—a waterfront rehabilitation plan—that included the needed physical plant upgrades—was already approved and financed, and the city manager signed a formal Letter of Intent *guaranteeing* that the city would meet all of the requested obligations, the last of the deal conditions was satisfied.

The Hudson family bought the business the next day, Aug. 18, 2000.

• • •

Although the *M/S Scotia Prince* provided amenities like a casino, dining and entertainment, the Hudsons immediately upgraded it. To add value to the ferry experience and to move the brand up-market, they spent $2 million on the ship's unused upper deck, adding a Tiki Bar & Grill, two hot tubs, a disco, an exercise area and a sun deck; below, new additions included a new salon and massage spa, photo shop and laboratory, doctor's office with fully equipped infirmary and a remodeled and expanded tax & duty-free shop. Existing cabins were refurbished and new passenger cabins and suites were added. Many new safety and navigational aids were installed, and they expanded the number and variety of vacation packages offered.

The Professor brought in top-notch senior management staff with whom he'd worked before, including John Hamill as chief operating officer, attorney Bob Schrader as general counsel and his son, Mark Hudson, as vice president of finance and marketing. The new management team created a solid foundation for success, developing an entirely new ethos that included new ways to report on programs, new methods to measure success and and new ways to encourage responsive management from the wharf to the call center, from the engine room to the bridge.

Hudson also promoted from within and brought in new people with the talents necessary to reorganize the business to meet 21st Century market demands. His employees were well educated, competent and willing to work at a demanding level year-round.

John Hamill moved into an office in the far corner of the terminal in 2001. Another take-charge guy, Hamill had worked with the Professor in Florida; before that he was in charge of all buildings and IT facilities at Miami International Airport.

Hamill bought an apartment in Portland, although his family remained in Sarasota. He saw them whenever he could during the summer operating season and was mainly home in Florida during the winter. When in Portland, he worked constantly, spending long hours in the terminal.

Bob Schrader had an office just down the hall. Now the legal eye of Scotia Prince Cruises, he too had worked with the Professor in Florida, where he'd been a prominent member of Florida's second largest law firm.

After just a few months, Schrader, who'd achieved 99.9 percentile ranking in his college and law school test scores, started having trouble remembering things. His back began bothering him and his allergies too, and he had frequent doctors' appointments for a variety of problems.

At the same time, the Professor noticed that John Hamill was often too tired to finish the 12- to14-hour days required to run the company during high season. Hudson was concerned and increasingly frustrated when these top-tier professionals began to lose focus and to make errors. He heard anecdotally about alarming symptoms in his other employees; they were taking sick days more frequently than ever before, and their health problems not only defied diagnosis, but didn't go away.

In spring 2001, Hudson rented a house near Portland for his family so he could work directly with the management team. He spent many hours every week checking reports, projections and financial results, making corrections constantly. Months went by, with errors and cognitive impairment now features of his daily life. Why are they making so many mistakes, he wondered? Worse yet, the same errors were being made over and over again, as if this staff had lost the ability to learn new things.

It must be the stress, Hudson thought. Long hours without a break create problems for everyone, even his dedicated staff; when you added separation from their families, of course things would be tough.

Or was there something else? Hudson was having his own short-term memory lapses. He didn't like to admit it, but he was irritable and more critical of shortcomings in others than he'd ever been as well. He was tired but not sleeping, putting on weight and becoming clumsy. Must be his age; perhaps he was "losing his touch." Why had he ever given up that nice retirement on Biscayne Bay?

That fall new accounting and reservations management staff were hired. By January 2002, however, these departments and their ancillary functions were moved from the terminal building to Yarmouth, Nova Scotia because the terminal was such an unpleasant place to work. Also, it had become quite clear that the city would not meet its 2002 deadline to complete the promised new facilities.

• • •

Portland's International Marine Terminal building is located behind a semi-porous sea wall in the harbor, and rests on what was once an old railway embankment. Rather than construct the building on pilings, the original builders added crushed stone and cribbing to the embankment, put down a reinforced concrete floor slab over that, and then built the terminal building on top of the whole thing. This "floating" construction method met city building codes at the time, and tends to stand up well over time unless it's disturbed.

The terminal building is constructed of a steel skin over two-inch by four-inch wood framing, with insulation in the space between the metal outer skin and the interior wallboard. There's no interior vapor or outer condensation barriers. The corrugated steel skin is simply attached to plywood panels. The cellulose in plywood is the food that mold needs to thrive.

Every morning the sun heats up one side of the steel structure, warming the seaside moisture-laden air in the wall spaces until it turns into vapor. As long as the skin and the wall cavity stay warm, the water stays in its vapor form, but as the sun moves overhead, one wall moves into shadow and begins to cool, while another wall begins to receive the solar "gain" and begins to heat up. As the metal skin cools further, moisture condenses inside the wall against the cooling steel, trickling down a corrugated valley to the bottom of the wall cavity, where a welcome mat of damp, moldy plywood awaits it.

Unfortunately, the building's outer wall has no flashing, weep holes or drainage sites, errors made too commonly in other cavity-wall construction like brick and block, where thermal gain inevitably causes condensation when subsequent cooling occurs. Water sitting in the bottom of a *hidden* wall cavity creates a haven for mold.

The terminal is a cavernous structure; its top story is an open attic that provides a safe roost for hundreds of pigeons, which in turn provide the building with an ongoing supply of guano. Mold isn't the only environmental agent that can make people sick, although guano wasn't the source of the illnesses in the staff. It didn't matter though, because the landlord, the City of Portland, didn't bother to investigate either problem.

Who knows when the mold silently spreading inside the walls of the terminal started to make people sick?

• • •

In a strategy common among ship owners, Hudson planned to add a winter route to the ship's itinerary, maximizing the *M/S Scotia Prince's* year-round income potential. He thought a new route between Tampa, Fla., and the Yucatan Peninsula would be popular. It wouldn't take much to modify the ship for the waters of the Gulf of Mexico and its longer voyages, and even with competition from endless cruise liners leaving from Miami, Fort Lauderdale and Tampa for the Caribbean and the Yucatan, a Tampa-to-Mexico ferry route made sense. The route could cater to U.S. travelers who wanted a cruise/drive Mexican vacation, as well as for Mexican vacationers who wanted to visit Florida.

In the winter of 2002–2003, several Scotia Prince Cruises executives spent a few months in Mexico and Florida to look after the route. Hudson feared his team would make far more mistakes in a new, challenging business environment, but happily, they didn't. In fact, things went surprisingly well, and no one really worried about the group's ongoing ill health and the mistakes. Except Matthew Hudson.

Up north, 2003 brought a struggling economy, new restrictions on international travelers, the Iraq war and SARS, a major problem in Canada. As a result, not many people planned to visit Canada. Scotia Prince Cruises had done well financially in 2001, even after 9/11, and 2002 had been solid as well.

The City of Portland had repeatedly formally agreed that no other ferry would compete with the *M/S Scotia Prince* on the Portland-Yarmouth route; in turn, they required that Scotia Prince Cruises never operate a competing ferry service from anywhere else in New England to Nova Scotia.

Nonetheless, that agreement didn't stop the owners of a high-speed catamaran ferry running between Bar Harbor and Yarmouth, Nova Scotia, from discussing an expansion into Portland in local newspaper articles. Legal questions complicated the issue, and many doubted the Portland-Yarmouth market could support both ships, but the Cat's owners have continued to express interest in bringing the high-speed ferry to Portland.

At this same time, local newspaper editorials asked whether Portland had outgrown the ferry agreement; some even questioned whether the city should continue to honor the existing agreement between the city and Scotia Prince Cruises, *the agreement on which the Hudsons had based their multi-million dollar investment.*

Don't forget that in 2003 the cruise industry was suffering. Portland newspapers estimated the number of cruise ships docking in their harbor in 2003 would drop 50 percent, causing as much as a $20 million loss to the local economy.

On September 29, 2003, a Category Two hurricane hit Nova Scotia. The storm brought a lethal storm surge nearly five feet above normal, causing an estimated $100 million in damages and eight deaths. U.S. and Canadian news shows reported that Hurricane Juan was one of the most devastating storms to hit Nova Scotia in more than a hundred years, further scaring visitors away.

There was a bright side, though: The catamaran, with its much shorter trip across the protected Bay of Fundy, was forced to cancel, but the larger monohulled *M/S Scotia Prince* sailed, arriving and departing Yarmouth on time.

Meanwhile, travel to Nova Scotia from the U.S. was down, too. By the end of 2003, the province had 20 percent fewer visitors from the previous year, although Scotia Prince Cruises increased its share of the market. Despite the slump in 2003, the efficiencies brought in by Hudson earned the company a profit and new capital investment continued. The tough market conditions continued into early 2004 but then picked up. By August, sales were excellent.

• • •

By summer 2004, Hudson knew that the problem with his management team wasn't just that that they were "losing their touch" or "getting old." There was something *wrong*. He suspected that the mysterious respiratory illnesses, chronic fatigue, unusual joint complaints and the cognitive problems his management team suffered from might *all* be related to exposure to the indoor air of the terminal building.

Hudson had definitively noticed that his own health was deteriorating as well. He was out of breath, he wasn't sleeping well, he was chronically thirsty, and most troubling of all, he was forgetting little things that he had in mind only moments before, and he wasn't finding words that he'd always known. And that bizarre taste! For a man whose wife is a fantastic cook, having everything taste like he was sucking on copper pennies was a real setback.

He was receiving regular medical care from some of America's finest physicians: nothing had changed in his life except where he lived and worked. Nothing on his short list of medical problems explained the deterioration.

By this time, every senior manager was ill or acting strangely. Bob Schrader got lost driving to the office in Portland from his home in New Hampshire. John Hamill was irritable and forgetful, always achy and tired. So were all the others who worked inside. The crew on the ship was fine, and everyone else who had never worked in the terminal building was well.

One of the port crew, a rock singer well-known locally, became ill working in the warehouse and couldn't sing well or even breathe normally. He was hospitalized for a pneumonia-like illness, though no diagnosis was made. Being away from the terminal and the wharf helped, but his symptoms re-emerged when he returned to work.

Meanwhile, Hudson's plan to be a "non-executive" had fallen completely by the wayside. He again stepped into the role of hands-on CEO, because all of his senior staff seemed to be suffering from a slow mental and physical deterioration.

The loss of people had been going on for a year. Port Captain Jay Frye had abruptly retired in June 2003 after "severe arthritis" made it impossible for him to work. Jay's job for twenty-five years was simple: he was responsible for everything regarding the ship in port and her highly choreographed 60-minute turnaround —twice a day, every day, May to October. If you needed to know about anything and everything operational, the answer was, "Ask Jay." He was there night and day.

Since he'd signed on with the Scotia Prince in 1978, he'd kept himself in top physical condition, but in the fall of 1997, both knees suddenly swelled up like balloons. By November, arthroscopic surgery was necessary; surgeons found massive inflammation, but no structural problems.

But the building had been the cause of health problems before. Jay's wife, Sadie Anne, an employee of the Nova Scotia government's Tourism Department, worked in the terminal building from 1995–2003. That same winter of 1997–1998 that Jay suffered from inflamed knees, she and Jay suffered through terrible sinus problems. Like Jay, Sadie Anne was hard-working; she worked until two days before her daughter was born at the end of June 1998. Following her summer maternity leave, though, she felt better.

Her symptoms returned within days of her return to work in the terminal building in October 1998. In those days, no one thought the building could make anyone sick. Back then, they'd have been shocked to find that it was mold illness that started to slow both Jay and Sadie Anne down, that eventually led to Jay taking early retirement in 2003. As we now know the inflammatory arthritis that finally defeated him was *caused* by exposure to mold.

By the summer of 2004, Hudson had fully realized that quite a few others were suffering health problems or acting oddly and unlike themselves: for example, Hudson's vice president of hotel operations stopped returning calls or acknowledging emails. Another vice president exhibited irrational personal and work-related behavior, then suddenly resigned, and the company's trustworthy, reliable and hard-working director of reservations began to suffer from terrible headaches and often had to work from home.

When Hudson sent three of his senior managers for sleep apnea testing because they seemed like zombies, all three were diagnosed with the condition, which causes the body to stop breathing as many as hundreds of time each hour. That was strange because none had the obesity and extra soft tissue in the neck that would lead to obstruction of air inflow while sleeping.

What was behind all these weird illnesses? Was something was going on in the building? What was the unifying factor? Professor Hudson began to do some research online.

In August, the company celebrated when the ship passed its latest semi-annual inspection by the Centers for Disease Control and Prevention (CDC) with flying colors—of a possible 96 points, the ship and its crew scored a perfect 96. In fact, the ship's CDC scores had increased steadily since the Hudsons had bought the company in 2000. The perfect result was the highest in the company's history.

How could it be? The ship's staff was perfect, but those who worked in the terminal were falling apart.

• • •

When Hudson began to read about Sick Building Syndrome, the symptoms exactly matched those of his co-workers. How could he have missed such an obvious group of clues? Why didn't he see what was right in front of him?

Finally, he saw the obvious: many previously well people exposed to a building with longstanding water intrusion were ill, but the others in the company who didn't have the same exposure were not ill.

The reality of the possible source of the illness shocked him. Mold! Could the terminal building contain molds that can cause immunosuppression, cancer and all the rest? And what about that Maryland doctor who said molds activated the innate immune response inappropriately? If the building had molds—and it seemed so obvious now that it did—was this mold *the* threat that caused chronic fatigue and all of the other symptoms that plagued his staff and his company?

Who could confirm what he suspected? He asked John Hamill to hire an environmental consultant. A technician came and found significant amounts of mold in the building in both air and bulk samples. The numbers were astonishing: they showed high levels of toxin-forming fungi growing throughout Scotia Prince Cruises' sections of the building.

When Hudson read more online about what amplified mold growth could do to people, his heart sank. He saw clearly why his senior staff had mysteriously turned into people who were in such a fog some days that they could barely tie their own shoes.

He couldn't sit still and do nothing. Mold was taking from him his most precious business resource, his friends and co-workers.

What he did, and what Bishop May did, needs to be done by anyone who confronts mold growing in water-damaged buildings.

The next morning, Monday, Aug. 23, Bob Schrader delivered copies of the lab results to the mayor and city manager of Portland. John Hamill had already closed Scotia Prince Cruises' portions of the building, although the city kept open the parts used by federal agency employees and the passengers being processed inbound. In fact, as I write this epilogue in mid-January 2005, they've never closed those areas. Attempted remediation? Yes. Cross-contaminated? Undoubtedly, and re-contaminated them too. But closed them? No.

John Hamill was experienced at emergency operations. Some years ago, when he was facilities manager at Miami International Airport when Hurricane Andrew hit, he'd been in charge of getting the airport working again almost overnight. So it was no surprise that by that same afternoon, he had tents set up in the parking lot adjacent to the terminal building. Scotia Prince Cruises removed all of their employees from the building, refused to let the public use it and immediately began operating out of the tents, processing passengers and vehicles there from Aug. 23, 2004 until the end of their season in late October.

Hudson needed good advice about mold quickly. He hired the most knowledgeable mold experts in the U.S., including Dr. Richard Lipsey of Florida. Environmental technicians gathered more samples, which were sent to a top-flight laboratory to evaluate.

Dr. Lipsey quickly told Hudson that the spore counts were dangerously high to people and that the directive to keep an unsuspecting public as well as his staff out of the building was the right decision. To make sure, Hudson brought in other internationally known experts for second and third opinions; the existence and the seriousness of the mold problem was confirmed by each.

No one was happy about operating from the tents in parking lot in a post-9/11 international travel era. But was there any other way to safely do business? Hudson had a brand to protect, a company to keep afloat—literally and figuratively—and employees who needed him to make the right decision. And what about the thousands of tourists who'd made reservations and who were counting on the *M/S Scotia Prince* in 2004? Hudson simply couldn't let anyone down.

Armed with irrefutable evidence, from August 23 on, Professor Hudson repeatedly gave copies of his growing sheaf of reports to city officials. "The premises aren't safe," he told them. "Please fix them, *now*."

Instead of saying what readers know *should* have been said, the city began the maddening "deny-and-obscure" dance we've seen all too often in *Mold Warriors*.

On Aug. 23, the city's port director told two Scotia Prince execs that they were "overreacting." On Sept. 1, a city official said in a press release that they'd "concluded that the public waiting areas at the terminal were safe for public use," and that "the results of testing in the portions of the building used by U.S. Customs, Immigration and Border Patrol indicated that levels of mold were low in these areas."

"Low." What did that mean? Did that mean only mold-susceptible patients—a whopping 24 percent of the population—would be sickened by being exposed to that level of toxigenic fungi? Somehow, Hudson wasn't reassured by the public relations spin.

On Sept. 17, the city's position shifted a bit; a city representative told the media, "We don't believe there are any air quality concerns over there," but admitted that, "there had been water damage from the (heating and air conditioning) system." Furthermore, the city intended to make repairs that included "replacing damaged carpeting, walls and ceilings, and should be completed next week."

In that same media release, city officials said, "they still believe Scotia Prince Cruises is overstating the threat posed by mold" and that the city "still believes passengers could safely use the building."

You haven't heard the whole bizarre story yet.

Curiously, on Sept. 25, 2004, City Manager Joseph Gray ordered the terminal closed after an engineering report showed its western end could collapse from a moderate seismic event, an intense snow load on the roof or a significant wind storm. He said in a written statement, "I have taken this extraordinary measure because it is imperative that no one be at risk of any injury using any city facility."

Three days later, an assistant city manager said, "The building, or at least part of it, could be safe to occupy again within a week."

"Based on what data is the building shown to be safe?" ask Mold Warriors. And naturally, silence is the only answer.

Things are no different in Portland, Maine than in Hampton Bays, N.Y., or Accomack County, Va. If the public doesn't cry out in protest over the blatant blarney, then the administration can sweep the problem away, under their spore-laden rugs.

The Scotia Prince operational season mercifully came to a close in late October 2004; senior Portland staff finally had a little time off to pull themselves together. A local Portland newspaper reported "a steep drop in ridership," and said that "the potential for new competition and its aging home on the Portland waterfront had tested" Scotia Prince Cruises, and that the business would "end its season with questions about the future of its market and its business."

Did anyone besides Hudson question the possibility of long-term effects on his staff that the mold exposure might have on their future health, their ability to work as they had before or even, in Jay's case, to work at all?

(Author's note: As of mid-January 2005 the Sept. 17, 2004 promise from Portland officials that "repairs ... should be completed next week" hasn't happened. Scotia Prince Cruises still has no building and no idea if it can sail in 2005. By now it should have been into its third month of marketing this season. If the city turned over *safe* facilities today, the company would be five months behind in preparing for a six-month season.)

• • •

The last insult was when Hudson's son Mark, a new father himself, developed strange, serious, unexplained illnesses, after he began to spend time in the terminal building, including leakage of cerebral spinal fluid without any known cause.

When Mark's mold evaluation labs came back in November 2004, showing abnormally high MMP9, ADH, and ACTH, very low VEGF and MSH too low to measure, his father was upset. Like Carol Anderson's horror about *causing* her daughter's illness by bringing her into the Ritz Carlton, Matthew Hudson was deeply troubled about having brought his son into a situation that made him very ill.

City officials wrote to Scotia Prince Cruises that same month to tell them they planned to immediately demolish most of the company's leased premises—not because of mold or bird guano, but because the *roof was unsafe*. The "safe" public areas would continue to be remediated. In December 2004, city contractors started to remediate the "safe" federal areas, and they're still doing that work a month later, as I write this.

• • •

After consulting with Dr. Lipsey and other mold experts, Hudson realized he needed an East Coast treating physician. The Professor took recommendations from his experts, and then he asked for my credentials. Shortly after an involved internal decision-making process, he decided to send a group of his staff to Pocomoke for my complete evaluation in early November. That group included Jay and Sadie Anne Frye, Bob Schrader, John Hamill, Mark Hudson and nine other Scotia Prince workers.

The Fryes, Bob Schrader and John Hamill were key figures in unraveling the mold illness caused by exposure to the terminal building.

Before the first group arrived in Pocomoke, November 5, 2004, Hudson told me, "I don't care what it takes or what it costs; I want our folks returned to health. If mold exposure isn't their issue, tell me. If it is, tell me what to do to correct the health problems. If you can prescribe therapies that will help, let's begin treatment right away."

Let's face it: I didn't expect the health histories of the fourteen terminal building patients to be any different from any other group of mold illness sufferers. Regardless of the source or the identified species of mold, the illness is virtually identical in genetically susceptible, primed and exposed people. Whether it's massive amounts of *Stachybotrys*, or just enough *Aspergillus* garnished with *Penicillium* to cause a primed mold patient to react, the illness involves toxins, cytokines, hormones, growth factors, autoantibodies, opportunistic, colonizing bacteria and abnormalities in the innate immune response.

Although this group had a presentation typical of mold illness, they were different in an important way. In every group, there's always someone who *isn't* sick, and the majority usually don't have mold-susceptible genes. Not this group ...

And mold illness follows the same principles in all its victims; the illness *isn't dose-related*. In a typical group, there will be a distribution of illness that matches their genotypes. My examinations of each of the Scotia Prince workers showed the same absence of other possible causes and the same twenty total symptoms, on average, that I see in countless other cohorts of mold-sickened patients. *But all of them were ill!*

Curiously, a couple of the workers (including the Professor) had a normal visual contrast test, despite plenty of symptoms. But overall, based on what I saw, I expected their labs to hold no surprises. They would be run-of-the mill mold patients' results.

But there were ***nine*** surprises:

1. the highest percentage of workers with genetic susceptibility to mold from any one building cohort I've seen since I started my biotoxin work in 1997;
2. the highest average MMP9 in any cohort;
3. the highest C3a, (over *10,000*), in the more than 800 sick people in our C3a database (normal is less than 146) and a level of 500 is alarming;
4. the highest osmolality;
5. the highest interleukin-1-beta;
6. the worst elevation of ADH and ...
7. ACTH (and both in one person!);
8. the highest VEGF (which means the *worst* VEGF-resistance!) and
9. the lowest erythropoietin, undetectable.

And these were the abnormal lab results from people who worked in a building that City of Portland officials called "safe." I wonder what they'd consider "dangerous?" Don't put my friends in there!

There was another baffling problem. If the design and construction of the walls of the terminal were the only cause of the mold growth, we might expect the illness to be very longstanding —twenty years or more. But it wasn't.

Why didn't Jay get sick until 1997? He'd worked countless hours in that building since 1978. Sadie, too; she'd worked for Nova Scotia Tourism since 1995. Of course, they'd blamed most of her difficulties on pregnancy, then on having the baby, then on being the mother of a young child. And the work that had piled up in her absence stressed her, too.

Interrupted sleep was expected for a new mother, so of course she'd be tired. No one asked about her light sensitivity and problems with memory? Well, hadn't she gained, "just a few pounds?" Surely, the weight gain was the cause, right? But after both Fryes left the terminal building in summer 2003 their symptoms didn't go away. In November 2004, both had sky-high MMP9, with Jay's VEGF level world class.

What had happened in 1997?

• • •

The first symptoms of the mold illness seemed to suddenly appear in summer/fall 1997. What was new then?

The bridge! That was when the State of Maine drove pilings right next to the terminal building to build the new Casco Bay Bridge. When they were driving in those pilings, the building and the ground it was on shook so badly that the soils under the building caved in. The "floating" structural floor slab of steel and concrete broke, and one end of the building slumped almost a foot. After that, water came up from below the floor every time it rained.

During his medical evaluation in November 2004, Jay remembered clearly when the work crew had arrived from Rhode Island that summer. They'd drilled holes in the structural slab and sprayed foam grout to attempt to fill the new voids that had appeared under the floor, hoping to correct some of the structural engineering and water-intrusion problems that appeared after the pilings were installed.

But the Rhode Island contractors never fixed the gaps in the floor, except where the foam might have filled in. And no one fixed the roof. The mold in the walls, now "shaken and stirred" from their wall cavity, released their mycotoxins and their spores almost as if they were sending a shower of yellow rain into the HVAC system. In late spring 1997, suddenly everything changed invisibly, as trillions of chemical nano-bombs, spread by the HVAC, colonized the newly moist areas and attacked anyone near them.

• • •

"Dr. Shoemaker," Hudson asked me, "How can we scientifically confirm that the terminal building made us sick? My Internet reading tells me that the Environmental Protection Agency considers a facility to have Sick Building Syndrome when 20 percent of the people have symptoms. But almost *all* of our people do. Your work shows that 25 percent of the adult population will be susceptible to mold. Yet our group is full of what you call the 'dreaded genotypes,' and mold-susceptible genotypes dominate the rest. Why is our group full of these particular genotypes?"

"Professor," I replied, "You hired these folks and you didn't know their genotypes. The fact that they're all susceptible must be a coincidence. I don't know of any link between mold-susceptible genotypes and work groups. So I can't comment on *why* all but one of your workers has susceptible genotypes. But now that we know your staff is highly susceptible, exposed and ill, you need to know how to approach mold illness."

I recommended that the group try the repetitive exposure protocol, duplicating causation, as we've shown in hundreds of patients and hundreds of buildings before.

"Let several of your employees improve with treatment, then show they are still well after they stop cholestyramine (CSM), but are away from the building," I suggested. "Then allow them to go back into harm's way, by going back into the terminal. Since there's certainly risk, maybe just John Hamill and Bob Schrader should become test subjects, after providing informed consent, of course. If they agree (Author's Note: they did), then after they're healed with CSM, and not ill anywhere else off CSM, send them into the facility without CSM for three days. Watch them become ill within hours after they go back into the building, and re-test them in three days to prove the mold and the indoor air is causing the illness. The symptoms will recur, visual contrast scores will fall and the repcat labs will confirm what you know to be true. Then we fix the illness again."

I could feel the Professor thinking faster now.

"What about the cognitive issues?" Hudson asked. "John, our President; Bob, our general counsel; and my son Mark—I know their mental capabilities have become degraded because I deal with them every day. I catch their mistakes often but I'm certain I don't catch them all. They know it and I know it—and as I've recently discovered—some professionals they each deal with outside the company know it, too. But the deficits can be so subtle, who else is going to believe they're mentally damaged?"

"Professor, the cognitive issues in your workers are no different from mold exposure than with *Pfiesteria*," I assured him. "Whether the patient is a high-school dropout or a fully accredited academic researcher, biotoxin illnesses cause measurable deficits in formal neurocognitive functioning."

"Well, then" he decided, "We'll have both John and Bob get neurocognitive testing before and after exposure, like the kind you mentioned in *Pfiesteria: Crossing Dark Water*."

"Matthew, no one has ever done hyperacute neurocognitive testing that quickly," I pointed out. "The tests can take up to ten hours. There's no way you can arrange for the same testing in one facility to be done twice in five days on the same two people. I'd worry that the patients might be able to learn how to do better from test day one to test day two. Wait a minute, let's think this out. OK, these guys aren't normal. They're *neurotoxic*. They *won't be able to* learn and assimilate."

That might have sounded like a challenge, but it wasn't. We'd been able to show neurocognitive *improvement* with therapy in few patients previously after treatment, but no one had never tried to show *loss* of faculties based on just three days of re-exposure because of the technical difficulties involved in getting the testing done. I had no proof that we could show measurable cognitive deficits in just three days, even though I see the problem constantly. If Hudson could get the tests done, proving rapid worsening in neurocognitive testing, well, we might make some neurotoxin history here.

• • •

Hudson developed a unique plan for his two senior managers. After their baseline studies were completed in Pocomoke, they would finish their first three-week treatment with cholestyramine (CSM), record symptoms and have more labs done. Then they'd stay away from the terminal—a known source of exposure to toxigenic elements—and stay off CSM for about seven days. A normal life without toxin-binding medicine! Then they would go to Maine, do the labs again and take the neurocognitive tests. Quest Labs in Maine, New Hampshire and Florida all agreed to send the samples to the Quest Lab in Baltimore, Md., where Dr. Bill Meyer has set up an efficient series of labs for biotoxin patients.*

---

*Thus Baseline first assessment; then CSM for 3 weeks, then second assessment labs; then 7 days normal—no CSM no exposure to the IMT; then third assessment labs and first neurocognitive evaluation; then still no CSM & into the IMT for 3 days in December; then immediately second neurocognitive and fourth assessment labs followed by CSM and ongoing treatment to resolve the secondary and tertiary effects of the poisons that had been in their systems since originally going to work in the IMT.

Finally, these two volunteers would go into the deserted terminal building for three days, noting any symptoms and taking the proverbial "hostage-with-newspaper" pictures of each other to prove they were there, then they would have neurocognitive tests done again and immediately go to Pocomoke for repeat testing and treatment.

Hudson had insisted the logistics of the protocol could work, and *they did*. In the end, he and I had data that no one in the world had. The repetitive exposure protocol worked well, as it always does. An extensive set of labs confirmed what I predicted. And the neurocognitive tests were thick icing on the proof—the terminal building had made those people sick.

Imagine proving prospectively that an excellent brain was *turned into mush* in three days by breathing the air in a moldy building! The next time a mold defense apologist says, "OK, we concede that *maybe* mold can cause some sniffles and a cough," I'll ask him if he would be so kind as to read a neurocognitive report. After reading what happened to Bob, I'll bet he's more likely to concede some more.

Three weeks of CSM started to make John Hamill and Bob Schrader better. Without CSM and without the building, they were improved and stable. Put back into the terminal, they both had a return of symptoms almost immediately. Not to mention the mush brains.

Professor Hudson had commissioned a team of health professionals from all over the East Coast to put together medical data that proved one thing: Portland's International Marine Terminal made people sick, took away their higher-level brain functions and damaged the innate immune response system. I have the data, the Professor has the proof, and now the sacrificial story has been told.

• • •

Now we need to work on healing John Hamill, Bob Schrader and the rest of the Maine cohort. We need to start the outreach process to other tenants of the terminal and to current and former employees, too. Hundreds, maybe thousands of federal employees have breathed in whatever bioaerosols were in the air inside that building for the past 35 years. Who knows how many people have been exposed overall?

Will the government of Nova Scotia cooperate to protect its employees? Quite a few were exposed since 1997. What about Customs and Border Protection staff? Scotia Prince Cruises warned the customs employees on several occasions, but in September 2004 city officials told them their premises were "safe," so they continued to occupy the building.

Mold illness isn't benign; denying it exists or ignoring its health effects threatens all of us in ways we might not suspect.

As I write this epilogue in mid-January 2005, the same landlord has been undertaking a "remediation" project since December in those "safe" Customs and Border Protection premises. From the reports given to me, the effort seems poorly designed, doesn't deal with water intrusion and is already causing cross-contamination. Maybe just telling this story will spur people to demand that they receive due medical process. Returning their health may be the first step in resolving the legal issues as well.

Although Matthew Hudson acted with extraordinary speed and efficiency to protect his people and his clients, and he even tried several times to protect federal employees who work in the same building, sadly, some of his group of 14 have the dreaded multiple-susceptibility genotypes; it might be too late for them to return to a full, normal life. And although many of the Scotia Prince senior staff members have improved after taking CSM, they certainly have a marked increase in the potential for future illness from mold exposure, solely caused by the terminal building.

They're "primed."

Maybe we'll receive the funding we need to satisfy the FDA that MSH replacement is safe and effective. I hope MSH research will be completed when the Maine cohort needs more help, as I believe they will.

And the City of Portland cannot plead ignorance of mold issues: they knew about Stachy, as they'd demolished a local school, Jack Elementary, in 2000 because of its "bad roof" (not to mention aerosolized Stachy at a reported 100 colonies per cubic meter inside the building.) The terminal building had an aerosolized Stachy count of 5,436 colonies per cubic meter! Bulk samples taken from one of the "safe" *public waiting rooms* had:

Stachybotrys 4.5 million colony forming units per gram (CFU/g);
Acremonium strictum of 2.1 million CFU/g;
Aspergillus sydowii 2.2 million CFU/g;
Aspergillus versicolor 2.2 million CFU/g; and
Penicillium chrysogenum 1.2 million CFU/g.

These weren't the only molds in the public areas, but they're all toxigenic. In all, Scotia Prince Cruises testing showed more than 35 different molds in the terminal building plus bacteria and their toxins—every one of them capable of causing adverse human health effects. The word toxigenic can sound empty, as sampling is too expensive and can be subject to technical errors if the lab isn't first rate. Hudson had PK Jarvis look for toxins in the public areas, the *"safe parts"* of the building. They found massive amounts of *ten* different fungal toxins!

What about those Scotia Prince officers and managers who won't make a 100 percent recovery, despite everything that we do for them? Will they ever again be able to work in such a responsible, high-level managerial capacity?

How about John Hamill and his high PAI-1 and MMP9? Will he become another David Selby, suffering a premature heart attack that someone else will incorrectly blame on donuts or sleep apnea? Or maybe he'll have an unusual neurologic event that's first called a stroke and then a seizure. I'm terribly afraid that we won't be able to reverse all of the abnormalities generated by the inflammatory response to mold toxins (among other biotoxins in the terminal), or be able to prevent the potential for innate immune response-generated heart attacks in the future.

What about Jay and Sadie Anne's daughter Bridgette, who often visited them at work? Or Tony's wife, Cheri, who used to wait in his office for an hour for a ride home each day and who now exhibits symptoms? Or Vinnie's baby daughter Mikayla, who played on the floor in his office? That office had a *Stachybotrys* airborne spore count *fifty-four times higher* than that of the now-demolished Jack School.

The ultimate victims of the terminal mold disaster are likely to be those who have already incurred pain, suffering and loss of quality of life. Some of them will not make complete recoveries, and all of them will be especially vulnerable to other environmental toxins for the rest of their lives.

It's likely that some of the visitors to the terminal with susceptible genotypes, even those traveling on a single occasion, will have been affected. Will any of the more than *one million* passengers since spring 1997 recognize the symptoms and take the Visual Contrast Sensitivity test on-line at http://www.chronicneurotoxins.com? What about the many repeat clients of Scotia Prince Cruises, some who travel 10 or more times each year? If they fail the VCS, will they want to pursue further the *source* of their health problems?

Professor Matthew Hudson became a Mold Warrior when he decided to protect his staff, the federal employees and the public, regardless of the cost to him professionally or personally. Early on, the landlord accused Hudson of making false claims in order to gain a financial advantage for his company. In fact, just as you would expect, although the company's sales had risen to become better than they'd been in years, after the mold was discovered, sales fell precipitously. It was only after Scotia Prince

Cruises set up tents in the parking lot and asked its employees and the public to do business in them that the company took a financial hit.

Among the victims of this debacle also may well be the taxpayers and voters of Portland. The City Fathers were at best reckless and at worst negligent in not properly repairing the building in 1997, in not providing proper maintenance over the years, and in not responding with due diligence in August 2004 when Bob Schrader hand-delivered fungal lab results to the mayor at his law firm, as well as to the city manager and assistant city manager at City Hall.

Will the politicians or the taxpayers be held responsible for personal injury suffered at the terminal building?

And what about the 111,000 *Aspergillus-Penicillium* colonies per cubic meter in the *air* in the warehouse office? Not 100 like the Jack School—111,000! Will some of the women who worked at the International Marine Terminal building be unable to carry pregnancies to term or be unable to give birth to *healthy* children because of their exposure to mold in the terminal building?

When I think of the ignorance and the disregard for human health that the landlord showed by ignoring such a mold problem, and the potential to harm tens of thousands, including children and the unborn, I'm angry, incredulous, speechless.

This book will have to do the speaking for me, for the Maine cohort, and for the millions of other mold victims who are out there suffering and praying for things to change.

• • •

APPENDIX 1

Biotoxin Symptoms

- Fatigue, weakness
- Muscle ache, cramps, unusual pain (ice-pick, "lightning bolt")
- Headache, can be confused with migraine
- Sensitivity to bright light, tearing (or lack of tearing), blurred vision, redness
- Chronic sinus congestion, cough, short of breath
- Abdominal pain (often labeled IBS), diarrhea, often secretory
- Joint pain, enthesopathy, morning stiffness; migratory, rarely true arthritis
- Cognitive impairment, recent memory, assimilation of new knowledge, abstract handling of numbers, word finding in conversation, confusion, difficulty sustaining concentration, disorientation, "brain fog"
- Skin sensitivity to light touch
- Mood swings, appetite swings, sweats, often at night, difficulty with temperature regulation
- Numbness, tingling, often non-anatomic, vertigo, metallic taste
- Excessive thirst, frequent urination, sensitivity to static shocks (doorknobs, car handles, light switch plates, kisses)
- Impotence, menorrhagia

APPENDIX 2

Biotoxin Lab Order Sheet

Ordered	Test	In	Lab to Use	Spec	Code #	DX Codes
	HLA		Lab Corp	Yellow, refrig	012542	279.10, 377.34, 279.8
	MSH		Lab Corp	Lav - freeze	010421	253.2
	Leptin		Quest	SST-freeze	84657N	253.2
	ADH		Quest	Lav - freeze	31260P	253.5
	Osmo		Quest	SST - freeze	26922P	253.5
	ACTH		Quest	Lav - freeze	21410P	255.4
	Cortisol		Quest	SST -freeze	11281X	255.4
	DHEAS		Quest	SST -freeze	21915R	MALE 257.2 FEMALE 256.39
	Testosterone		Quest	SST -freeze	29868W	MALE 257.2 FEMALE 256.39
	Andro-stenedione		Quest	SST -freeze	18549	MALE 607.84 FEMALE 256.39
	ESR		Here	Lav		
	TNF		Specialty	Green freeze	3294	416.9
	PAI-1		Lab Corp	Blue freeze	146787	437.6
	CRP		Lab Corp	SST - refrig	006627	378.54
	Lipid with Phenotype		Lab Corp	SST - refrig	033886	272.0
	CBC		Lab Corp	Lav - refrig	005009	285.0
	CMP		Lab Corp	SST - refrig	322000	780.79

	GGT		Lab Corp	SST - refrig	001958	250.00
	Anticardiolipin		Lab Corp	SST - refrig	161950	710.0
	Nasal Culture		Esoterix	Rm temp	609980	478.21
	MMP-9		Quest	SST- freeze	41865	340
	VEGF		Quest	Lav - freeze	14512X	416.9, 253.2, 710.0
	Myelin Basic Protein		Specialty	SST - freeze	1056	340
	Antigliadin		Lab Corp	SST - refrig	163402	579.0
	IgA		Lab Corp	SST - refrig	001784	579.0
	TTG IgA		Lab Corp	SST - refrig	164640	579.0
	Erythropoietin		Quest	Red - freeze	22376P	285.9
	Immune Complexes {C3d; C1q binding} (Profile # 82086)		Quest Baltimore	SST freeze SST freeze	C3d 11218X C1q 52282P	279.8
	+ immune complexes {C2, C4, C4d, C3a} (Profile #82118)		Quest Baltimore	SST freeze SST freeze Lav freeze 4ml plasma/ 4ml serum	C2 44842P C4 44982E C4d 4943N C3a 42003	279.8
	+ C3a {factor B, C3} (Profile #82100)		Quest	SST freeze SST freeze	factor B 46706P C3 44859W	279.8
	IgE		Lab Corp	SST - refrig	002170	493.01
	Lyme WB		Lab Corp	SST freeze	163600	088.81
	IL-1B		Quest	SST- freeze	1757N	716.99
	TSH		Lab Corp	SST - refrig	004259	244.8
	Saves					
	Saves					
	Saves					

APPENDIX 3

Biotoxin Lab Definitions

Dear Lab Staff:

We have sent our mutual patient in for some lab tests that might not be ones you see everyday. In the past, some labs have done the wrong tests when they were not sure what we wanted. Here are some definitions. Call us before you guess at what a test is.

HLA—immune response genes; (yellow-refrigerate) **LabCorp only**

Alpha MSH—melanocyte stimulating hormone (pre-chilled lavender; add 0.5 cc Trasylol, *immediate centrifuge*, freeze) **LabCorp only**

DHEA-S—(SST-refrigerate or freeze) **Any lab**

TNF—tumor necrosis factor alpha (SST, freeze) **Specialty only**

PAI-1—plasminogen activator inhibitor-1 (blue, freeze) **LabCorp, Esoterix**

Anticardiolipin—(SST, refrigerate or freeze) **Any lab**

Antigliadin—(SST, refrigerate or freeze) **Any lab**

MMP-9—matrix metalloproteinase 9 (freeze, SST) *immediate centrifuge* **Quest only**

VEGF—vascular endothelial growth factor- (lav, freeze) **Quest only**

MBP—myelin basic protein (SST, freeze) **Specialty only, no exception**

Erythropoietin—(SST, Refrigerate) **Any lab**

TTG-IgA—Tissue trans-glutaminase IgA (SST, refrigerate) **Any lab**

Complement assays as follows: USE BALTIMORE CODES

- **C3a**—activated 3rd component of complement (lav, freeze) **Quest, Baltimore.**
- **C3d**—immune complex (serum, freeze) **Quest, Baltimore**
- **C1q**—complement immune complexes (serum, freeze) **Quest, Baltimore**
- **C2**—complement C2 (serum, freeze) **Quest, Baltimore**
- **C4**—complement (serum, freeze) **Quest, Baltimore**
- **C4d**—fragment, EIA (lavendar, freeze) **Quest, Baltimore**
- **Factor B**—properdin factor B (serum, freeze) **Quest, Baltimore**

Please note the Quest complement assays need to be sent to the Baltimore lab (aren't the same as other sites across the country).

Send all complement samples, MMP-9, VEGF frozen in dry ice, overnight, next day delivery, to:

Quest Diagnostics Incorporated
1901 Sulphur Spring Road
Baltimore, Maryland 21227

APPENDIX 4

Shoemaker Protocol

1. Take a decent environmental exposure history. Don't forget that 80% of all mold patients will tell you there is no visible mold in the sites that make them ill. Look for ciguatera fish poisoning! The absence of a positive blood test for Lyme doesn't rule *out* the diagnosis when the lab is a large commercial outfit. The presence of a positive blood test for Lyme from a small "Lyme lab" doesn't rule *in* the diagnosis. If the symptoms fit and no confounders are found, begin therapy. 8% of all biotoxin patients will have a false negative VCS; < 1% will have a false positive. Do the VCS.
2. The first of several steps in treatment begins with CSM, taken four times a day on an empty stomach. Kids over 3 years old and less than 120 pounds get 60 mg/kg/dose for each of three doses a day. Deal with the predictable gastrointestinal problems from CSM. The first several days of CSM can be difficult. Push fats and oils in diet, especially if you are slender (low leptin). Hang in there!
3. After 30 days, record symptoms and VCS. Next eradicate MRCoNS if patient is MSH deficient. Use topical Bactroban, applied each inside surface of nostrils three times a day. Work a little bit of the cream or ointment deep into the aerobic area of the nose. Add rifampin, 600 mg in the AM with food and one other oral antibiotic, depending on sensitivities of the organism.
4. Eliminate gliadin if there are antibodies to gliadin.

5. If ACLA are highly positive, consider using low dose heparin, 5000 units twice a day.
6. If VEGF is low or if VEGF and erythropoietin are mismatched (both made by genes linked to each other), use VEGF therapies in Phil Ness chapter.
7. Correct pituitary/peripheral hormone problems. Maybe all you need to do is supplement DHEA to slowly feed androgen manufacturing processes. Maybe you'll need to supplement with low dose DDAVP to correct ADH deficiency. Be cautious with volume depleted patients!! Don't use corticosteroids unless you have to!!
8. If VEGF doesn't rise, despite all your efforts, consider high dose Omega three monounsaturated fish oils and Actos in combination to push PPAR gamma. No amylose!!
9. If nothing works, consider use of low dose erythropoietin replacement, after informed consent, of course. 8000 units taken two times a week with 81 mg of aspirin; monitor symptoms and hemoglobin.
10. Trental is a weak anti-cytokine, but it can increase capillary hypoperfusion and reduce brain fog.

There are many more elements to treatment of biotoxin illness in the complete text of Mold Warriors. These are just a start. Always monitor VCS, symptoms, VEGF, leptin C3a and MMP9. Don't forget the most common reason for failure of the protocols is failure to use the meds correctly, with ongoing mold exposure the second most common. If you have an MARCoNS in your deep aerobic spaces in your nose, and you'll only know by doing an API-STAPH culture (you can insist on it!!), CSM won't help much and MSH won't rise until the organism is eradicated.

Always expect a mold susceptible genotype person to relapse with re-exposure. Teach mold avoidance!

Over the years, the most common errors in failure to diagnose have been the attempt to, "save on a few tests," by not looking at all the Biotoxin Pathway. Sure, the tests are technically demanding. Sure, the tests must be done right to yield reliable information.

But you already knew that healing these complex illnesses was complicated before you read this far. We are learning new features of the Biotoxin Pathway continuously. Who knows what breakthroughs we will have next year, especially if a large organization backs us in bringing MSH replacement to the millions who need it every day.

Ritchie C. Shoemaker MD
11/21/04

APPENDIX 5

HLA DR by PCR Registry

HLA Rosetta Stone

	DRB1	DQ	DRB3	DRB4	DRB5
Multisusceptible	4	3		53	
	11/12	3	52B		
	14	5	52B		
Mold	7	2/3		53	
	13	6	52A,B,C		
	17	2	52A		
	18*	4	52A		
Borrelia, Post Lyme Syndrome	15	6			51
	16	5			51
Dinoflagellates	4	7/8		53	
Multiply Antibiotic Resistant Staph epidermidis (MARCoNS)	11	7	52B		
Low MSH	1	5			
No recognized significance	8	3,4,6			
Low-risk Mold	7	9		53	
	12	7	52B		
	9	9		53	

APPENDIX 6

HLA DR by PCR Rosetta Stone

1. Look at the LabCorp report. There are five categories of line entries: DRB1, DQ, DRB3, DRB4 and DRB5. Each individual will have two sets of three alleles, unless the DRB1 is 1, 8 or 10. Those patients will only have a DQ and won't have DRB 3,4,5. Each individual with a DRB1 other than 1, 8, or 10 will have a DQ and one other allele from DRB 3,4,5. If you are expecting to find two entries in, say DR 3,4, 5, but only find one, the patient is homozygous for that allele and only one allele will appear on the PCR. Don't be worried that the numbers and letters look impossible to understand. They just have to be translated.
2. You will translate these categories into B1, DQ, 52 (A, B, C), 53 and 51 respectively. If the translation were easy and made sense, you wouldn't need this!
3. The numbers and letters in each of the 5 categories are given to an excessive amount of detail. Write down only the first two numbers in each line.
4. When you see 03 as one of the two genes, an allele, for DRB1, rewrite it as 17.
5. Record the entries in B3 by converting the 01 to A, the 02 to B and the 03 to C; this will give you 52A, 52B and 52C, respectively.
6. B4 is 53.
7. B5 is 51.

8. Record the genotypes in two columns, one each representing each parent, using the templates in the appendix.
9. When the International Language on HLA DR changes again, we might have to change the Rosetta Stone again. I hope not.

APPENDIX 7

Biotoxin Timeline

Let's Look at Biotoxins

- First time *Pfiesteria*, Pocomoke River Md, NC, Virginia, Florida 1997
- Expand to include different dinoflagellate, ciguatera 1998
- Add fungal species (SBS) 1998, learning disability from toxins
- BG algae and Lyme 1999,—TNF, PPAR gamma and cytokines
- Brown recluse and apicomplexans 2000,—environmental acquisition of obesity and diabetes
- Endogenous sources 2000, PAI-1 and atherosclerosis; *Chattonella*
- Fibromyalgia and Chronic Fatigue 2001
- MSH, coag neg Staph 2001—downstream hormones; metallic taste
- 2001–2 Anthrax vaccine cases and Post anthrax syndrome
- MMP9 2002—link to MS; HLA susceptibility
- Myelin basic protein antibodies, VEGF, antigliadin, anticardiolipins 2003
- Babesia GPI, Staph Exotoxin A and B, 2003
- Dr. John Ramsdell, NOAA and brevetoxin, 2003

APPENDIX

Biotoxin Research Timeline

- 2004 Biotoxin Pathway
- Wingspan, autoantibodies and genomics
- Reduced VO2 max, anaerobic threshold
- Acute and chronic sustained activation of alternative pathway of complement
- Stabilization of C3a; link to ASCVD

APPENDIX 8

Paradigm Conference VCS Slide

**Reacquisition of Sick Building Syndrome:
VCS Deficit in Initial Illness,
Resolution After Cholestyramine Therapy,
Stable without Re-exposure,
Deficit Reacquisition with Re-exposure,
Second Resolution with CSM Therapy**

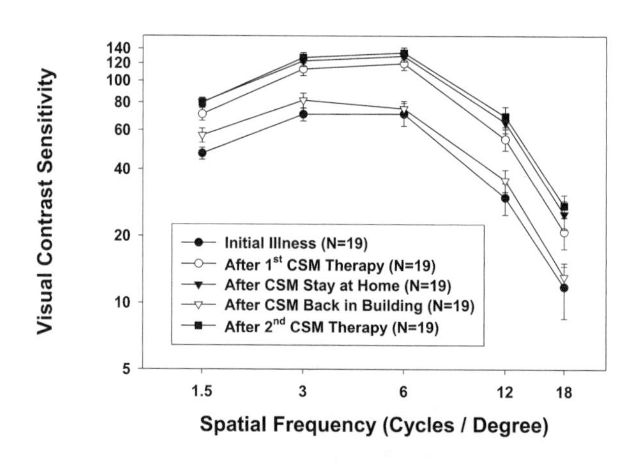

Mean # Symptoms Out of 26 Assessed:	1. Initial Illness = 14.9 2. After 1st CSM = 1.6 3. Stable at Home = 0.8 4. Re-exposure = 9.5 5. After 2nd CSM = 1.3

APPENDIX 9

Johanning Conference VCS Slides

Controls Versus Initial Illness:

Occupational Exposure, Residential Exposure,
Consultation but No Treatment

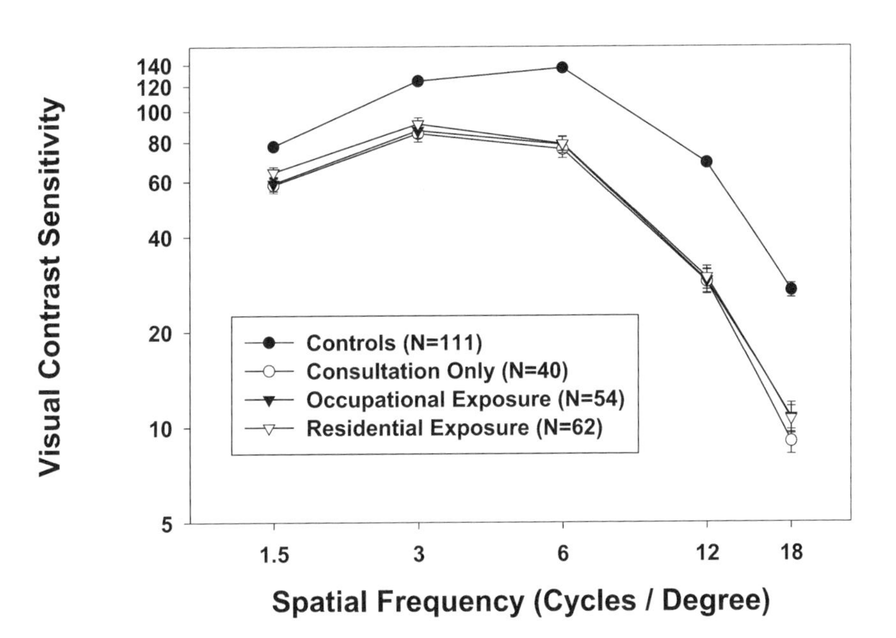

Symptoms	
Control	1.7
Consult	20.3
Occupational	17.9
Residential	18.5

Controls Versus Initial Illness:

Fungi Genera Identified, Visible Evidence of Fungi Only,
Water Damage Evidence Only

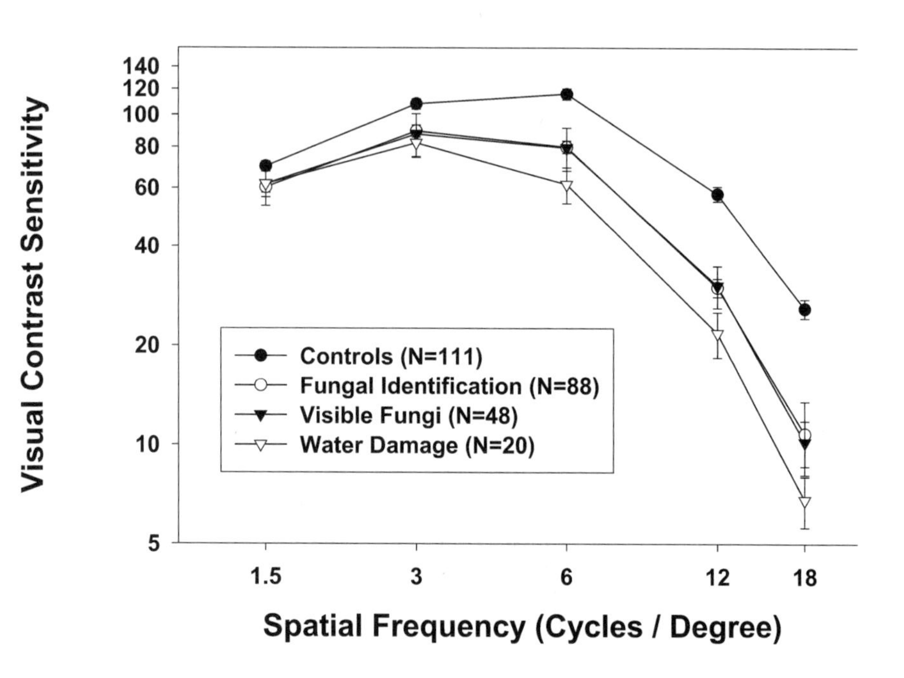

Symptoms	
Control	1.7
Fungal ID	18.8
Visible	18.3
Water Damage	17.1

Fungal Identification Cohort: Time Series

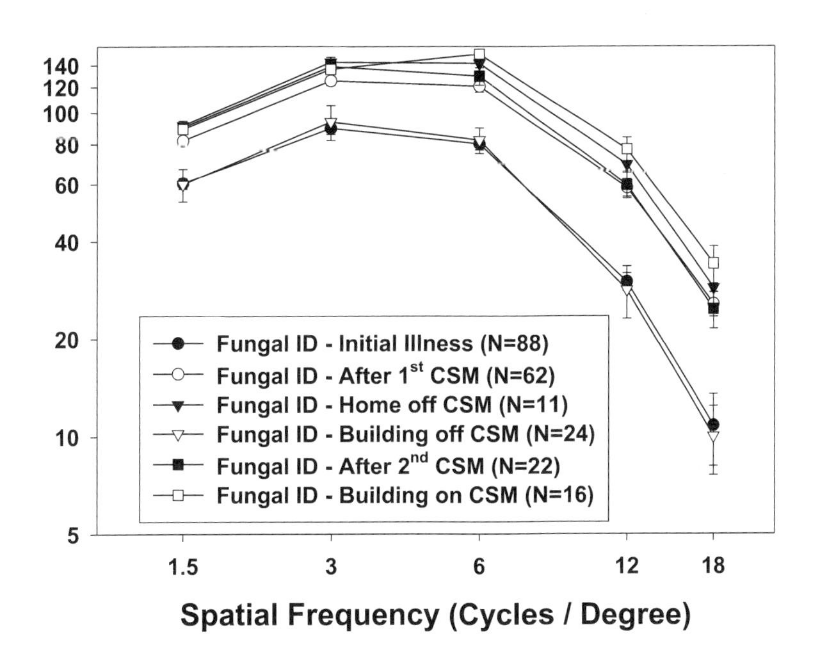

Symptoms	
Control	1.7
Initial	18.8
AC-1	2.6
HOC	1.5
BOC	10.0
AC-2	1.4
Prophyl	1.7

Occupational Cohort: Time Series

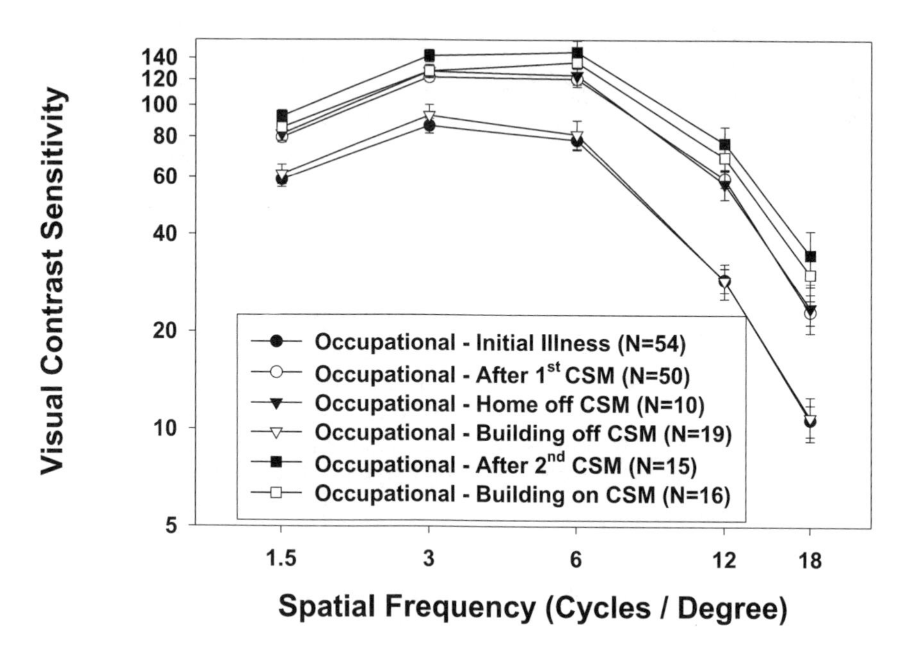

Symptoms	
Control	1.7
Initial	17.9
AC-1	2.3
HOC	1.1
BOC	6.5
AC-2	2.3
Prophyl	1.6

Residential Cohort: Time Series

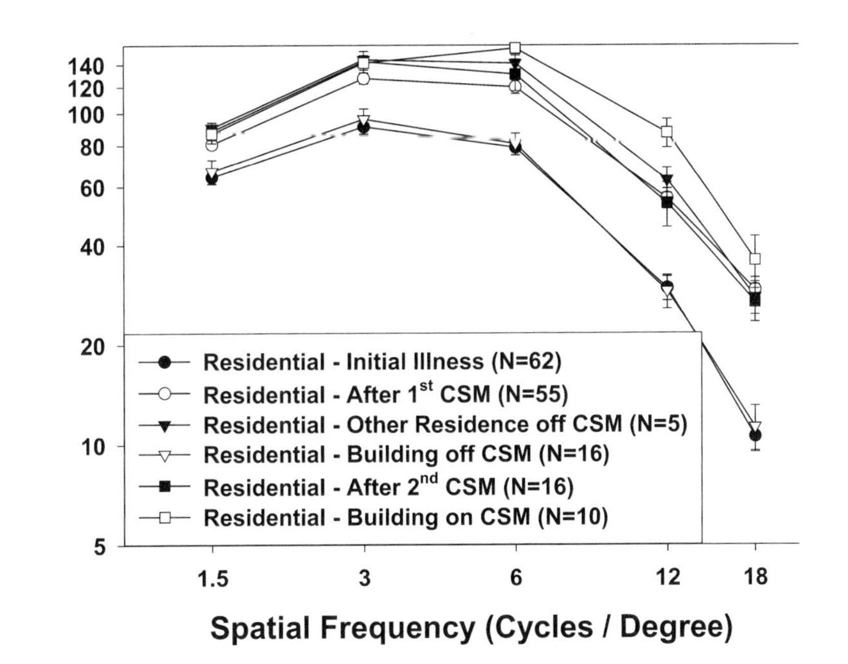

Symptoms	
Control	1.7
Initial	18.5
AC-1	3.1
OROC	2.0
BOC	8.7
AC-2	1.1
Prophyl	2.1

APPENDIX 10

Hampton Bays VCS Slides

Visual Contrast Sensitivity for All Participants Screened

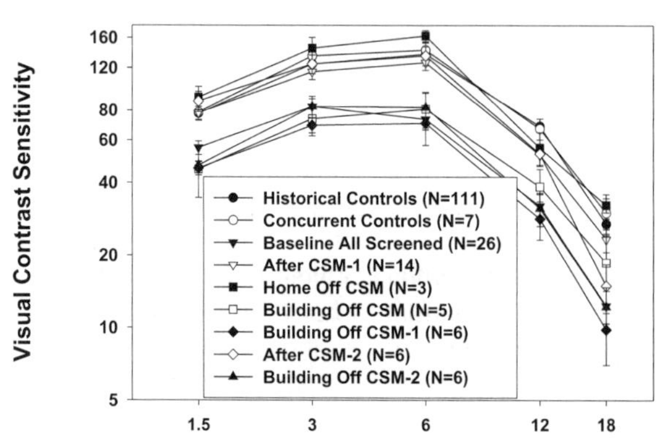

Symptom Means & Standard Error of the Means

Historical Controls = 1.7 + 1.1
Concurrent Controls = 2.7 + 0.4
Baseline All Screened = 20.2 + 1.2
After CSM-1 = 3.0 + 0.7
Home Off CSM = 1.0 + 0.0
Building Off CSM = 11.6 + 1.6
Building Off CSM-1 = 15.2 + 2.8
After CSM-2 = 1.5 + 0.6
Building Off CSM-2 = 11.3 + 1.7

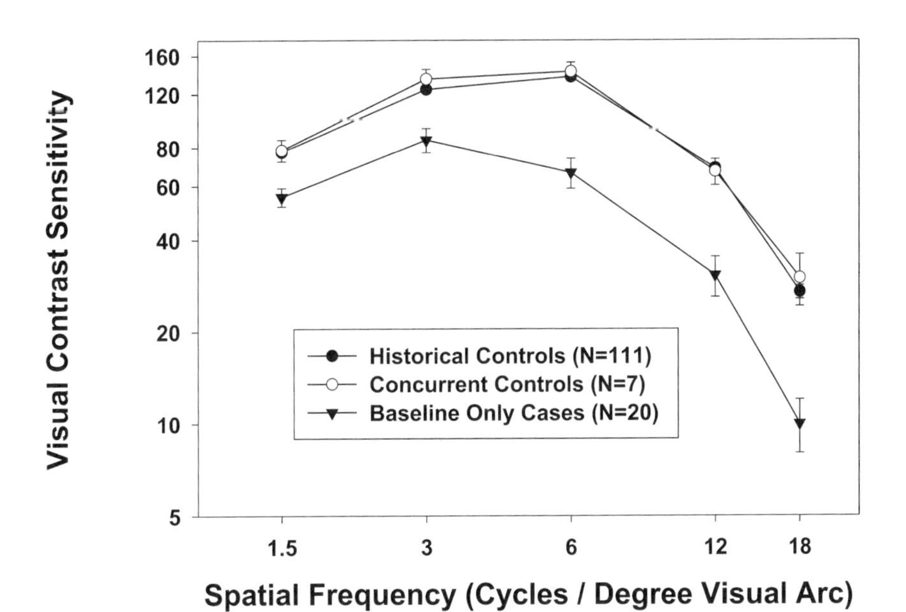

Symptom Means &
Standard Error of the Means

Historical Controls = 1.7 + 1.1

Concurrent Controls = 2.7 + 0.4

Baseline Only Cases = 19.3 + 1.7

Cases that were Re-exposed Once after Treatment

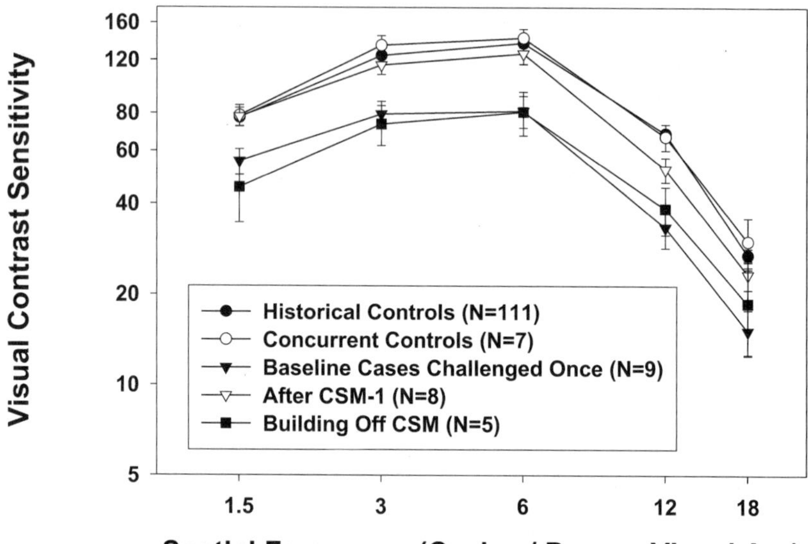

Symptom Means &
Standard Error of the Means

Historical Controls = 1.7 + 1.1

Concurrent Controls = 2.7 + 0.4

Baseline = 20.9 + 1.3

After CSM-1 = 3.0 + 0.7

Building Off CSM = 20.9 + 1.3

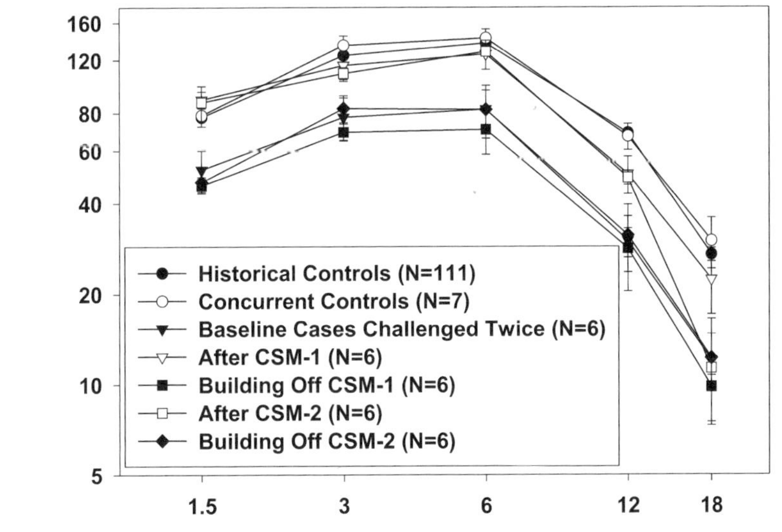

Symptom Means & Standard Error of the Means

Historical Controls = 1.7 + 1.1

Concurrent Controls = 2.7 + 0.4

Baseline = 23.5 + 1.6

After CSM-1 = 3.3 + 1.1

Building Off CSM-1 = 15.2 + 2.6

After CSM-2 = 2.0 + 0.7

Building Off CSM-2 = 11.3 + 1.7

APPENDIX 11

No Amylose-Diet

No Amylose Diet
0-0-2-3

No Skipping meals. The starvation response burns protein.

Adequate protein. 6–8 oz. Final cooked weight.
  Note: This diet is HIGH carbohydrate and HIGH protein

Lactose (milk) and fructose (fruit): OK
Learn the differences in carbohydrates
Artificial sweeteners: OK

Avoid:
- Glucose—may be disguised as dextrose and sucrose
- Amylose—you must learn this simple, short list too
  - Bananas—only forbidden fruit
  - All foods which grow beneath the ground (only onions and garlic are OK)
- The cereal grains are the biggest problem foods
  - Wheat, rice, oats, barley and rye
  - Corn has a natural inhibitor of amylose—OK!

No low-fat, corn syrup, maltodextrin (hidden sugars)

Careful with cheeses and yogurt: fat counts

Spices and condiments: OK

Popcorn and baked corn chips ok, tortilla chips always safe
  Corn chips may have sugar in spicy coating

No cereal, no chocolate, no fast foods, no regular soft drinks, no commercial fruit juices, squeeze your own

Diet soft drinks: OK
Caffeine drinks (coffee/tea): OK

Don't forget: moderation in eating includes eating good food, prepared well, presented pleasingly and eaten with gusto, not gluttony.

APPENDIX 12

Tape Lift Sampling Protocol

Tape Lift Mold Samples

1. Wash your hands first, or use gloves to do sample.
2. Find an area of visible mold in your environment where there is water intrusion or high humidity.
3. Wipe a piece of clear (see-through) Scotch tape over the mold until all of the adhesive stickiness is gone.
4. Do not fold specimen.
5. Place inside sealable plastic baggie.
6. Write a check for $35.00, payable to P&K Microbiology **FOR EACH SPECIMEN**.
7. Enclose the check and baggie in second plastic container. Clearly identify the location of the sample.
8. Mail to my office at 500 Market Street, Suite 102, Pocomoke City, MD 21851.
9. I will create a change of custody and forward the specimen to P&K.
10. Specimens collected improperly or not pre-paid will be discarded.
11. A tape lift sample is not the same as a building survey done by a qualified indoor hygienist, but it often tells us if we need to look into the possibility of amplified mold growth as the source of illness.

APPENDIX 13

IgE and Mold Illness

IgE by Diagnosis

	Control	*Mold*	*Asthma*	*Nasal Allergy*
N=	129	216	24	29
Average	34.4	42.2	574.8	397.2
<=10	54	80	0	0
>235	2	4	18	19

APPENDIX 14

MSH Information

MSH Fact Sheet

Q: What is alpha melanocyte stimulating hormone (MSH)?

A: MSH is an anti-inflammatory, regulatory hormone made in the hypothalamus. It controls production of hormones, modulates the immune system and controls nerve function, too. It is made when leptin is able to activate its receptor in the proopiomelanocortin (POMC) pathway. If the receptor is damaged by peripheral immune effects, such as the release of too many pro-inflammatory cytokines, then the receptor doesn't work right and MSH isn't made. Leptin controls storage of fatty acids as fat, so MSH and leptin are a major source of interest.

Q: Isn't this damage to the leptin receptor just like what we see in insulin resistance?

A: Actually, the mechanism is almost identical. A lot of our obesity and diabetes is inappropriately blamed on overeating when the real problem is damage to these regulatory receptors.

Q: What does MSH do?

A: MSH sits as the central hub of a series of important effects. MSH controls hypothalamic production of melatonin and endorphins. Without MSH, deficiency creates chronic non-restful sleep and chronic increased perception of pain, respectively. MSH deficiency causes chronic fatigue and chronic pain. MSH also controls many protective effects in the skin, gut and mucus membranes of the nose and lung. It also controls the peripheral release of cytokines; when there isn't enough MSH, the peripheral inflammatory effects are multiplied. MSH also controls pituitary function, with 60% of MSH deficient patients not having enough antidiuretic hormone. These patients will be thirsty all the time, urinate frequently and often will have unusual sensitivity to static electrical shocks. 40% of MSH deficient patients won't regulate male hormone production and another 40% won't regulate proper control of ACTH and cortisol.

Q: What illnesses are associated with MSH deficiency?

A: Any illness that begins with excessive production of pro-inflammatory cytokines will usually cause MSH deficiency. This is the basic mechanism that underlies damage caused by exposure to biologically produced toxins neurotoxins (biotoxins) made by invertebrate organisms, including fungi (molds), dinoflagellates (ciguatera and *Pfiesteria*), spirochetes (Lyme disease), blue-green algae (*Cylindrospermopsis* in Florida and *Microcystis* all over the world) and bacteria, like anthrax. Nearly 100% of the patients who have Chronic Fatigue Syndrome (CFS) will have MSH deficiency. Do we know that all CFS is due to biotoxins? No.

Q: OK, if something is going on in the body that causes inflammation, like exposure to toxins made by mold in Sick Building Syndrome, and the immune system is cranking out these proteins, cytokines, that are great for our health when they are released in the right amount at the right time, but harmful when too many are made at the wrong time, why don't we just fix the cytokine response and watch MSH get going again?

A: Good question. We are looking at an incredibly small area in the hypothalamus, one in which there is a real risk of permanent damage from cytokines to blood flow to this pathway. And who is to say that the vital receptors aren't destroyed by too much attack for too long? Once MSH production is damaged too much, it is too late.

Q: So, if the toxin and cytokine illness goes undiagnosed, or someone says the illness doesn't exist, like what we have seen for a long time with mold and in Lyme disease, there are going to be people whose MSH supply just dwindles down to nothing.

A: Right. These patients are miserable. They live, but there is no life. They are given Oxycontin or Neurontin or Elavil or Xanax, for example, but nothing really helps. Families are destroyed, careers are ruined, financial resources are poured into tests that mean nothing and therapies that hopefully won't cause harm, because they never help. Even worse, because MSH levels are rarely measured, many doctors look at the MSH deficient patients as if they are making up the illness, making up the pain to get drugs or looking for disability. And lots of them end up on disability, costing our society not just the unnecessary expense but also costing us the lost productivity.

Q: What is the answer for those people who are MSH deficient? Why don't we just give them MSH? It should be so simple, like giving insulin to a diabetic who needs it, right?

A: Should be, but it isn't. The FDA is real particular about giving potent hormones like MSH to just anybody. Just look at cortisone. A little bit is necessary for life, but too much will kill you. MSH can't be given to people until animal toxicity studies are done, showing safety, and that costs a lot of money, even if the paper work can be done.

Q: Why don't all the big drug companies jump on this? If what you are telling me is that there are a large number of people who will need daily replacement of MSH, and by the way, it looks like our supply of Sick Buildings that will generate MSH deficient patients won't dry up any time soon, then there should be a huge market for somebody.

A: No doubt about that. There are companies looking at MSH as we speak for weight loss, skin pigment changes and as an anti-depressant, but no one is looking at MSH in chronic, fatiguing illnesses.

Q: So, if you can raise the money to prove MSH is safe, then what?

A: Researchers at the NIH, like Dr. Robert Star, would likely be willing to help with the experiments in humans. I'm sure there are other academic researchers who would come forward, but the problem is that the experience of physicians with MSH is so limited. Certain researchers, like Dr. James Lipton of Zengen and Dr. Star, who know MSH, know what a huge area of medicine is involved with MSH. But the average endocrinologist, for example, just doesn't deal with MSH deficiency. We will use my database of nearly 2000 patients with illnesses associated with MSH deficiency to develop a double blind, placebo controlled, cross-over clinical trial after we have done a titration study to prove how much MSH is needed, given in what route, for how long, with what side effects and with what adverse reactions over time. When our work demonstrates what I feel it will, then suddenly, there is hope for a large number of chronically suffering people of all ages.

Q: We'll wait to see what happens. Good luck to you and your research.

APPENDIX 15

CRBAI Donation Form

CRBAI Donation
Center for Research on Biotoxin Associated Illnesses
500 Market St. Suite 103
Pocomoke, MD 21851

Please help us with our research so that we may help others. We will send our new publication, Biotoxin Pathways, to all contributors of $25 and more.

Enclosed is a check for ________, payable to CRBAI.

If you would like to direct your donation to specific areas of research, please specify below:

If you would like to specify patient stipends for your donation, please tell us:

Please tell us what projects you would be willing to support with additional donations:

If you would like to use a credit card to donate, please give us a phone number where we can reach you, as we would rather not have credit card numbers passing through the mail:

Many thanks for all your help. We will win the battle to find solutions to chronic fatiguing illnesses!

Ritchie. C. Shoemaker MD
410-957-1550 phone
410-957-3930 fax
www.chronicneurotoxins.com

APPENDIX 16

Book Ordering Information

Books:

Mold Warriors $24.95

The inside story on how the battle to confirm mold illness was won

Desperation Medicine $16.00

The primer on biotoxin illness, this is where you start

Lose the Weight You Hate $21.00

Weight loss, cholesterol, environmental acquisition of diabetes, recipes and more

Pfiesteria: Crossing Dark Water $14.00

Here is the beginning, including science for sale

To order:

Please send check for the book(s) chosen with a shipping/ handling fee of $5.95 per book and $1.50 for each additional book to:

Ritchie C. Shoemaker, M.D.
500 Market Street, Suite 102
Pocomoke City, MD 21851

You may also call the office at (410) 957-1550 and order over the phone or may place an order electronically at www.chronicneurotoxins.com.

FAQ

25 Frequently Asked Questions On Mold Illness

1. I'm sick. How do I know if I'm sick from toxins in a water-damaged building?

Mold Warriors' careful scientific definitions will help you figure out whether you have the symptoms of neurotoxic illness. A series of biological markers—including the gene susceptibility HLA DR test and the visual contrast sensitivity (VCS) test—will also help you sort out whether the illness could be related to mold toxins, such as from exposure to water-damaged buildings. The VCS test has been used for 40 years to assess the potential for exposure to biologically produced neurotoxins. Also, evaluating your specific lab tests will help you and your physician confirm the diagnosis. These labs include hypothalamic hormones, MSH deficiency and abnormal regulation of pituitary hormones, antidiuretic hormone and concomitant measure of osmolality; ACTH and simultaneous cortisol, VEGF, cytokines, complement and measurement of MMP-9.

2. How do I know if the mold growing on the ceiling in my living room is dangerous? Same question for the mildew on my books?

Sampling molds is incredibly easy. Making a "tape-lift," rubbing a piece of clear Scotch tape over the mold, is the first step. Send the tape to a reputable lab for identification. P and K Microbiology charges $35 for the service. Contact your physician to find out more details, as most labs won't accept samples from patients.

3. How will my doctor know if I'm sick from neurotoxins?

You will have many health symptoms, VCS deficits and abnormalities in labs from the Biotoxin Pathway. Any physician can order the lab tests and look for the markers that will diagnose neurotoxic illness. If you meet the diagnostic definition and your exposure is simply to mold, then treatment is begun with cholestyramine. If you have Lyme disease, you'll need antibiotics first, and then pretreatment with Actos before cholestyramine is begun.

4. Is the mold illness contagious?

No, exposure is necessary.

5. I took CSM for my neurotoxic illness but I'm not any better. Why not?

Cholestyramine (CSM) therapy is just the first step in a series of steps that are followed sequentially. Do one intervention at a time! All biotoxic illness patients, for example, follow one month of CSM therapy before beginning the second step. If there's a biofilm-forming, multiply antibiotic resistant coagulase negative Staph colonizing your nose, that organism must be eradicated. If there is auto-immunity present, those complicating factors must be identified and changes initiated. If there are reductions of levels of VEGF, that also needs to be identified and treated. There are some people, especially those with 4-3-53 and the 11/12-3-52B genotype, who unfortunately, often won't get better. We're using new therapies for these patients and hope to have results to report shortly.

6. My doctor said the blood tests Dr. Shoemaker recommends are too expensive and too new. What should I do?

The tests aren't new; they have been in use for a long time. The blood tests that we use are standard in commercial laboratories and insurance reimbursable. The illness that you have has already defied diagnosis, and the standard tests used, such as a CBC or simple metabolic profiles, don't show any abnormalities. The blood tests listed at the end of this book will profile your illness and show you what's wrong. In fact, *not* doing those blood tests and *not* knowing what's wrong is what's *really* too expensive!

7. I took the VCS test online but my doctor doesn't know how to read it.

The eye test available at this link http://www.chronicneurotoxins.com is a screening test; it's designed to assess your potential for exposure to biologically produced neurotoxins. The test itself does not diagnose the illness; what we're looking for is the presence of a distinctive number of errors that will signal to your physician that additional work up is necessary. You can sign up for an interpretation of your VCS test at the chronicneurotoxins website.

8. How can I best individualize my neurotoxic illness treatment?

Every patient has a unique illness profile. Do the complete symptom recording, VCS and labs to profile your illness. Although biotoxin illness causes common problems in people, ultimately, like a tailored dress or suit, your treatment will be unique.

9. A lot of my co-workers are sick but their doctors say the problem is allergies and fibromyalgia. How can we tell if it's really a neurotoxic illness?

We've done a number of building studies where we go on site and screen everyone in the building, using visual contrast and symptom questionnaires. If at least 25% of the patients have the distinctive pattern of Sick Building Syndrome symptom abnormalities, then we do some additional testing. If there are correlations between visual contrast deficits and those with symptoms, the likelihood of neurotoxic illness exceeds 90%. The laboratory tests will those with define the illness. The best way to know if there really is a problem in the workplace is to follow the 5-step repetitive exposure protocol. When one person is ill, we call him a complainer; two people working in the same area are a conspiracy; and three a cohort. Find a couple of buddies to go through the battles with you; the action can get rough.

10. I find myself unable to follow your suggestions. I just cannot follow through as I could have done 10 years ago. What should I do?

These illnesses are chronic; they don't leave on their own. There's no question that proper approach to treatment requires discipline, time and personal commitment. If you're not able to take medication, there's no way you'll improve. I suppose it's not necessary to understand the illness, but without understanding, the patient doesn't have the "power" to control his own health as much as possible. Cognitive issues are almost always present in the illness. Ask family or friends to help you, take *Mold Warriors* to your physician for help, or contact info@moldwarriors.com or info@chronicneurotoxins.com.

11. What can I do if my doctor is unable to read my HLA results?

At *first* glance the HLA wording is difficult to grasp. But we're very excited to offer a special key—follow the Rosetta Stone in the appendix—to help you interpret HLAs and sort out which genotype you have. After you've done a few of these, it'll seem like second nature.

12. I filed a lawsuit against my landlord because of mold, but my attorney said he was worried about the strength of testimony from defense medical experts, so he didn't want to bring a personal injury claim. What can I do to strengthen my case?

You'll run up against a seemingly unbeatable array of experts routinely hired by insurance companies in court cases. Fear not. The strength of your case is based on documenting your illness with symptoms and lab markers; the most unbeatable evidence is your response to treatment and re-exposure in the 5-step repetitive exposure protocol. Given that as those who testify against you aren't treating physicians and have no knowledge of this illness, you certainly can be proven to be right.

13. I improved when I took CSM, but my MMP-9 stayed high and I still have many neurologic abnormalities. They said I had MS, then Parkinson's and now they don't know. What advice do you have for me?

I hope that you were able to document your lab abnormalities *before* you started on therapy. I hope that we know what your potential for exposure is. If the problem is simply elevated MMP-9, I would strongly suggest that you also have a test of myelin basic protein (see Chapter 14) to understand better what the MMP-9 is doing in your brain. If MMP-9 alone is your problem then simply using the no-amylose diet and taking Actos will quickly lower that level.

14. What proof do you have that mold causes CFS?

Direct patient care and multiple studies. In my referral practice, I often see people with a prior diagnosis of chronic fatigue syndrome. Many of those patients have an illness that's simply caused by mold in their basement, bathroom and/or attic. The problem here is that when a diagnosis of chronic fatigue is made according to CDC guidelines, no biological markers were used and no exposure histories were recorded. By expanding what we understand about chronic fatiguing illnesses, eventually we'll show that true chronic fatigue syndrome has multiple causes, one of which is mold.

15. Why do you attack the CDC and consensus panels of medical experts repeatedly, just like you did in Desperation Medicine*? Why don't you think the CDC acts in our best interests?*

The CDC is one of the strongest federal organizations with a huge influence on current public health thinking and medicine. My conflict with the CDC is I see that much of the extraordinary work done by their excellent research scientists is then "spun" by managers whose priority is CDC image, or merely a concern about appearances and public opinion. The CDC isn't an organization that will take the initiative readily, unless there's a clear public health mandate. It's clear the CDC is failing to act to defend the public in mold illness, Lyme Disease and Chronic Fatigue Syndrome. And that's a national disgrace.

16. I have one of the "dreaded" genotypes and low MSH, but I don't feel ill. What advice do you have for me?

We'd probably like to include you in our study of people we're following to see what affects future symptoms. MSH deficiency alone doesn't create symptoms. MSH controls the innate immune responses. Clearly, with your high susceptibility, exposure to biologically produced toxins puts you at a tremendous risk to develop illness because you don't have the protection from MSH. It's been our experience that patients who have low MSH and the dreaded genotype need to be treated rapidly if biotoxin exposure occurs. So I would suggest that you have frequent testing to assess your health status, especially VCS, C3a and MMP-9.

17. Why are MARCoNS difficult to eradicate?

Most of theses organisms make biofilms, which are near-impenetrable barriers for antibiotics to cross. Also, their presence reduces our white blood cells' ability to leave mucus membranes and attack these colonizing organisms. This community of organisms' capability to rapidly make antibiotic resistance factors is extraordinary. The new age in microbiology must take into account biofilm formation and how it affects human illness.

18. Are other physicians treating mold the way you do?

Fortunately, yes, we have a large number of physicians who are using my protocols and beginning to collaborate. It won't be long before we have national standards from treating physicians that would lead the way for those in academia who have scientific credentials, but limited access to primary care in biotoxin illnesses. The biggest shift in physician attitudes is likely to follow the beginning of legal actions *against* treating doctors for failure to diagnose.

19. Why is VEGF deficiency a major problem for your patients who don't show improvement?

VEGF controls delivery of oxygen and nutrients in capillary beds. Cells that aren't receiving necessary increases in oxygen and nutrients don't function normally. Please see Chapter 17 for the additional discussion about the incredible importance of VEGF. It's necessary to measure a VEGF level repeatedly to follow treatment.

20. Autoantibodies are a big deal in your Biotoxin Pathway, but they also occur in people without biotoxin illnesses.

The presence of an autoantibody in your blood doesn't necessarily mean you have an autoimmune illness. The blood test is just a part of a clinical evaluation. The study of autoantibodies is still an emerging part of medicine, and we're only starting to explore why biotoxin patients commonly have significant autoantibodies. Even basic autoimmune issues are still unclear. For example, science doesn't yet understand how people develop antibodies to gliadin, a protein found in gluten, which is not part of our "self," yet ends up being treated as "self" by the immune response.

21. My blood tests show all these antibodies—IgA, IgG and IgM—to fungi. My doctor says these tests help him diagnose and treat.

Be careful with the analysis of antibody tests. Changes in antibody levels have been thought to be helpful in many illnesses, but that idea has never been confirmed. For example, antibiotic therapy bears no clear role in antibody levels in Lyme Disease.

22. Why am I now so sensitive to various chemicals, perfumes, petrochemicals, inks, fumes from computers and copying machines?

So-called Multiple Chemical Sensitivity (MCS) is still a commonly seen clinical condition without much in the way of a confirmatory diagnostic literature. I have treated more than 500 patients with MCS, and have developed the condition myself, but *I still haven't found anyone with MCS who didn't begin their illness with mold exposure*. My illness responded nicely to medications; many don't. Following a long recent discussion with Dr. Martin Pall, we're looking at changing levels of citrulline, an amino acid made when nitric oxide interacts with another amino acid, arginine, after an exposure adequate to initiate MCS symptoms. If we're able to create MCS symptoms, it could implicate changes in nitric oxide metabolism in MCS patients. But no one knows why MCS occurs.

23. Why do so many school kids get nosebleeds after they're exposed to mold?

This is a difficult question to answer, since noses bleed after many common events. If you watch people in a car, waiting at a stoplight, you'll routinely see men pick their noses and women fluff their hair. Kids in a classroom are always at their noses. So nosebleeds, commonly caused by nasal irritation and nasal trauma, both from humidity and contact, are a common event. In a group of People Who Denied Nose-Picking, With Unusual Bleeding, including nosebleeds, after mold exposure, we have found hyperacute changes in C3a affecting changes in clotting. We don't have the number of kids studied before exposure to answer the question, "do physiologic responses to mold/mold toxins cause nosebleeds?" properly yet. But we're working on it.

24. If I'm exposed to mold and sick from it, what's the first thing I should do?

Plan to get out before it's too late! If you remain sick after you're no longer exposed, don't be surprised. That's a marker for toxin illness.

25. I'm planning to spend thousands of dollars to remediate my home. Is this a waste of money?

Often it is. Remediation should mean removal of *all* sources of biotoxin illness. That's often impossible. Make sure you have some back-up financial protection in case you get sick with re-exposure to your "remediated home." Since merely leaving a fraction of spores or biotoxins behind can make a home "sick," many find their remediation was unable to prevent a return of their illness

GLOSSARY

a priori—a term from philosophy, that which is known before hand.

Acquired immune response—antibody formation by special a B lymphocyte after an HLA-tagged antigen is presented to the B lymphocyte by naïve T cells (see dendrite cells).

ACTH—adrenocorticotrophic hormone, released by anterior portion of pituitary. It stimulates release of cortisol and other hormones by the adrenal gland. Early in MSH deficiency, high levels of ACTH compensate for loss of MSH regulatory control. ACTH often is dysregulated in the face of MSH deficiency; measure simultaneously with cortisol.

Actos—*pioglitazone*, one of the thiazolidinedione group of drugs. FDA approved to help with treatment of diabetes, it is also has anti-inflammatory effects by blocking the cytokine nuclear receptor activity and thereby lowering elevated levels of leptin, MMP9, PAI-1. It may have benefit in treatment of multiple chemical sensitivity (MCS) if the MCS is diagnosed early. Actos is one of the most useful of the medications of the Biotoxin Pathway.

ADH—antidiuretic hormone made by posterior portion of pituitary; retains free water from kidney, also called vasopressin. Relative reduction, or dysregulation, is common in mold patients and is seen in approximately 60% of patients with MSH deficiency. Measure simultaneously with osmolality.

Adhesion molecules—compound made by endothelial cells that cause white blood cells in the blood to "stick" or adhere to one location. Adhesion molecules are part of the innate immune response as they're released when TNF, IL-IB, interferon and other cytokines, as well as anaphylatoxins from complement, bind to receptors on endothelial cells. Adhesion molecules provide a link from innate immune response activation and atherosclerosis.

Allele—an alternative word for a gene. There will be two copies of alleles of each of the immune response genes HLA DR. An allele is one member of a pair of genes on the same location of a chromosome.

Alternaria—common indoor mold, not thought to be a toxin-former, but can cause allergy symptoms.

Alternative pathway of complement—the series of proteins activated by antigen only, without antibody, particularly important in illnesses caused by biotoxins in HLA-susceptible individuals.

Amerospores—general classification for small, round spores that can't be identified accurately by direct microscopic examination. Toxin-forming organisms that are responsible for production of amerospores include *Acremonium, Aspergillus, Penicillium*, and *Trichoderma*. Look for the location of the sample to help with identification.

Amplified mold growth—indoor growth of mold in water damaged buildings in the absence of a significant group of predators. Usually accompanied by production of secondary metabolites.

Amylose—Amylose-free diet—the diet that recognizes that

dietary restriction of the group of carbohydrates that have the complex plant starch, amylose, will prevent the rapid rise of blood sugar after a meal. Amylose-containing foods that must be avoided are wheat, rice, oats, barley and rye; bananas; and vegetable, that grow below the ground. See, *Lose the Weight You Hate*, Ritchie C. Shoemaker MD, 2002.

Anaerobic threshold (AT)—the level at which oxygen is no longer delivered in capillary beds to muscle cells. Low AT is commonly seen in mold patients.

Anaphylatoxin—name often given to C3a and C5a. These compounds can activate smooth muscle spasm, promote release of adhesion molecules by endothelial cells, increase vascular permeability and cause release of cytokines and inflammatory compounds from mast cells and basophils. Their link to clotting systems may explain the increased amount of bleeding problems seen clinically in mold patients.

Androgens—male hormones released following stimulation by anterior pituitary hormones called gonadotrophins. We measure total testosterone, DHEA-S (not DHEA) and androstenedione. MSH controls pulsatile release of gonadotrophins. Abnormalities in androgens are found in 40% of patients with MSH deficiency.

Anterior pituitary—anatomical division of pituitary.

Antibody—compound made by acquired immune response designed to eliminate specific molecule, recognized as "not-self," by lymphocytes.

Antigen—specific molecule that stimulates antibody response.

Arthrinium—fungus found on plants. It's rarely found growing indoors.

Ascospores—a general classification for spores produced by sexual reproduction and can include *Aspergillus* and *Penicillium*. Often identified from fungi growing in damp areas.

***Aspergillus/Penicillium*—like**—*Aspergillus* and *Penicillium* spores are often indistinguishable. Commonly called Asp/Pen.

Assay—measure of activity of a biological compound.

Atherosclerosis—hardening of the arteries. Results from delivery of oxidized LDL cholesterol *across* endothelial cell membranes, *through* the supporting basement membrane and *into* the matrix of artery walls. There, LDL is then engulfed (innate immune response) by macrophages, initiating inflammation. As the deposits of LDL cholesterol increase in size, exerting pressure from the outside of the blood vessel, the internal diameter of the blood vessel is reduced. Over time, the deposits of cholesterol can erode into the blood vessel wall ("rupturing of a plaque"), causing an acute obstruction of blood flow. If the blood vessel's internal diameter is reduced too much due to the growth of plaque from the outside in the heart or the brain, we become at risk for heart attack and stroke. Also called arteriosclerosis.

Attention deficit disorder (ADD)—diagnosis often given to children with difficulty maintaining concentration, overactivity and learning disorder. Often misdiagnosed when mold exposure isn't considered.

Aureobasidium—common outdoor, soil-dwelling fungus. When found indoors, it's a marker for moisture accumulation. When a patient talks about "black mold" growing on shower curtains, tile grout, windowsills and fabrics, expect to find Aureo and *Cladosporium*.

Autoantibodies—antibodies produced by the acquired immune response to our own tissues. Autoantibodies to nuclear material ("false positive ANA"), cardiolipins and myelin basic protein are common in mold patients. Antibodies to gliadins, a small protein component of gluten, also commonly found in mold patients, aren't technically the same as autoantibodies but are included in that group for discussion purposes. Curiously, those patients with wingspan greater than weight (normally less than 10% of patients) comprise 80% of the group of patients with autoantibodies.

Basidiospores—a group of molds commonly found in gardens, leaf litter and forests. Can cause dry rot.

Benzodiazepine—anti-anxiety medications.

Bernoulli's equation—mathematical formula that associates pressure and flow within a tube.

Biomarker—a test that reflects biological activity of an illness. Some biomarkers are static, like a HLA DR genotype. Others are called dynamic, in that they change with disease activity.

Bioterrorism—use of biologic agents as weapons against civilians.

Biotoxin exposure history—an environmental history that looks to associate exposure to an area with resident biotoxin forming organisms with illness.

Biotoxin Pathway—the main theory of biotoxin illnesses. The Pathway includes the dynamic interaction of genetics, cytokines, autoantibodies, vascular growth factors, central hypothalamic hormones, pituitary hormones, peripheral hormones and opportunistically growing microbes.

Biotoxin-associated illness—illness caused by exposure to biotoxins.

Biowarfare agents—biologically produced toxins that can be made into weapons to be used against troops or civilians.

Building Related Illness (BRI)—term often used to attempt to describe illnesses found in multiple members of the exposed group. Can apply to chemical illness, irritant illness or toxin illness. Term should be discarded.

CA-1—acronym for unnamed *Babesia*-like organism found in California.

Case definition—an organized approach to what we mean when we say someone has a diagnosis or is a case of mold illness.

Causation—the logical effect of process A on element B to produce result C, such that a given result C is seen when B is exposed to A.

C3a—a by-product of activation of the third component of complement. C3a can continue to activate C3 when the alternative pathway of complement is activated, creating a high risk for chronic fatigue and persistent symptoms in mold patients (see *anaphylatoxin*).

Celiac disease—a systemic illness with multiple symptoms associated with antibodies to a specific compound, endomysial antibody, an IgA, as well as to gliadin.

CDC—Center for Disease Control and Prevention.

The Center for Research on Biotoxin Illnesses—Our non-profit 501-c-3 research organization. Please help us with a donation!

Centromere—the constricted region near the center of a chromosome that has a critical role in cell division.

Chaetomium—toxin-former commonly found on damp sheet-rock.

Chimera—According to Larrouse World Mythology, a classical mythological monster with a fire-breathing goat's head, the fore-quarters of a lion, and the hind part of a dragon. Killed by the Coninthian hero, Bellerophon, on his winged horse, Pegasus. Also a statue near the Rayburn Office Building.

Cholestyramine (CSM)—non-absorbable anion-binding resin used to lower cholesterol for over 40 years. This compound has multiple quaternary ammonium side-chains that present a net positive charge with a radius of approximately 1.4 Angstroms. The physicochemical properties of CSM enable it to bind to a diverse variety of toxin molecules, including biotoxins, as well as many others.

Chronic Fatigue Syndrome (CFS) or Chronic Fatigue and Immune Dysfunction Syndrome (CFIDS)—The clinical syndrome defined by a consensus panel in 1994 that includes a series of symptoms, especially fatigue, felt to represent a distinct clinical entity.

Cigarette science—another way to say, "smokescreen." In the argument over whether or not exposure to cigarettes could cause cancer, defenders of the tobacco companies were hired to publish research findings that would say that nothing in tobacco smoke has ever been shown to hurt anything.

Ciguatera—the dinoflagellate toxin illness caused by eating fish that are predators on tropical reefs. The illness is a classic biotoxin-associated illness, one of the few that comes from eating toxin-contaminated food.

Cladosporium—common outdoor mold that often grows indoors on damp wallboard and fabrics. Not as common in areas of water damage that dry out intermittently.

Classical pathway of complement—the series of proteins activated by a combination of antigen and antibody joined to make an "immune complex." Classical pathway activation is commonly seen in diseases of autoimmunity, especially lupus, but is also highly important in eradicating organisms causing infectious disease.

Cognitive—related to thinking.

Cohort—epidemiologist's word for groups of patients with similar exposures.

Complement—innate immune first responder. Thirty different proteins, receptors and inhibitors. Third most common group of proteins in blood. Highly associated with immune response to acute exposures. Chronic elevation of alternative pathway increases C3a; a major factor in chronic health symptoms.

Compounding pharmacist—A pharmacist who is capable of making, mixing or otherwise compounding individually made prescriptions not otherwise made available by major manufacturers. Compounded prescriptions include topical preparations for those who cannot take pills, and flavored medicines to make it easier for children to take.

Confounding illness—another illness that can produce similar symptoms as those in the illness under consideration.

Consensus panel—groups of individuals, usually authoritative scientists, brought together by a large body, usually governmental, for the sole purpose of considering a problem and developing an opinion that can be used by the larger body for policy. Ideally, this group could be the purest form of sharing of thought. In reality, the direction of the panel discussion is usually directed by the larger organization for the purpose of "managing" the public perception of what is true about the given scientific question.

Controls, control group—people without exposure who are used in scientific studies to compare to the actual effects of exposure on cases. If exposure causes something to occur, the effect won't be seen often in controls, but will be seen often in cases.

Cortisol—powerful anti-inflammatory steroid hormone released by adrenal upon stimulation by ACTH. Must measure simultaneously with ACTH to assess possibility of dysregulation.

CRP—c-reactive protein. An often ordered marker of inflammation initiated by release of interleukin-6. Innate immune response element.

Cytokines—proteins that are the effective elements of immune response. Three main functions of cytokines are part of life/death control: recognition of foreign antigens, recruitment of WBC response and activation of "killer" functions.

Cytokine nuclear receptor—controller of copying (transcription) of a group of genes found in DNA of fat cells, endothelial cells and white blood cells. When activated by NFkB, these cells manufacture the proteins and cytokines that regulate the immune response to foreign invaders (antigens).

Dendrite cell—"professional" antigen presenting cell that engulfs antigen (innate immune response) before processing it and presenting the antigen for antibody formation (acquired immune response).

DHEA-S—important androgen (male) hormone.

DNA—the genetic code.

Ehlers-Danlos Syndrome—a group of patients with unusual physical features characterized by long arms and wingspan compared to height. There are several different types.

EKG—12 lead electrocardiogram.

Endorphins—the natural opiates of the brain. These compounds regulate pain reception.

Endothelial cells—cells that line blood vessels.

Epidemiologist—a scientist who studies the presence of disease in populations.

Epicoccum—Rarely felt to be a toxin-former. Often found on continuously damp materials.

Erythropoietin—"epo" an "anti-cytokine" protein that serves several functions. It's used clinically to increase red blood cell production in bone marrow of patients receiving chemotherapy, those with HIV, those with renal failure and those who must increase red blood cell mass before surgery. Epo supplementation usually provides marked clinical improvement in mold patients who have refractory symptoms and persistent low VEGF, despite treatment.

Exotoxins—both A and B destroy MSH, perpetuating MSH deficiency, that in turn permits ongoing colonization of MARCoNS in deep spaces of nose.

Expert medical testimony—opinion in a legal case that comes either from one with knowledge, either from practice or education, not available to the everyday person that could help a jury sort out contradictory testimony.

Fibromyalgia—my pet peeve diagnosis. The diagnosis named as a result of a 1988 consensus panel that was trying to sort out why so many people had chronic, multisystem health effects. This one is the most egregious of the consensus panel opinions that abuse common sense.

Five-step, repetitive exposure protocol—a stylized study design that enables one to determine causation of illness in mold cases.

Fungal contamination—growth of fungi indoors in water damaged buildings.

Fungi—group of organisms that lack chlorophyll and blood vessels that include yeasts, smuts, molds and mushrooms.

Fusarium—common indoor and outdoor fungus. Often a toxin-former, especially in areas where broad spectrum fungicides are used. Indoors, it causes damping off of house plants. Look for it in water reservoirs for humidifiers and drip pans. *One species, Fusarium oxysporum schlecht,* can release cyanide into its root zone.

Gastrointestinal—referring to the stomach and intestines, as well as liver, gall bladder and pancreas.

Genetic susceptibility—predetermined group of immune-response genes represented by HLA DR. Particular toxin illnesses are highly linked to presence of particular genotypes. While the genes don't make patients ill, exposure to toxins will result in illness essentially only in those with the "susceptible" genotype. The HLA link to *lack* of antibody formation is a bridge from the innate immune response to the acquired immune response.

Genomics—the study of the functions and interactions of all the genes in the genome, including their interactions with environmental factors.

Genotype—a person's genetic makeup; his or her DNA sequence.

GGTP—important liver enzyme that converts glutathione. Found along small bile ducts.

Gliadin—protein found in gluten consisting of 18 amino acids. The toxic element in gluten. Some evidence exists to suggest that gliadin can be incorporated into wall of small intestine, acting as a "self" molecule.

Gluten, gluten intolerance—a protein found in wheat, barley, rye and oats.

Haplotype—a group of nearby alleles that are inherited together.

Health assessment—a systematic review of elements of health.

Hemolysins—MARCoNS protein consisting of 22 amino acids that disrupts red blood cell membranes, releasing iron necessary for growth of bacteria living in biofilms.

Heterozygous—having two different alleles at a specific gene locus.

HLA DR—groups of immune-response genes that control attachment of HLA "tag" to antigen in dendrite cells.

Homozygous—having two identical alleles at a specific gene locus.

Hyphae—part of the vegetative mass of molds.

Hypothalamus—the master gland of endocrine system. Exerts influence on mood, appetite, sweating, temperature regulation. Has multiple mechanisms to affect changes in nerve function as well. Main source of MSH production.

Heat, Ventilation and Air-Conditioning (HVAC) system—the system that warms and cools many of our homes and buildings.

IBS—Irritable bowel syndrome. Commonly diagnosed condition abdominal pain with alternating constipation and diarrhea. Also called spastic colon. Patients who have mold as a cause for their "IBS," will often have more trouble with constipation from CSM than those without IBS. The diagnosis is made following consensus opinion that multiple unexplained gastrointestinal symptoms are caused by emotional factors and benign, but uncomfortable responses to food and environment. Usually wrong.

IL-IB—Interleukin-1-beta; one of the cytokines that causes increased inflammation. This compound is frequently elevated in blood of mold patients.

IL-6—Interleukin 6; pro-inflammatory cytokine released as part of recognition of foreign antigens and after activated by TNF and IL-1B.

Independent medical exam—exam performed at the request of a third party in medical claims disputes. Often, the third party is an insurance company and often the IME physician is one who makes his living providing an assessment of the medical claim for the insurance company. Patients are often surprised to find the report of the IME examiner bears little resemblance to what is actually present in the case. If you have to go through an IME, take a video camera and a tape player, as well as a significant other. You are provided with a list of your rights before the IME. Be prepared to be outraged by what the IME physician does in his opinion. Demand that the IME physician be available for corrections.

Infectious disease—illnesses caused by growth of organisms growing inside or on us. As opposed to colonization, when an organism doesn't invade from its secure site in or on us, an infectious disease will include growth beyond boundaries.

Innate immune response—primordial set of "first responders" includes dendrite cells, macrophages, toll receptors, complement and cytokines.

Intima—the endothelial cell and basement membrane of blood vessel walls.

JAK, Janus kinase—one of the innate immune response enzyme systems that helps activate receptors for insulin and leptin.

Leptin—hormone made by fat cells. It's also a cytokine that participates in regulating body mass, and storing fatty acids efficiently. It also activates production of MSH, which in turn regulates much of the immune response from the innate immune system. High leptin and low MSH are the markers for obesity caused by toxins. With high leptin in a biotoxin patient, weight loss is incredibly difficult.

Linkage disequilibrium—the nonrandom association in a population of alleles at nearby loci.

Lupus—systemic lupus erythematosis is an autoimmune illness characterized by multiple organ involvement and presence of antibodies to our own nuclear material. Full discussion of SLE is beyond this glossary.

Lyme Disease—the infectious and biotoxin illness caused by carriage of the spirochete, *Borrelia burgdorferi*, following a tick bite. Both over-diagnosed and underdiagnosed, a significant contributor to chronic fatiguing illnesses.

Macrophages—white blood cells that can engulf antigens, immune complexes and release cytokines. They are activated by cytokines (innate) and immune complexes (acquired).

Matrix—the collection of connective tissue molecules between the intima of blood vessels and smooth muscle cells.

Melatonin—a compound that regulates restorative restful sleep. When MSH is low, supplemental melatonin is usually ineffective in restoring normal sleep.

Memnoniella—Not clearly shown to be a consistent toxin-former indoors. Often grows on cellulose.

Mildew—superficial growth of fungi on organic surfaces.

MMP9—matrix metalloproteinases, including #9, are enzymes that effect the tissues underneath blood vessel walls (subintimal space). MMP9 is particularly adept at delivering inflammatory elements out of blood vessels and putting them where they shouldn't be in brain, nerve, lung, muscle and joint. If you see high MMP9, look for toxins and cytokines.

Mold—fungi capable of digesting organic matter.

Mold facies—a red rash on the face, probably due to release in skin of complement or VEGF, or maybe both, due to mold exposure. Usually more common in fair-skinned patients, especially blondes, and far more common in women than men. Topical preparation of CSM helps, but nothing is a better cure than prevention of exposure.

Molecular mimicry—the concept that a compound made by one organism can share structural features of molecules made by other organisms such that the antibody made recognizing one compound will result in recognition of the second compound as well.

MSH—alpha melanocyte stimulating hormone. A small peptide hormone (13 amino acids) that regulates nearly all aspects of the innate immune response. Made in the hypothalamus, it controls nerve, hormone, cytokine function, skin cells (keratinocytes) and mucus membrane defenses, as well as controlling production of endorphins and melatonin.

Multiple sclerosis (MS)—a demyelinating illness of the brain.

Multiply antibiotic resistant coagulase negative staphylococci—once benign, but no longer, these colonizers of the skin and nose of MSH-deficient patients make biofilms that let them differentiate and live as a community of bacteria. In a biofilm colony, individual MARCoNS may variously make antibiotic resistance factors, hemolysins (small proteins that help the bacteria obtain iron) that activate cytokine responses and exotoxins that split and inactivate MSH. When a mold patient has a biofilm-forming MARCoNS and low MSH, clinical improvement will not occur until the MARCoNS is eradicated. Will be resistant to methicillin in 60% of isolates.

MBP—antibodies to myelin basic protein.

Mycotoxin—biologic toxin made by molds, fungi.

Myelopathy—an illness caused by disease of a nerve.

NFkB—the nuclear factor that rapidly activates the cytokine nuclear receptor. NFkB is normally sheltered by a protecting protein, IkA, in the cytoplasm of a cell. NFkB is released from IkA by a second messenger sent from the cell surface membrane after a toxin binds to a surface Toll receptor.

Occupational exposure—exposure to biotoxins while at work.

Osmolality—measure of salt and water content of the blood. Loss of free water into urine in ADH-dysregulated patients results in higher than expected osmolality. We use osmolality as an indicator of dehydration. Measure simultaneously with ADH.

Pandora's Box—In Greek mythology, Pandora was the first woman on earth. The gods endowed her with many talents; gave her beauty, music, persuasion, and so forth. Hence her name: Pandora, "all-gifted." When Prometheus stole fire from heaven, Zeus took vengeance by presenting Pandora to Prometheus' brother. Pandora had a jar which she was not to open. Impelled by curiosity, Pandora opened the jar, and all evil contained escaped and spread over the earth. She quickly closed the lid, but the jar's contents had escaped, except for one thing which lay at the bottom, and that was Hope.

Paradigm—the example of a whole. A small element of mold illness represents much of the entirety of biotoxin illnesses.

PEAS—Possible Estuarine Associated Syndrome; acronym coined by a consensus panel convened by the CDC to study Pfiesteria Human Illness Syndrome.

Pfiesteria—the estuarine toxin former that started my career in biotoxin illnesses. These dinoflagellates, from one of the oldest groups of living creatures, killed fish and made people sick in the Pocomoke (Md) River in 1997. When the State of

Maryland said no one was made ill by *Pfiesteria*, my introduction to the Appearance of Good Science was initiated.

PFT—pulmonary function test.

Phenotype—the clinical presentation or expression of a specific gene or genes, environmental factors, or both.

PIC—proinflammatory cytokine.

Post-Lyme Syndrome—grouping of symptoms and physiologic abnormalities persistent in a patient with Lyme disease after adequate treatment with antibiotics. Found invariably in patients with HLA 15-6-51 or 16-5-51.

Posterior pituitary—anatomical division of pituitary.

Pseudoscience—junk science, see cigarette science.

Pulmonary stress test—supervised aerobic exercise, monitoring oxygen delivery, oxygen consumption and cardiac function. This test yields clinically important measures like V02 max and anaerobic threshold (AT).

Punitive damages—extra money awarded in legal cases, usually as a result of attitude or behavior of the defendants.

Pustules—small areas of clogged pores on the skin that have small growth of bacteria.

Remediation, remediation plan—the process of safely removing mold contamination from inhabited spaces.

Residential exposure—exposure to toxigenic fungi at home.

Rheumatologists—medical specialists in illnesses caused by inflammation.

Rosacea—the distinctive facial rash seen in people over 25 that often looks like acne. Often it is due to mold exposure.

Rusts—rusts are plant pathogens. Not a human health threat (yet!).

Sampling program—A systematic plan to test for the presence of toxigenic fungi, actinomycetes and bacteria in water damaged buildings. Look at the sampling plan to see if it could possibly be adequate to answer a simple question: Are there toxigenic organisms in the air in this building that can make me sick?

Selective serotonin reuptake inhibitors (SSRIs)—antidepressants that work by deactivating mechanisms that remove serotonin from neural pathways in the brain.

Sick Building Syndrome (SBS)—one of the most common names for the illness acquired from working in moldy or water-damaged buildings.

Smuts/Myxomycetes—parasitic plant pathogens. Myxomycetes are occasionally found indoors, but they're not something I worry about.

Spore—the reproductive structure of a mold.

Stachybotrys—The "black mold" of media hype. Needs lots of humidity, has sticky spores that rarely are found in air samples, but if they are in an air sample, look out!

Stealth toxin—a toxin whose presence is undetected.

Sub-intimal space—the area of matrix and smooth muscle cells in blood vessel walls.

Toll receptor—named for the German word roughly translated as "Eureka," Toll receptors are part of the innate immune response team that provide rapid response after they're activated by components of bacteria, viruses and toxins circulating in blood.

Torula—when you find yeast indoors, look for wet cellulose.

TNF—tumor necrosis factor alpha—one of the cytokines that increases inflammation. It's rarely elevated in mold patients but is often elevated in patients who have the Post-Lyme syndrome.

21st Century Medicine—the new approach to inflammation that includes assessment of innate immune responses.

Ulocladium—Ulocladium is a rarely toxin-former, often found on sheetrock and fabrics.

Unidentified conidia—fungal spores indicating growth, but without unique identifying characteristics.

VCS test—Visual Contrast Sensitivity test—the neurotoxicological test that shows a distinctive group of deficits in biotoxin illness patients.

VEGF—vascular endothelial growth factor—probably one of many growth factors. Exerts extraordinary control of blood flow in capillaries. High levels of VEGF stimulate new bloods vessel growth (angiogenesis); role of VEGF in cancer and heart disease is a hot research topic. Low VEGF is common in biotoxin patients. Affects NFkB and cytokines.

VEGF resistance—high VEGF levels in the face of low V02 max and low anaerobic threshold (AT). The VEGF receptor is not responding to VEGF, despite high levels in blood. Must distinguish this condition from high VEGF without resistance by performing pulmonary stress test.

VO2 Max—a measure of the upper limits of oxygen delivery from capillaries to the entire body during exercise. Low VO2 max is commonly seen in chronically short-of-breath mold patients.

WA-1—Acronym for as yet unnamed Babesia-like organism found in the State of Washington.

WBC—white blood cell.

INDEX

B

C

F

G

H

M

N

O

P

R

S

T